SEMICONDUCTOR DEVICE-BASED SENSORS for GAS, CHEMICAL, and BIOMEDICAL APPLICATIONS

SEMICONDUCTOR DEVICE-BASED SENSORS for GAS, CHEMICAL, and BIOMEDICAL APPLICATIONS

Edited by
FAN REN & STEPHEN J. PEARTON

CRC Press
Taylor & Francis Group
Boca Raton London New York

CRC Press is an imprint of the
Taylor & Francis Group, an **informa** business

CRC Press
Taylor & Francis Group
6000 Broken Sound Parkway NW, Suite 300
Boca Raton, FL 33487-2742

First issued in paperback 2017

ISBN 13: 978-1-4398-1387-4 (hbk)
ISBN 13: 978-1-138-07539-9 (pbk)

Visit the Taylor & Francis Web site at
http://www.taylorandfrancis.com

and the CRC Press Web site at
http://www.crcpress.com

Contents

Preface

According to a 2007 report entitled "Sensors: A Global Strategic Business Report," the global sensor market grew at an annual average rate of 4.5% between 2000 and 2008 and is expected to reach US\$61.4 billion by 2010. The US\$11.5 billion chemical sensor market represents the largest segment of this global sensor market. This includes chemical detection in gases, chemical detection in liquids, flue gases and fire detection, liquid quality sensors, biosensors, and medical sensors. Semiconductor-based sensors fabricated using mature microfabrication techniques and/or novel nanotechnology are one of the major contestants in this market. Silicon-based sensors remain dominant due to their low cost, reproducibility, and controllable electronic behaviors, and the abundant data for chemical treatments on silicon oxide or glass. However, the Si sensors are generally unable to be operated in harsh environments—for instance, high temperature, high pressure, and corrosive ambients—so the application areas of silicon-based sensors are still limited. Wide-bandgap Group III nitride-based compound semiconductors are very good alternative options to replace silicon because of their many advantages, for example, high chemical resistance, high temperature operation, high electron saturation velocity, high power operation, and blue and ultraviolet optoelectronic behaviors.

For the bio- and medical-sensing applications, disease diagnosis by detection of specific biomarkers (functionally or structurally abnormal enzymes, low-molecular-weight proteins, or antigen) in blood, urine, saliva, or tissue samples has been established using several approaches, including enzyme-linked immunosorbent assay (ELISA), particle-based flow cytometric assays, electrochemical measurements based on impedance and capacitance, electrical measurement of microcantilever resonant frequency change, and conductance measurement of semiconductor nanostuctures. ELISA possesses a major limitation in that only one analyte can be measured at a time. Particle-based assays provide a spotlight for multiple detection by using multiple beads, but the whole detection process takes more than 2 hours and is not practical for bedside detection. Electrochemical devices have attracted attention due to their low cost and simplicity, but significant improvements in their sensitivities are still needed for use with clinical samples. Microcantilevers are capable of detecting concentrations as low as 10 pg/mL but, unfortunately, suffer from an undesirable resonant frequency change due to viscosity of the medium and cantilever damping in the solution environment. Nanomaterial devices so far have provided the best option toward fast, label-free, sensitive, selective, and multiple detections for both preclinical and clinical applications. Examples of electrical measurements using semiconductor devices include carbon nanotubes for lupus erythematosus antigen detection, compound semiconducting nanowires, In_2O_3 nanowires for prostate-specific antigen detection, and silicon nanowire array detecting prostate-specific antigen, carcinoembryonic antigen, and mucin-1 in serum for diagnosis of prostate cancer.

The book provides a forum for the latest research in semiconductor-based sensors for gas, chemical, bio, and medical applications. It features a balance between original theoretical and experimental research in basic physics, device physics, novel materials and device structures, process and system, because with sensors the transformation of research into products is key. The main objective of the book is to provide the reader with a basic understanding of new developments in recent research on semiconductor-based sensors. This book is written for senior undergraduate and graduate students majoring in solid-state physics, electrical engineering, and materials science and engineering, as well as researchers involved in the field of sensors for gas, chemical, bio, and medical applications.

In this book, we bring together numerous experts in the field to review progress in semiconductor and nanomaterial-based sensors. We feel that this collection of chapters provides an excellent introduction to the field and is an outstanding reference for those performing research on different sensor applications. Most of the following subfields, as well as others at the cutting edge of research in this field, are addressed: nanomaterial-based sensors, organic semiconductor-based sensors, GaN-based sensor arrays for nano- and pico-fluidic systems for fast and reliable biomedical testing, metal oxide-based biosensors, MSM-based optical sensors, wireless remote hydrogen sensing systems, InN-based sensors, GaN MOS-based gas sensors, and AlGaN/GaN chemical and medical sensors, and Si-MOS-based sensor, nanomaterial-based sensor. Chapter contributors are the following: Dr. Cimalla and his collaborators on AlGaN/GaN sensors for direct monitoring of nerve cell response to inhibitors; professors Gwo and Yeh on InN-based sensors; Dr. Chang and co-workers on AlGaN/GaN HEMT-based sensors for biomedical applications; Professor Heo and his collaborators on ZnO thin film and nanowire-based sensor applications; Dr. Anderson from the Naval Research Laboratories and his collaborators on reviewing wireless remote sensing; Dr. Landheer and his collaborators from National Research Council in Canada on bioaffinity sensors based on MOS field-effect transistors; and professors Xie and Qi on MEMS-based optical chemical sensors.

Fan Ren and Stephen J. Pearton
University of Florida
Gainesville

The Editors

Fan Ren is distinguished professor of chemical engineering at the University of Florida, Gainesville. He received his Ph.D. in inorganic chemistry from Polytechnic University in Brooklyn, New York. His current research interests include semiconductor-based sensors, Sb-based HBTs, and nitride-based HEMTs. He is a fellow of the APS (American Physical Society), AVS (American Vacuum Society), ECS (Electro-Chemical Society), and IEEE (Institute of Electrical and Electronics Engineers).

Stephen J. Pearton is distinguished professor and alumni chair of the Materials Science and Engineering Department, University of Florida, Gainesville. He has a Ph.D. in physics from the University of Tasmania and was a postdoctoral fellow at the University of California–Berkeley prior to working at AT&T Bell Laboratories in 1994–2004. His research interests are in the electronic and optical properties of semiconductors. He is a fellow of the APS, AVS, ECS, IEEE, MRS (Material Research Society), and TMS (The Metallurgical Society).

List of Contributors

Travis J. Anderson
Naval Research Laboratory
Washington, DC

C.Y. Chang
Department of Materials Science and Engineering
University of Florida
Gainesville, Florida

Yuh-Hwa Chang
Institute of Nanoengineering and Microsystems
National Tsing Hua University
Hsinchu, Taiwan

Byung Hwan Chu
Department of Chemical Engineering
and Department of Materials Science
    Engineering
University of Florida
Gainesville, Florida

Irina Cimalla
Technical University Ilmenau
Institute for Micro and Nanotechnologies
Ilmenau, Germany

Volker Cimalla
Fraunhofer Institute for Applied Solid State
    Physics
Freiburg, Germany

M.J. Deen
Department of Electrical and Computer
    Engineering
McMaster University
Hamilton, Ontario, Canada

G. Dubey
Steacie Institute for Molecular Sciences
National Research Council of Canada
Ottawa, Ontario, Canada

Michael Gebinoga
Technical University Ilmenau
Institute for Micro and Nanotechnologies
Ilmenau, Germany

Shangjr Gwo
Department of Physics
National Tsing Hua University
Hsinchu, Taiwan

Young-Woo Heo
School of Materials Science and Engineering
Kyungpook National University
Daegu, Korea

Yu-Liang Hong
Department of Physics
National Tsing Hua University
Hsinchu, Taiwan

W.H. Jiang
Institute for Microstructural Sciences
National Research Council of Canada
Ottawa, Ontario, Canada

D. Landheer
Institute for Microstructural Sciences
National Research Council of Canada
Ottawa, Ontario, Canada

Vadim Lebedev
Fraunhofer Institute for Applied Solid State
    Physics
Freiburg, Germany

Hong-Mao Lee
Department of Physics
National Tsing Hua University
Hsinchu, Taiwan

Jenshan Lin
Department of Materials Science Engineering
and Department of Electrical and Computer
Engineering
University of Florida
Gainesville, Florida

G. Lopinski
Steacie Institute for Molecular Sciences
National Research Council of Canada
Ottawa, Ontario, Canada

W.R. McKinnon
Institute for Microstructural Sciences
National Research Council of Canada
Ottawa, Ontario, Canada

D.P. Norton
Department of Materials Science and
Engineering
University of Florida
Gainesville, Florida

Stephen J. Pearton
Department of Electrical and Computer
Engineering, Department of Chemical
Engineering, and Department of Materials
Science and Engineering
University of Florida
Gainesville, Florida

Vladimir Polyakov
Fraunhofer Institute for Applied Solid State
Physics
Freiburg, Germany

Z.M. Qi
State Key Laboratory of Transducer
Technology
Institute of Electronics
Chinese Academy of Sciences
Beijing, China

Fan Ren
Department of Chemical Engineering
University of Florida
Gainesville, Florida

Andreas Schober
Technical University Ilmenau
Institute for Micro and Nanotechnologies
Ilmenau, Germany

Yen-Sheng Lu
Institute of Nanoengineering and Microsystems
National Tsing Hua University
Hsinchu, Taiwan

M.W. Shinwari
Department of Electrical and Computer
Engineering
McMaster University
Hamilton, Ontario, Canada

N.G. Tarr
Department of Electronics
Carleton University
Ottawa, Ontario, Canada

Yu Lin Wang
National Tsing Hua University
Hsinchu, Taiwan

H. Xie
Department of Electrical & Computer
Engineering
University of Florida
Gainesville, Florida

J. Andrew Yeh
Institute of Nanoengineering and Microsystems
National Tsing Hua University
Hsinchu, Taiwan

1 AlGaN/GaN Sensors for Direct Monitoring of Nerve Cell Response to Inhibitors

I. Cimalla
Technical University Ilmenau, Institute for Micro
and Nanotechnologies, Ilmenau, Germany

M. Gebinoga
Technical University Ilmenau, Institute for Micro
and Nanotechnologies, Ilmenau, Germany

A. Schober
Technical University Ilmenau, Institute for Micro
and Nanotechnologies, Ilmenau, Germany

V. Polyakov
Technical University Ilmenau, Institute for Micro
and Nanotechnologies, Ilmenau, Germany

V. Lebedev
Fraunhofer Institute for Applied Solid State Physics, Freiburg, Germany

V. Cimalla
Fraunhofer Institute for Applied Solid State Physics, Freiburg, Germany

CONTENTS

1.1   INTRODUCTION

Mammalian neuronal cells represent one of the most sophisticated and complex signaling and processing systems in any living organism. Many diseases, such as Alzheimer's and Parkinson's disease or amyotrophic lateral sclerosis, have their genesis in disturbed processes on nerve cells. The study of the mechanisms of the physiological changes in nerve cells, which are induced by stressors such as neuroinhibitors or analgesics, represents one way to obtain helpful information for understanding such neural disorders and opens the way for the development of treatments. However, new drugs for efficient treatments of such diseases are the products of a long development process, the first step of which is often the discovery of a new enzyme inhibitor. In the past, the only way to identify such novel inhibitors was a "trial and error" principle: screening huge libraries of compounds against a target enzyme and studying the response in the hope that useful results would emerge. This "brute force" approach is to some extent successful and has even been extended by combinatorial chemistry approaches capable of producing large numbers of novel compounds quickly, as well as by high-throughput screening technologies to rapidly screen these huge chemical libraries for useful inhibitors. However, despite the increasing demand for high-throughput functional screening methods in the areas of environmental protection, toxicology, and drug development [1–5], these methods are not yet efficient in the case of cellular measurements [3,5,6].

In pharmacology, it is important to study the side effect spectrum of a given compound; thus, the application of complex functional tests at the whole-organism level are necessary [7,8]. To assess the global toxicity of the wide variety of chemicals that are possible drugs, research on functional cells using biosensors could be more effective than pure physicochemical methods [9]. In recent years, the application of whole-cell biosensors for toxin detection, drug screening, and recording of cell action potential has developed from an academic principle to a widely accepted method [9–19]. Already in the early 1950s, Hodgkin and Huxley studied the electrical properties of a single cell by impaling the cellular membrane with carefully designed glass microelectrodes. With this procedure, the action potential of the neuronal cells was recorded, and the presence of voltage-gated ion channels has been demonstrated [20]. Two decades later, further improvement of this method led to the development of the patch-clamp technique by Neher and Sakmann (1976) [21]. In this case, the glass electrode is placed close to the cell membrane, and a low-impedance electrical contact to the cell's interior is established [22]. Using this whole-cell configuration, voltage-gated ion-channel ensembles or even the function of single ion-channels on the cell membrane are studied. However, this novel and nowadays widely accepted method for medical, pharmaceutical, and physiological research has some disadvantages. The patch-clamp procedure is complicated to handle, invasive, and invariably destroys the cell after measurement. It is also limited to observing only a few cells at the same time, and long-time measurements are not feasible as the cells are subjected to a high degree of stress during a patch-clamp measurement, which makes an observation over several hours impossible.

Almost at the same time as the development of the patch-clamp system, a planar microelectrode array for the recording of extracellular electrical activity of cells cultured in vitro was designed by Thomas et al. (1972) [23]. The system was created using gold-plated nickel electrodes of known spacing and size on a glass substrate passivated with patterned photo resist. A glass ring was fixed to the substrate with wax to form a culture chamber. Using Thomas' techniques, eventually with slightly improvements, numerous researchers have utilized microelectrodes arrays to examine a wide variety of cell types under different conditions. In this mode, the first recording of extracellular electrical responses from explanted neural tissue was reported by Gross et al. (1977) [24]. Only a few years later (1980), Pine proposed to combine two developed methods: a glass micropipette inserted through the cell membrane and a microelectrode array, in order to record at the same time the intracellular and extracellular signal, respectively [25]. This was an important simultaneous experiment for the validation of the extracellular technique and a calibration of the recorded extracellular signal for comparison with the large quantity of available data on intracellular recordings in the literature. Pine was also the first to replace the organic photoresist passivation layer with silicon dioxide in order to use standard silicon chip processing technologies. Based on the structure and techniques mentioned earlier, the study on cultured cells was continued by numerous research groups examining many different cell types over the years [26–30].

In 1991, Fromherz et al. [12] launched a new era of biosensing by utilization of an integrated field-effect transistor (FET) as the sensing element instead of the previously used bare metal electrodes. The extracellular signal from single dissociated cells, which were manually placed on the transistor, was recorded. A glass micropipette was inserted through the cell membrane and used to stimulate the cell and monitor the intracellular voltage. Later, Fromherz used similar systems to explore the physical characteristics of the neuron–silicon junction using an array of transistors positioned below the neurons [31] as well as capacitive stimulation of the neuron through a thin oxide layer [32].

The basic sensing element is the pH-sensitive FET proposed by Bergveld in 1970 [33], which initiated a large number of publications devoted to ion-sensitive FETs (ISFETs) based on established silicon technology. A historical overview, which also highlights future challenges and possibilities, is given by Bergveld [34]. Silicon-based biosensing ISFETs (BioFETs) were further classified as shown in Figure 1.1.

Despite good prospects and promising demonstrations of these devices, the fact that they are based on silicon technology, while being beneficial for mass production, impairs their long-term

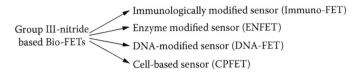

FIGURE 1.1 Classification of Bio-FETs by Schöning and Poghossian. (After Schöning, M. J. and Poghossian, A., 2002, *Analyst*, 127, 1137; Schöning, M. J. and Poghossian, A., 2006, *Electroanalysis*, 18, 1893.)

chemical stability and thus limits the areas of application [37]. This triggered the search for new functional and passivating surface layers, which has not yet led to entirely satisfying results or to commercial available devices employing BioFETs.

At the beginning of this decade, a novel ISFET based on the group III nitrides (GaN, AlN, and alloys) was reported. Group III nitrides exhibit a number of superior properties compared to Si or conventional III-V semiconductors, which makes them promising candidates for biosensors [38–40]. In 2003, Steinhoff et al. reported the first application of these ISFETs for pH sensing [41]. Two years later, the same authors reported the recording of action potential from heart muscle cells using sensor arrays and demonstrated the superior electrical characteristics of ISFET based on group III nitrides compared to similar devices based on silicon technologies [19].

For future applications, the basic understanding of the recorded signals is still a crucial precondition. In the present study, ISFETs based on an AlGaN/GaN heterostructure are further characterized with respect to the coupling to nerve cells in order to enable the application in CPFETs. Specifically, the following topics are addressed:

- The biocompatibility and impact of the typical device processing steps on the reaction of living cells are evaluated.
- The neuroblastoma-glioma hybrid NG 108-15 nerve cell line as a cellular system for the extracellular sensors is employed.
- The applicability and reliability of the nerve cell-to-FET hybrid system are investigated.

This work is organized as follows: Section 1.2 gives a short introduction to sensors employing GaN-based FETs. Section 1.3 summarizes the fabrication steps along with the basic electrical characterization of the ISFETs. Special attention is paid to the stability and the biocompatibility of the sensors. Section 1.4 describes the experimental conditions for the cell–transistor coupling experiments, including the employed cell line and the dosed neurotoxins. Section 1.5 summarizes the calibration experiments to characterize the sensor sensitivity to different ions. Based on these first coupling experiments, Section 1.6 describes the monitoring of the reaction of the NG 108–15 nerve cell–FET sensor hybrid system in terms of drug screening by application of different neurotoxins to the cells. Finally, these results are analyzed in Section 1.7 by self-consistent simulation of the heterostructure employing the site-binding model, and by an estimation of the ion flux between the nerve cell and the FET sensor from the measured signal variation.

1.2   AlGaN FIELD-EFFECT SENSORS

Optoelectronic devices and high-frequency transistors are the most considered and investigated application fields for group III nitride semiconductors. Especially, the two-dimensional electron gas (2DEG) confined in AlGaN/GaN heterostructures has demonstrated extremely promising results for the fabrication of high electron mobility transistors (HEMTs) with applications in high-frequency power amplifiers [42]. Simultaneously, optical and electronic properties of group III nitride-based micro- and nanostructures have been proved to be extremely sensitive to any manipulation of surface charges. For example, the exposure of unpassivated AlGaN/GaN HEMTs to water vapor causes a significant depletion of the polarization-induced 2DEG, resulting in an unwanted variation of the electronic device performance. Profiting from the research on HEMT devices and utilizing the basic HEMT structure, chemical sensors for gases and liquids were developed that received increased attention in the last five years (for details, see the reviews by Eickhoff [39] and Pearton [40]). Due to their wide band gaps and strong bond strengths, this material system has a very good chemical stability as seen in their resistance to wet etching procedures [43,44], and their biocompatibility [45,46]. In combination with low noise due to the wide band gap [19] and high sensitivity to changes of the surface charge [38], group III nitrides are considered to be one of the most promising materials for biosensor devices [47]. Moreover, the optical transparency in the visible part of the spectra opens up the possibility of combining standard optical and electrical measurements. With respect to the construction of integrated devices, the monolithic integration of chemical sensors with group III nitride optoelectronics for combined spectroscopic analysis, transistors for on-chip signal processing, and surface acoustic wave devices [48] for analog signal filtering is possible. Wireless sensor networks using radio frequency identification (RFID) have also been suggested [40].

In a conventional high electron mobility transistor, the 2DEG forms the channel between drain (D) and source (S) ohmic contacts and is modulated by a top gate. On an uncovered gate, the surface potential, and thus the 2DEG density, is affected by the presence of charged species from the environment. Although the strong response on charges was first termed a parasitic effect that had to be controlled for the optimization of HEMTs, the potential for sensor applications was demonstrated shortly afterward and represents the basic idea of realizing ion-sensitive field-effect transistors (ISFETs). These particularities of AlGaN/GaN FETs in electrolytes are essential for their operation as biosensors.

In contrast to Si-based conventional ISFETs, the AlGaN/GaN heterostructure utilizes the piezoelectric and spontaneous polarization properties of the material system [49,50]. As a consequence, the AlGaN/GaN heterostructure exhibits a captivating simplicity in its structure without the need for additional passivation films or elaborate designs while retaining or exceeding the sensitivity, noise characteristics, and chemical and thermal stability of Si-based ISFETs. Leaving a bare GaN surface by omitting the gate metallization, a modulation of the channel current I_D by ions was first observed by Neuberger et al. [38]. Ions were generated by a plasma spray device and directed to the surface. The change in the surface charge by the incident ions affected the 2DEG density directly. Applying water as well as different organic solvents to the surface of the same structure illustrated that I_D is influenced by the dipole moment of the molecules in the liquid [51].

In AlGaN/GaN electrolyte gate field-effect transistors (EGFETs) as a special configuration of ISFETs, the open gate area is directly exposed to an electrolyte. This EGFET is the basic element of all AlGaN/GaN-based biosensors. Instead of the fixed gate voltage, a reference potential U_{ref} is applied to the electrolyte–oxide–semiconductor system via a reference electrode that is dipped into the electrolyte [52] (Figure 1.2).

A positive drain voltage UDS drives a drain current ID parallel to the AlGaN–GaN interface. Transistor action is possible since an additionally applied reference voltage U_{ref} shifts the Fermi level with respect to the conduction band of the undoped GaN layer. Due to the charges on the open gate surface (donors in the AlGaN layer are not occupied), most of the voltage drop is across this AlGaN layer, thus establishing a quasi-insulating barrier between the open gate and 2DEG. Depending on the reference voltage, the triangular potential well at the interface is raised or lowered in energy and the channel is emptied or filled. This corresponds to an alteration of the drain-source current. For a large enough reference bias, at the threshold voltage U_T, the 2DEG region completely

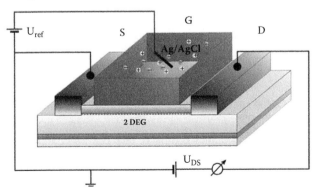

FIGURE 1.2 Schematic view of an AlGaN/GaN EGFET. The metal gate is replaced by an electrolyte bath chamber contacted with an Ag/AgCl reference electrode.

depletes and the current channel is pinched off. The drain current in the nonsaturated region (below pinch-off) is

$$I_D = \beta \cdot \left(U_{ref} - U_T - \frac{1}{2} U_{DS} \right) \cdot U_{DS}, \tag{1.1}$$

where β is a parameter depending on the geometry (the channel width to length ratio W/L), the mobility μ of the electrons in the 2DEG, and the gate insulator capacitance per unit area C_{ox}:

$$U_T = E_{ref} - \Psi_o + X_{sol} - \frac{\phi_{AlGaN}}{e} - \frac{Q_{SS} + Q_{ox} + Q_B}{C_{ox}} + 2\Phi_F. \tag{1.2}$$

In the case of an ISFET, the gate voltage is the voltage at the reference electrode $U_{ref} = U_{GS}$. The threshold voltage also contains terms that reflect the properties of interfaces between the liquid and the gate oxide on one side, and the liquid and the reference electrode on the other side. The work function of the gate metal in a conventional MOSFET is here replaced by the reference electrode potential relative to vacuum E_{ref}. The vacuum level can be calculated with the electrode potential relative to the normal hydrogen electrode (NHE) plus the value of the Fermi level relative to the vacuum states ($U_{NHE} = 4.7$ V [53]). For the widely used Ag/AgCl electrode, a value of $U_{ref} = 4.9$ V is calculated. The interface potential at the gate oxide–electrolyte interface is determined by the surface dipole potential of the solution χ_{sol}, which is a constant, and the surface potential ψ_0, which results from a chemical reaction, usually governed by the dissociation of oxide surface groups. In this case, the threshold voltage is given by

$$U_T = E_{ref} - \Psi_o + X_{sol} - \frac{\phi_{AlGaN}}{e} - \frac{Q_{SS} + Q_{ox} + Q_B}{C_{ox}} + 2\Phi_F, \tag{1.3}$$

where Q_B is the depletion charge in the AlGaN; Φ_F, the Fermi potential; φ_{AlGaN}, the material work function; Q_{SS}, the charges in surface states at the GaN surface; and Q_{ox}, the fixed oxide charge.

All terms are constant except ψ_0, which is responsible for the sensitivity of an ISFET to ions. The influence of the ion concentration on ψ_0 can be explained by the site-binding model, which was introduced by Yates et al. in 1974 [54]. It proposes that atoms in the surface layer of metal as well as semiconductor oxides act as amphoteres when in contact with an electrolyte. That means that, depending on the H$^+$ (OH$^-$) concentration in the electrolyte, they can release protons into the electrolyte (acting as a donor) and thus are negatively charged, form neutral OH sites, or bind protons from the electrolyte (acceptor), resulting in a positive surface charge.

$$\text{donor: M–OH} \leftrightarrow \text{M–O}^- + \text{H}^+ \tag{1.4a}$$

$$\text{acceptor: M–OH} + \text{H}^+ \leftrightarrow \text{M-OH}_2^+ \tag{1.4b}$$

These surface reactions depend on the acidity and the alkalinity constant of the oxide groups, as well as the concentration of H$^+$ in the electrolyte. In the case of a high concentration of H$^+$ (low pH), the M-OH groups tend to accept a proton instead of releasing one; thus, most of them act as acceptors, and the oxide surface becomes positively charged. In contrast, if the concentration of H$^+$ is low (high pH), most of the M-OH groups release a proton, and the surface charge becomes negative. To determine the total change in surface charge, the sum over all surface sites N_s has to be considered, which depends on the material (e.g., $N_s(\text{SiO}_2) \sim 5 \times 10^{14}$ cm^{-2}, $N_s(\text{Al}_2\text{O}_3) \sim 8 \times 10^{14}$ cm^{-2}, $N_s(\text{Ga}_x\text{O}_y) \sim 9 \times 10^{14}$ cm^{-2}) [55]. These changes in surface charge due to the change of pH in the electrolyte directly affect the surface potential ψ_0.

For AlGaN/GaN-based sensors, the first reproducible and quantitative results for pH sensing were reported in 2003 by Steinhoff et al. [41]. They compared different transistor structures with respect to their pH response and found that the thin surface oxide layer forming upon exposure to atmosphere is sufficient for a linear response in the range from pH 2 to 12. X-ray photoelectron spectroscopy (XPS) analysis of as-deposited GaN revealed the almost immediate formation of a thin oxide film at the surface [56]. In contrast to the established Si-based ISFETs, neither thermal oxidization nor a specific ion-sensitive oxide layer (e.g., Ta_2O_5 or Al_2O_3) is needed. Thus, although the site-binding model [54] was developed for Si-ISFETs [53,57,58], it was also proposed to explain the pH response of AlGaN/GaN-ISFETs. According to this model, amphoteric hydroxyl groups are bonded to Ga surface atoms (Ga-OH) when in contact with aqueous solutions, and can be protonated (Ga-OH_2^+), neutral (Ga-OH), or deprotonated (Ga-O^-) depending on the pH of the solution (Figure 1.3) as described earlier. This modifies the surface charge σ_s, which in turn influences the sheet charge density n_s in the 2DEG, that is, enhancing and depleting the 2DEG at low and high pH, respectively. The resulting response of the drain current for a sensor investigated in this work is shown in Figure 1.4.

For highly concentrated acidic or alkaline solutions, deviations from the simple site-binding model with stable amphoteric are observed. In addition, ISFET-based pH monitoring often has shown hysteresis and drift effects [60–62]. These can be attributed to instabilities of the used passivation of the contacts and of the group III nitride surface at high pH values. Thus, despite the well-known properties of a basic AlGaN/GaN heterostructure, technological issues and passivation strategies are very important to developing sensors for "real" applications. Thus, before discussing the sensing properties of AlGaN/GaN biosensors, the technological steps of ISFET fabrication as well as the results of a basic electrical characterization are described shortly in the following section. Comprehensive reviews covering further details of growth, properties, and applications of group III nitrides can be found elsewhere [43,63,64].

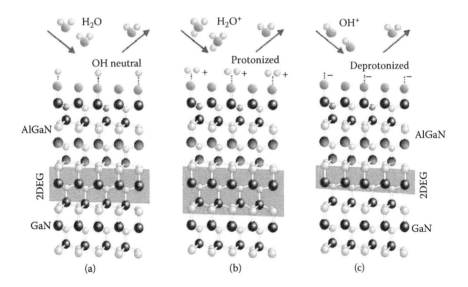

FIGURE 1.3 (See color insert). Hydroxyl groups on the surface of AlGaN/GaN heterostructures in (a) water, (b) acidic, and (c) alkaline solutions and the resulting impact on the 2DEG carrier density. (From Ambacher, O., and Cimalla, V., in *Polarization Effects in Semiconductors: From Ab Initio to Device Application* (eds. C. Wood and D. Jena), Springer, New York, 2008, p. 27.)

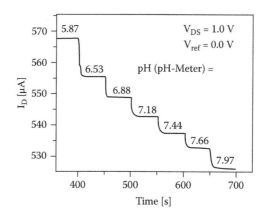

FIGURE 1.4   Drain current I_D versus pH value during a titration of KOH into HCl solution. (From Spitznas, A., 2005, Diploma thesis, Technical University Ilmenau.)

1.3   FABRICATION AND CHARACTERIZATION OF AlGaN/GaN SENSORS

1.3.1   GROWTH OF THE HETEROSTRUCTURE

The first AlGaN/GaN heterostructures for sensor applications were grown by plasma-induced molecular beam epitaxy (PIMBE) [38,41,51,65,66]. More recent studies [19,64–69], however, illustrate that most groups currently use the MOCVD technique. For this work, AlGaN/GaN heterostructures were grown on sapphire substrates using PIMBE [70] and MOCVD [71]. Sapphire substrates were preferred because of their high crystal quality and availability at relatively low cost compared with SiC and GaN substrates. Hence, for biosensors, optical transparency and chemical stability are important criteria for substrate selection. Moreover, sapphire is the most extensively used substrate for growth of group III nitrides, and the technology of growth is quite mature [63,72]. A thin AlN nucleation layer (10–150 nm) is usually grown to lower the dislocation density and thus obtain better crystal quality. This layer also defines the polarity of the following GaN layers to be metal-face. The GaN buffer (1–3 μm) is covered with an $Al_xGa_{1-x}N$ barrier with an Al content of 20%–30% and a thickness between 10 and 30 nm. To enhance its chemical stability, the device is usually topped with a 2–5 nm GaN layer.

Clearly, PIMBE and MOCVD growth are quite different methods, and therefore the layer parameters are different despite the nominal identical heterostructure design. Due to the lower growth temperature, PIMBE layers exhibit a higher number of material defects [73]. MOCVD-grown material has a higher structural quality; however, more impurities can be expected due to the high temperature growth at higher pressure (50–200 mbar compared to 10^{-8} mbar in MBE). Moreover, MOCVD-grown layers are more homogeneous, and due to the higher growth rate thicker buffer layers can be achieved. In order to characterize the influence of the different properties, series of AlGaN/GaN heterostructures were grown by both methods with varying thickness and Al content in the barrier (see, for example, Table 1.1 and Figure 1.7 for a series of samples for the investigation of pH-sensing properties).

1.3.2   SENSOR PROCESSING

Prior to further processing, the wafers are cleaned in acetone and isopropanol, and dried in N_2 flux. M mesa etching is performed by inductively coupled plasma (ICP) etching in Cl_2–Ar gas mixture [38] to laterally confine the active area of the ISFET sensor (Figure 1.5).

TABLE 1.1
AlGaN/GaN heterostructure parameters

	Sample	AlN nucleation	GaN buffer	AlGaN barrier[a]	GaN cap layer
PIMBE	I	180 nm	270 nm	$d_1 = 10$ (10.3)	2 nm
	II			$d_2 = 13$ (12.1)	
	III			$d_3 = 15$ (13.7)	
MOCVD	IV; V	20 nm	1200 nm	$d_1 = 10$ (8.1; 8.5)[b]	
	VI			$d_2 = 15$ (14.3)	2 nm
	VII			$d_3 = 20$ (18.4)	

[a] The barrier thickness values are calculated from growth rate and in the brackets from C–V measurements. The GaN cap layer is included in the measured value.
[b] Modified MOCVD heterostructure with Si-doped GaN interlayer.

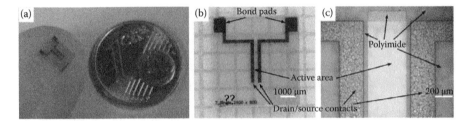

FIGURE 1.5 (a) Sensor chip after mesa etching and metallization, (b) single chip demonstrating the transparency in the visible range, (c) magnification of the active area with drain/source contacts and the polyimide passivation.

Ohmic contacts to the 2DEG channel are realized by employing a conventional Ti/Al/Ti/Au metallization scheme [74]. Dependent on the two growth methods PIMBE and MOCVD, it is necessary to have two different annealing temperatures. For the PIMBE-grown heterostructures, an annealing at 750°C for 60 s was found to be optimal, while MOCVD-grown heterostructures required a higher annealing temperature of 850°C for 50 s. The final sensor structures are shown in Figure 1.5.

The passivation of the sensor structure is a very important factor in achieving stability and reproducibility of the device in liquids such as electrolytes. It eliminates the possible electrochemical reaction between metal contacts and electrolyte. Therefore, the passivation must be mechanically and chemically stable in both acidic and alkaline solutions. In addition, for biosensors, layer biocompatibility is an important requirement.

For the ISFET biosensor, the polymer polyimide PI-2610 from DuPont proved to be the most promising passivation material. The polyimide material has been used already for different sensor applications [75]. It was shown to be stable in contact with biological cells and organic solutions [46] and facilitates cellular attachment as will be shown later.

The PI-2610 exhibits a desirable combination of beneficial film properties such as low stress, low coefficient of thermal expansion, low moisture uptake, high modulus of elasticity, and good ductility. The transparency in the visible light region is also a big advantage of this polyimide passivation material, which allows the examination of cells using inverse optical microscopy. The moisture uptake of 0.5% for the PI-2610 series is relatively low for high-temperature polymers [76]. The low moisture uptake is an important property for sensors working in electrolytes, to avoid memory effects by modification of the passivation depending on the environment. After a curing procedure at 350°C in nitrogen, the polyimide thickness is found to be about 1.2 μm, and the compact layer

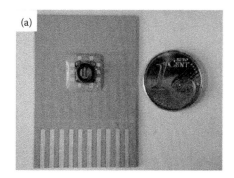 

FIGURE 1.6 (See color insert).   Encapsulation of the sensor chip: (a) on an LTCC frame, (b) on a conventional PBC frame.

covers all the sensor contacts and the areas between the sensors, leaving only the active transistor area and the bonding pads free.

Finally, for measurements in liquids, the sensors are encapsulated using two kinds of holders (Figure 1.6). The first variant uses a printed board circuit (PBC) holder with Cu metallization, where the chip is glued on using silicon rubber. The contacts are made with Ag paste or by a bonding technique using Au wires. This encapsulation can be realized very quickly, it is flexible (the sensor chip can be taken out and mounted again without being damaged), but has the disadvantage of instable contacts (using clamps), which can generate undesired noise effects and which have to be replaced from time to time. An improved encapsulation variant for the use of the sensor for longer time periods without renewed encapsulations employs a low-temperature cofired ceramics (LTCC) frame. This multilayered technique is based on glass or glass–ceramic composites with a sintering temperature of less then 920°C. Unfired tapes of this material have a typical thickness of 50 to 300 μm and are easy to structure by punching or laser cutting. The final result of the stacking, lamination, and sintering processes is a rigid substrate with high reliability and chemical resistivity [77].

1.3.3   Electrical Characterization of the AlGaN/GaN Heterostructure

After growth, the AlGaN/GaN heterostructures were appraised to evaluate their electronic characteristics, crystal quality, and surface properties. A summary of the different heterostructures (Figure 1.7, Table 1.1) is given in Table 1.2. In the MOCVD-grown heterostructures with the smallest barrier thickness (sample IV, V), an additional Si-doped GaN layer is included close to the AlGaN–GaN interface (Figure 1.7c). This doped interlayer enhances the sheet carrier concentration in the confined 2DEG and minimizes persistent photocurrent effects, which are often present in group III nitrides [71,78].

The electrical characterization of the selected heterostructures revealed that MOCVD-grown samples exhibit superior electronic transport properties. The lower electrical performance in PIMBE-grown samples is mainly caused by the higher dislocation density. This effect has important consequences for the application as chemical or biosensor, since such defects are preferentially etched and can be the origin of early breakdown for such sensing devices. Moreover, it has been shown that high dislocation densities in PIMBE-grown samples also limit the possibilities of cleaning in alkaline solutions, which is a required procedure for applications in the food industry [63]. On the other hand, PIMBE-grown heterostructures have been shown to exhibit better light stability [78,79]. In the case of MOCVD growth, the Si-doped interlayer was necessary to achieve a similar or even improved behavior.

TABLE 1.2
Results of the C–V and Hall measurements for the investigated heterostructures

Barrier [nm]	n_s [cm^{-2}]—C–V	n_s [cm^{-2}]—Hall	μ [cm^2/Vs]	R_{est} [Ω/]
		PIMBE		
10	4.78×10^{12}	5.06×10^{12}	563	2192
13	6.26×10^{12}	6.41×10^{12}	798	1221
15	6.88×10^{12}	8.38×10^{12}	700	1065
		MOCVD		
10	1.65×10^{13}	1.19×10^{13}	920	571.6
10	1.43×10^{13}	1.17×10^{13}	918	579.8
15	1.29×10^{13}	1.12×10^{13}	901	577.4
20	1.49×10^{13}	1.21×10^{13}	969	570.7

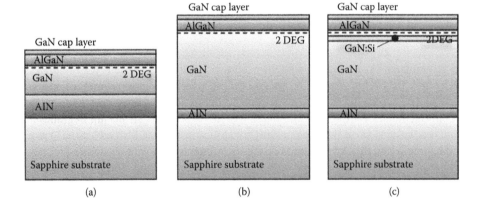

FIGURE 1.7 (a) PIMBE and (b) undoped MOCVD heterostructures for ISFET biosensors, and (c) modified MOCVD heterostructure with doped GaN interlayer.

1.3.4 TECHNOLOGY IMPACT ON THE SURFACE AND SENSOR PROPERTIES

The impact of typical device processing steps (KOH, HCl, HF wet chemical etching, SF$_6$ and Cl plasma etching) on the surface properties of group III nitride-based chemical sensors was investigated with emphasis on wetting behavior, chemical composition, biocompatibility, and electrical performance of the sensor [46]. KOH-free developer was used since KOH is known to attack group III nitrides [44,80,81]. The influence of KOH was investigated by etching the sample in pure solutions with a concentration of 15% for 30 s at room temperature and 70°C. Exposure to HF and HCl solution is standard cleaning procedure that produces the lowest coverage of oxygen and carbon on GaN sensor surface. The influence of 10% HF or 15% HCl baths for 5 min at room temperature was studied. ICP mesa etching in Cl$_2$–Ar gas mixture [82,83] was used to create 3D structures and to laterally confine the active area of the sensor.

For the biocompatibility tests, first the surface of the nitride heterostructure must be sterilized. Two different sterilization methods are explored: one is the autoclaving procedure in water vapor for 20 min at 121°C and the other is a room temperature method of sterilization by washing in ethanol for 20 min and drying in nitrogen. This second method is used later also for cell–transistor

TABLE 1.3
Summary of the impact of different surface treatments on the surface roughness (rams), the contamination with carbon, oxygen, and other elements (ions are Zn, Ca, K), the contact angle Φ, and the sheet carrier concentration in the 2DEG

	rms	C	O	Cont	φ	2DEG
F, dry etch	↘(+)	○	↗↗	F	↘↘	↗↗(++)
CI, dry etch	↗(−)	○	↗	(CI)	↘	
HF, wet etch		○	↘	(F)	↘↘	
HCI, wet etch		○	↘	(CI)	↘	
KOH, wet etch	↗↗(−−)	○	○	K	↘↘	↘↘(−−)
autoclave		↗↗	↗		↗↗	
CHO-K1		↗	↗	ions	↗	
HEK293 FT		↗	↗	ions	↗	
DMEM		↗	↗	ions	○	
FCS		↗	↗	ions	○	↘(−)

Note: Arrows up and down mean increasing and decreasing value, respectively; circle: no change; in the case of rms and 2DEG, + and − denote improving or degrading effect, respectively [46].

coupling experiments. As biological model systems, the widely distributed mammalian cell lines HEK 293FT and CHO-K1 are used. The results of these investigations are summarized in Table 1.3. Almost all of the investigated treatments can alter slightly the amount of surface contamination; however, only etching in a strong alkaline solution or in a dense chlorine plasma degrades the sensor behavior. These results prove the expected high stability of the AlGaN/GaN ISFET for chemical sensing in aggressive media. More details on the experiments are given elsewhere [46].

1.3.5   BIOCOMPATIBILITY OF THE SENSOR

A critical issue for the realization of cell-on-chip concepts is the survival of living cells on the sensor surface. As shown before, group III nitride-based ISFETs are chemically stable under physiological conditions, exhibiting promising properties for sensor applications in liquid and electrolyte environments. By way of comparison, silicon is known to be attacked by many biological important agents; however, an appropriate cell growth can be achieved on an oxidized surface. Other semiconductors require great efforts for passivation (e.g., GaAs [84]). In contrast, AlGaN-based alloys are chemically inert. First studies with rat fibroblasts (3T3 cells) have demonstrated good adhesion properties of the group III nitrides independent of the aluminum concentration in the $Al_xGa_{1-x}N$ alloy (x = 0, 0.22, 1), and slightly improved behavior after a pretreatment by oxidation [47]. Recent investigations using lactate dehydrogenase (LDH) on cerebellar granule neurons prepared from 7-day-old Wistar rats [85] clearly demonstrated that adhesion and growth of living cells on GaN is superior to silicon.

The biological model systems for the present work are the widely distributed mammalian cell lines HEK 293FT, which adhere only poorly and can be removed simply by cell medium flow, and CHO-K1, which are adherent fibroblastoid cells that can only be removed by gentle enzymatic treatment with trypsin. As a third cell line, neuroblastoma-glioma hybrid NG 108-15 nerve cells were used, which are the subject of the transistor–cell coupling experiments.

Cells were cultivated in Dulbecco's modified Eagle's medium-high glucose (DMEM) supplemented with 10% fetal calf serum (FCS) and 1% penicillin/streptomycin according to standards set by the American tissue-type culture collection (ATCC). The sterilized sample with 1 cm² sensor material was placed in wells of 6-well-tissue-culture plates. One control well without sensor material was used as a reference for growth normalization and cell morphology studies. Thereafter, 4 mL of prepared medium was added on every well and approximately 1×10^5 cells were seeded uniformly. After incubation for two days at 37°C in humidified 5% CO_2 atmosphere, the cells were analyzed by optical microscopy, removed from the well and sample surface, and counted with a Neubauer counter. In addition to the AlGaN/GaN heterostructures, n- and p-type silicon substrates were used for comparison. The surface was not modified using fibronectin or other organic materials, which are commonly used in order to improve cell adhesion and biocompatibility of different substrate materials.

The biocompatibility tests revealed that all cell lines were growing well on the AlGaN/GaN material (Figure 1.8a). Independent of processing steps, the growth on the silicon surface was substantially lower. The surface contamination by carbon, hydrocarbons, and metal ions [46] seemed to decrease the cell proliferation slightly. There is evidence that HEK 293FT cells are more sensitive since they show a reduced growth. The cell proliferation is also good on polyimide, which is used for contact passivation in the final sensor device, which is exposed to the media. Most important for the present work is that the NG 108-15 nerve cell lines also proliferates very well on the sensor surface, and the results are similar to the other two investigated cell lines. To further enhance the proliferation of cells on AlGaN, a surface modification with fibronectin can be applied as reported by Yu et al. [86]; however, in contrast to other semiconductors, it is not an essential precondition.

Similar to the discussed impact of different treatments on the electrical and surface properties of the sensors, the influence of these steps on the biocompatibility properties was investigated (Figure 1.8b) [46]. Obviously, the treatment of the sensor surface has no substantial influence on the proliferation of the cell lines. Thus, in contrast to silicon-based ISFETs, no complex modifications are needed to promote the adhesion and proliferation of cells on AlGaN/GaN-ISFETs, though fibronectin modification might be beneficial.

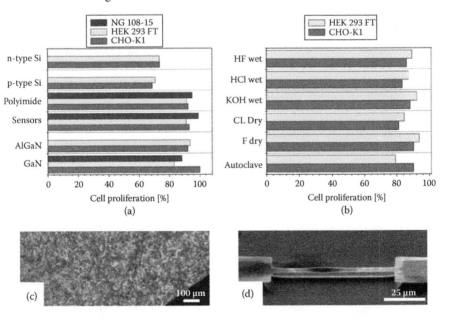

FIGURE 1.8 Cell population for three cell lines (HEK 293 FT, NG 108-15, and CHO-K1) after two days cultivation (a) on different substrates and (b) differently treated AlGaN/GaN heterostructures; (c) HEK 293FT cells on the active area of a planar AlGaN/GaN sensor; and (d) CHO-K1 cells on a 3D MEMS sensor [46].

1.4 MEASUREMENT CONDITIONS

One of the major tasks for biosensors is to monitor the ion fluxes and concentrations [87]. In electrolytes, most of the processes can be translated into changes of the H^+ concentration and, thus, to the pH value. In consequence, the pH sensors are a reliable base for biosensors in order to detect different reactions that take place in vitro or in vivo.

1.4.1 BUFFER SOLUTIONS

Solutions able to retain a constant pH regardless of small amounts of acids or bases added are called *buffers*. Their resistance to changes in pH makes buffer solutions very useful for chemical manufacturing and essential for many biochemical processes. Buffer solutions were used in this work in order to investigate the pH-sensing behavior in constant environments as well as to stabilize the pH in cell media during cell–transistor coupling experiments. The employed buffer solutions are summarized in Table 1.4.

1.4.2 CHARACTERIZATION SETUP

For basic characterization of the sensors in buffer solutions as well as for measurements with cells, two different measurement setups were used (Figure 1.9). In both cases, the electrolyte bath chamber was realized from PEEK (polyetherketone) by conventional precision engineering. The advantages of this material include high chemical stability, the very low conductivity, and simple process technology.

TABLE 1.4
The buffer solutions used in this work

Name	pK_a at 25°C	Buffer range	Molecular weight	Full name and molecular formula
HEPES	7.48	6.8–8.2	238.3	n-2-hydroxyethylpiperazine-N'-2-ethanesulfonic acid; $C_8H_{18}N_2O_4S$
TRIS	8.06	7.5–9.0	121.14	2-Amino-2-hydroxymethyl-propane-1,3-diol; $C_4H_{11}NO_3$
CS11	12	5–8		Citrate 25 mM + bis-tris-propane 75 mM; $NaH_2PO_4 + Na_2HPO_4$
Mettler-Toledo		4–10		

Source: Olmsted, J. III, and Williams, G. M., 1997, Wm. C. Brown Publishers, USA.

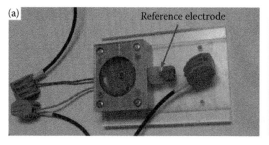

FIGURE 1.9   Measurement setup for AlGaN/GaN sensor characterization and cells research (a) on PBC, (b) on LTCC.

Due to its transparency, low cost, and good mechanical properties, Plexiglas was used as holder for the sensor chip, which is encapsulated on PBC (Figure 1.9a; for the sensor, see Figure 1.6b). Because of these properties, the same material was also used as the lid for the constructed electrolyte chamber in order to prevent evaporation processes and alkalization of the cell medium as a result of CO_2 consumption at room temperature. The disadvantage of Plexiglas material is its low conductivity, which does not permit ground contact or screening of the sensor chip from electromagnetic noise. Moreover, the possibility of electrical charging can influence the sensor signal by touching the lid during measurements.

A second measurement setup for the LTCC ceramic-encapsulated sensor chip (Figure 1.9b; for the sensor, see Figure 1.6a) is more electrically stable than the first one because it employs an aluminum holder for electrical screening and mechanically stable contacts.

Three types of reference electrodes can be used with these two systems:

- A double-cell glass reference electrode, InLab from Mettler-Toledo, which has an ARGENTHAL™ silver ion trap that ensures the electrolyte remains completely free from silver ions
- A homemade reference electrode with a 2 mm Ag/AgCl pellet incorporated in a gel from Fermentas [90] with 120 mM KCl solution
- A leakproof Amani reference electrode from Warner Instruments with a unique, chemically resistant, highly conductive (10 kΩ), and nonporous junction

Before any quantitative measurements are made, the sensors must be calibrated using the standard buffer solutions by recording the transistor characteristics, establishing its working point, and determining the sensor sensitivity at different pH values. A good pH sensor must show sensitivity to H^+ ions in accordance with the Nernst law. Based on this, properties of the sensor such as signal-to-noise ratio and drift behavior of AlGaN/GaN pH sensors are evaluated. Sensitivity to pH is the main parameter characterizing the sensors, and is defined as

$$S_{pH} = \frac{\Delta U_{ref}}{\Delta pH}, \qquad (1.5)$$

where ΔU_{ref} is the difference in reference potential for two buffer solutions with different pH's. For the used sensors, a sensitivity of up to $S_{pH} = 58.5$ mV/pH is achieved, which is in good agreement with the 59 mV/pH from the theoretical approximation using the Nernst equation.

1.4.3   NOISE AND DRIFT

In the measurement setup described earlier, several sources of noise appear. External noise sources (thermal, electronic, light) are discussed and quantified elsewhere [60]. Noise originating from the heterostructure and its interaction with the electrolyte has a level of about 100 nA. In the case of cell measurements, additional noise caused by the biochemical processes appears, which can decrease or increase the overall noise level: decreasing due to the stabilization of the surface chemistry, and increasing if the additional source is stronger. The stabilization effect occurs because the cells need a couple of days to proliferate on the active area. At the end, the noise level is decreased to 20–50 nA. In any case, these values of noise are acceptable if we take into consideration the fact that the sensor response to extracellular potential is between 1 and 15 µA. For the transistor–cell coupling experiments in this work, a glove box was used (Figure 1.10), which also provides special conditions for the measurements such as constant temperature, humidity, pressure, and atmosphere composition. In this glove box, the light sensitivity of the sensors can also be controlled [78].

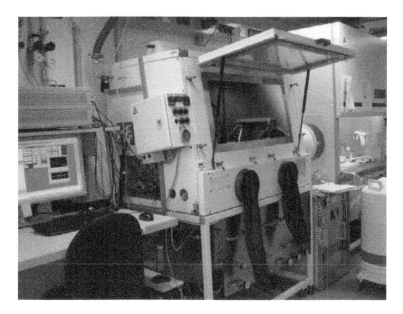

FIGURE 1.10 The measurement setup used to record the sensor signal: (right) glove box with sample loading from a laminar box, (left) the PC for monitoring the signals.

In addition to noise, a signal drift appears during the measurements, which is a common problem for all chemical sensors, and which complicates quantitative measurements. However, for the study of cell reaction to different inhibitors, the signal changes are expected to be small and the long preparation periods necessary for the proliferation of cells ensures a stabilization of the sensor signal as far as is possible. Moreover, most of the drift can be neglected due to the fast response of cells. Consequently, even if quantitative pH sensing cannot be ensured, the cell–transistor coupling can be studied.

1.4.4 CELL-BASED SENSOR—CPFET

With the mechanisms described in the preceding sections and the excellent stability, group III nitrides have good potential for biosensing applications. To achieve selectivity of the HEMT sensor device, further functionalization is usually necessary. Depending on the proposed application, several basic principles have been demonstrated to achieve selective gas, pH, ion, or biosensors. An overview of these possibilities is given in Reference 91. In contrast, for the cell–transistor coupling in a CP-FET employing an AlGaN/GaN heterostructure, no surface functionalization is necessary because of (1) the excellent chemical stability of the sensor surface (2) and the good biocompatibility, as will be shown in the following section.

The CPFET couples whole cells as a biorecognition element with a transducer device. It offers the possibility of studying the effect of drugs or environmental influences on the cell metabolism directly by measuring, for example, extracellular acidification or intra- and extracellular potentials. Although these cell–sensor hybrids suffer from some fundamental limitations, such as a short lifetime in the range of several days or a difficult and time-consuming preparation of the device, the fact that the direct response of a living system can be recorded in situ allows unique information to be obtained. Such cell biosensors are suitable for a variety of applications, such as drug screening in pharmacology, detection of toxins, and environmental monitoring.

Similar to the reports on silicon-based ISFETs [92], the acidification as a result of cell metabolism can be monitored using AlGaN/GaN ISFET [93]. An example is shown for P19 cells (embryonic mouse-teratocarcinoma stem cell) adhering to an unmodified open gate of the transistor. After one

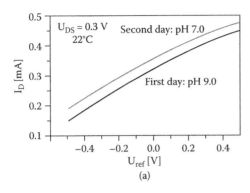

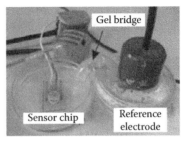

(a)

(b)

FIGURE 1.11 (a) The shift of transfer characteristics of AlGaN/GaN FET as a result of P19 cell acidification, (b) the measurement setup.

day under incubator conditions, the cell medium became acidic due to cellular activity. A corresponding change from pH 9 to pH 7 was measured by analyzing the shift of transistor transfer characteristic (Figure 1.11a).

A further development of the concept was demonstrated with cardiac myocyte cells of embryonic Wistar rats on ISFET arrays [19]. After 5 to 6 days in culture, a confluent monolayer of cells developed on the fibronectin-modified GaN surface with spontaneously contracting aggregates. The extracellular potential was recorded by measuring the drain-source current in constant-voltage mode, and the corresponding gate voltage Uref was calculated using the transconductance gm. Transistor signals were 100–150 ms in duration with an amplitude of 70 µV, firing at a stable frequency for several minutes. The signal shape was assumed to be determined by the K+ exchange; however, the exact reason for this signal shape remains to be clarified. Furthermore, the authors evaluated the dependence of the gate-source voltage noise on the frequency, compared it with silicon-based devices, and concluded that noise in Si devices under the same conditions is one order of magnitude higher.

Recently [86], a combined measurement setup using patch clamp and AlGaN/GaN ISFETs to detect the reaction of saos-2 human osteoblast-like cells to different concentrations of known ion channel inhibitors was shown. Quaternary ammonium ion (TEA) and tetrodotoxin (TTX), which block K+ and Na+ channels, respectively, are used to influence the ion currents through the cell membrane. The intracellular potential of the adherent cells on the fibronectin-modified open gate of the transistor was controlled using a patch pipette. For the measurements, a rectangular 90 mV pulse was applied to the cell, and the extracellular voltage as a function of time was recorded using the AlGaN/GaN ISFET. With increasing inhibitor concentration, the amplitude of extracellular voltage decreases and a complete blocking of the membrane channels was achieved with the addition of 20 mM TEA and 50 nM TTX.

From the experimental details, it can be concluded that AlGaN/GaN ISFETs are well suited for the construction of CPFETs owing to the inherent properties of group III nitrides, such as chemical stability and low noise. The biosensing capabilities of the AlGaN/GaN CPFET will be demonstrated in the next section on NG 108-15 (mouse neuroblastoma × rat glioma hybrid) nerve cells by adding different inhibitors.

1.4.5 NG 108-15 Nerve Cell Line

The used NG 108-15 is a fibroblast hybrid cell line derived by fusion of mouse neuroblastoma clone N18TG-2 with rat glioma clone C6BU-1. This cell line has become a widely used in vitro model system for studying neuronal functions [94–106]. When cultured in serum-containing medium, the cells proliferate well and exhibit significant motility. These cells can easily be differentiated [95–98], and it was observed that the cell properties after this process are greatly enhanced with respect to the following properties: the differentiated cells show increased activity of the membrane protein, which in the case of NG 108-15 expresses at least four major families of voltage-sensitive channels (including Na^+, Ca^{2+}, and K^+) [99–110]. These voltage-sensitive channels respond to a variety of ion channel blockers including tetrodotoxin [103–110], bradykinin [111], and to the drugs, which can produce cell apoptosis (cell death), such as staurosporine [112] and buprenorphine hydrochloride [113].

The cells are electrically active and can be induced to produce action potentials via injection of current [114]. However, spontaneous action potential (as is the case of cardiac cells) does not occur. If the cells are cultured in serum-free media, the electrophysiological and morphological properties are changed. Proliferation ceases, and the cells begin to produce neurites and other extensions, which are capable of forming synapses to other tissues and cells [114,115]. These cells exhibit features characteristic of neurons and are more likely to produce action potentials than their counterparts cultured in serum-containing media. This electrical activity can even be spontaneous under some circumstances. However, as with most neurons, the strength of this spontaneously generated extracellular signal is low, since only a small percentage of cells is active. For these reasons, NG 108-15 cells were not used in action potential studies. Instead, they were used for studies of cellular impedance, examining both motility issues as well as channel conductance changes [94].

Recently, cocultures of NG 108-15 cells and chick myotubes were reported, which grew together and formed functional neuromuscular synapses. Then, the nerve cell induced the upregulation of muscle AChE expression, which was persistent when the muscular activity was blocked by α-bungarotoxin [116]. Taking such studies into account, the counterpart of the coculture to NG 108-15 should be replaceable by a sensor in order to monitor cell activity. The good biocompatibility of the AlGaN/GaN sensor material favors its choice for a coupling with this nerve cell line for the recording of extracellular potential as response to inhibitors. For proper proliferation of the nerve cells on the sensor surface, it is necessary to create optimum conditions of cell cultivation, that is, good cell media and stable environmental conditions.

1.4.6 NEUROTRANSMITTERS–ACETYLCHOLINE

Nerve cells come in contact with each other and also with muscles and glands at junctions called *synapses*. Most synapses are chemical. Chemical synapses depend on specific chemicals called *neurotransmitters* to conduct the signal across the junction. The neurotransmitter binds to a receptor on the membrane on the second cell, and this produces a change in ionic conductance of the receptor or the signal is transmitted to the cytoplasm of the second cell, which is called the *second messenger system*. Between the membranes of the two cells is a gap of about 20–50 nm, the synaptic cleft. Because of this distance, no electrical signal transport is possible (in contrast to the electric synapse). In this case, the nerve signal is transmitted to the postsynaptic membrane by means of a neurotransmitter. One of the most common neurotransmitters for synapses between neurons outside the central nervous system as well as for the neuromuscular junction is acetylcholine (ACh).

The ACh molecules are confined in numerous synaptic vesicles situated near the terminus of the presynaptic axon (Figure 1.12). The presence of the nerve pulse causes large transient increases in the permeability of the presynaptic membrane to Ca^{2+} ions; thus, Ca^{2+} ions flow down their electrochemical gradient into the axoplasm. In the cell, Ca^{2+} ions cause the synaptic vesicle to move to the presynaptic membrane and fuse with it. In this mode, the ACh is released in less than a millisecond into the synaptic cleft. The ACh molecules diffuse across to the postsynaptic membrane of another cell, where they bind to specific receptor proteins.

When ACh binds to its receptor in the neuromuscular junction, a channel in the receptor opens for about a millisecond and approximately 3×10^4 Na^+ ions pass through into the cell and, simultaneously, the K^+ ions move outward. An enzyme called acetylcholinesterase (AChE) immediately begins to degrade the ACh into acetate and choline so that the resting potential of the postsynaptic

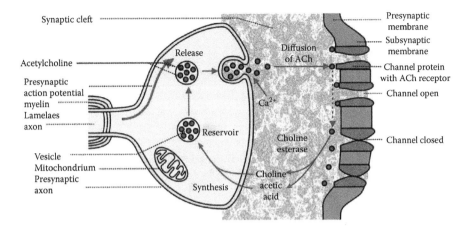

FIGURE 1.12 Transmission of a signal across a cholinergic chemical synapse. (After Vogel, G. and Angermann, H., 2002, Chemical Signals in Cells, Munich: Dt. Taschenbuch-Verl.)

membrane is rapidly restored [118–121]. The prompt degradation of ACh is absolutely essential. Without the effect of AChE, the ACh would continue its stimulatory effect until the ACh diffuses away, and control of nervous activity would quickly be lost. For example, the control of the ACh/AChE is an important task in diseases such as Alzheimer's and Parkinson's [121–123]. There are several organic compounds that act as inhibitors of AChE and are therefore potential neurotoxins. Such compounds are synthesized for use in pharmacological screening [123], as insecticides [124] in agriculture, and as nerve gases for chemical warfare. The effect of such types of neurotoxins on nerve cells will be studied in this work using AlGaN/GaN FET.

1.4.7 NEUROINHIBITORS

Once ACh has been released into the synaptic cleft and depolarization of the postsynaptic membrane has occurred, the excess ACh must be rapidly hydrolyzed. Without this process, the membrane cannot be restored to its polarized state, and further transmission will not be possible. The enzyme responsible for hydrolysis is AChE, and every substance that inhibits the AChE activity is potentially toxic.

An irreversible, potent AChE inhibitor is diisopropylfluorophosphate (DFP). It has a lethal dose (LD50) for rats of 4 mg/kg. DFP, a structural analogue of Sarin, is an oily, colorless or faint yellow liquid with the chemical formula $C_6H_{14}FO_3P$. It is used in medicine and as an organophosphate insecticide. It is stable, but undergoes hydrolysis when subjected to moisture, producing hydrofluoric acid [125].

Phenylmethanesulfonyl fluoride (PMSF), with the molecular formula $C_7H_7FO_2S$ (Figure 1.13), is a serine protease inhibitor, which has a weaker action than DFP. PMSF is rapidly degraded in water, and stock solutions are usually made up in anhydrous ethanol, isopropanol, corn oil, or dimethyl sulfoxide (DMSO). Proteolytic inhibition occurs when a concentration between 0.1 and 1 mM PMSF is used. The half-life time is small in aqueous solutions. The LD50 is about 500 mg/kg. Both DFP and PMSF are used successfully to inhibit the AChE in this work (Section 1.7).

A synthetic and reversible channel blocker, which is also used in this work, is amiloride (Figure 1.13), $C_6H_8ClN_7O$. It works by directly blocking the epithelial sodium channel (ENaC), thereby inhibiting Na+ reabsorption in the distal convoluted tubules and collecting ducts in the kidneys (this mechanism is the same for triamterene). This promotes the loss of Na^+ and water from the body, without depleting K^+. The compound is soluble in hot water (50 mg/mL), yielding a clear, yellow-green solution. Amiloride is freely soluble in DMSO. The amiloride was used in this work together with PMSF and DFP inhibitors to study the combined reaction of nerve cells and to identify the ions that contribute to the sensor signal.

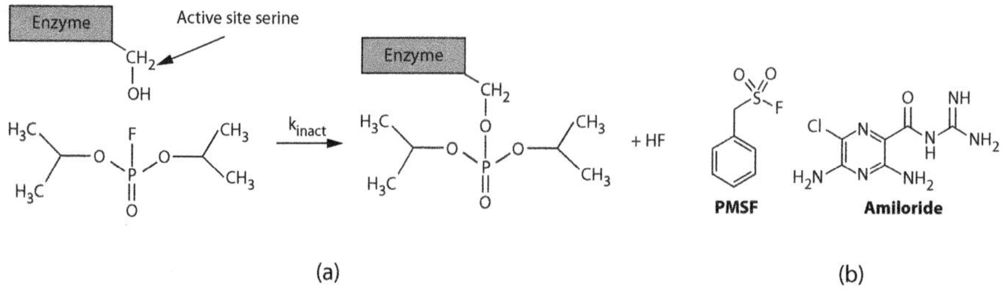

FIGURE 1.13   (a) Reaction of the irreversible inhibitor diisopropylfluorophosphate (DFP) with a serine protease, (b) chemical formulas of the phenylmethanesulfonyl fluoride (PMSF) and amiloride used in the present work as neuroinhibitors.

Consequently, all processes and cell reactions result in a change of specific ion fluxes across the cell membrane. The goal of this work is to establish a measurement technique for the changes in ion concentrations outside the cell as a reaction to neuroinhibitors by a planar sensor, on which the nerve cells are cultivated.

1.4.8 Cell Media

For in vitro cell growth, it is important to use a special liquid, that is, a cell media that contains well-defined substances that make cell growth and reproduction possible. For different cell types, it is necessary to use cell media with quite different properties in order to prevent any change of the cell function and the differentiation from the cell stem.

Cell media usually contain a high amount of different substances, which can influence the tests negatively. Nevertheless, for most of the cell media, it is necessary to add a buffer solution for pH adjustment and stabilization. From the point of view of pH stability, a highly buffered cell media would be beneficial, while for cell proliferation a concentration below 25 mM HEPES is preferable. As a compromise, an HEPES concentration as low as 25 mM was chosen for the further coupling experiments of NG 108-15 nerve cells with the AlGaN/GaN sensor. To improve the stability of the cell media pH, and to avoid an increase of Na$^+$ ions concentration in cell media, finally an HEPES TRIS 25 mM buffer solution was used.

For the cell culture technique, a stable pH value is a very important precondition [89,126]. A pH value of cell media below 7 or exceeding 8 can negatively influence cell proliferation as well as their reaction to different stimuli. A pH value of about 7.5 for cell media represents an optimum environment for cell growth. To control the pH value of the cell milieu, additional chemical substances are used that balance it around the desired value (similar to buffer solutions; see Section 1.5.3). Sodium hydrogen carbonate is such a substance; it can be added in cell media and acts not only as buffer but also as nutrition for plant cell culture. An increase in the CO_2 content in the milieu causes a decreased pH value. Because the cells need special conditions to proliferate (5% CO_2 and 37°C), the acid–alkaline equilibrium is maintained using sodium hydrogen carbonate in relatively large concentrations. The buffer system in the cell media thus has two components: $NaHCO_3$ and CO_2.

$$NaHCO_3 + H_2O \Leftrightarrow Na^+ + HCO_3^- + H_2O$$

$$CO_2 + 2\,H_2O \Leftrightarrow H_3O^+ + HCO_3^-$$

This reaction depends on the partial pressures in the environment. For this reason, the cells are kept in special incubators with well-defined atmospheric pressure, temperature, and CO_2 concentration. If the CO_2 concentration is at a low level, the OH$^-$ concentration in media is high and the pH is

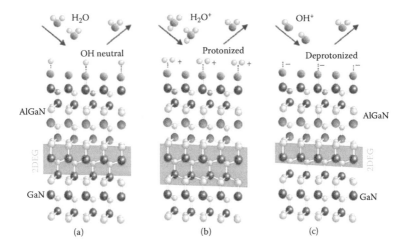

FIGURE 1.3 Hydroxyl groups on the surface of AlGaN/GaN heterostructures in (a) water, (b) acidic, and (c) alkaline solutions and the resulting impact on the 2DEG carrier density. (From Ambacher, O., and Cimalla, V., in *Polarization Effects in Semiconductors: From Ab Initio to Device Application* (eds. C. Wood and D. Jena), Springer, New York, 2008, p. 27.)

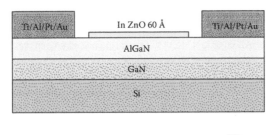

FIGURE 1.6 Encapsulation of the sensor chip: (a) on an LTCC frame, (b) on a conventional PBC frame.

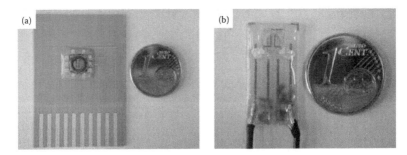

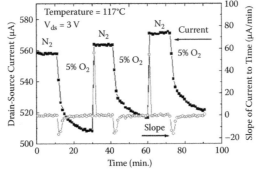

FIGURE 2.1 (Top) Schematic of AlGaN/GaN HEMT-based O_2 sensor. (Bottom) Drain current of IZO-functionalized HEMT sensor measured at fixed source-drain during the exposure to different O_2 concentration ambients. The drain bias voltage was 0.5 V, and measurements were conducted at 117°C.

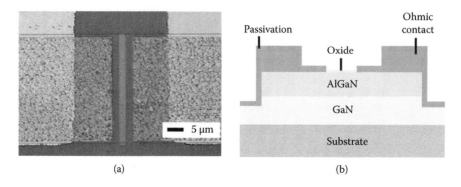

(a)   (b)

FIGURE 2.7   SEM and schematic of gateless HEMT.

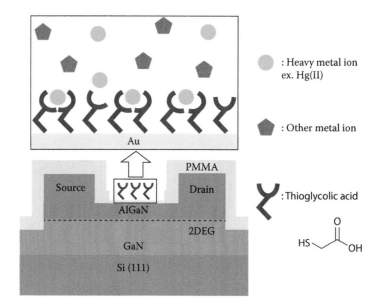

FIGURE 2.10   Schematic of AlGaN/GaN HEMT. The Au-coated gate area was functionalized with thioglycolic acid.

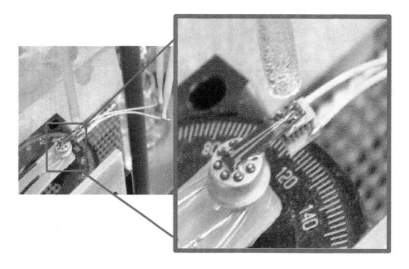

FIGURE 2.21   Optical image of sensor mounted on Peltier cooler.

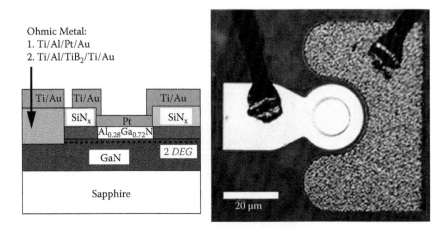

FIGURE 2.45 Photographs of integrated pH sensor (top) and receiver/transmitter pair (bottom).

FIGURE 3.1 Cross-sectional schematic of completed Schottky diode on AlGaN/GaN HEMT layer structure (left) and plan-view photograph.

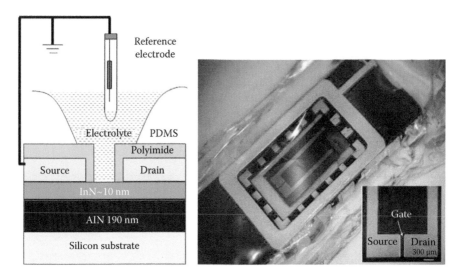

FIGURE 4.3 (a) Schematic cross-sectional structure of an ultrathin (~10 nm) InN ISFET. The native oxide at InN surfaces functions as the sensing material, which is in direct contact with the sensing environment. The adsorption of analyte ions onto the InN surface builds up a Helmholtz potential, resulting in an output current variation. The source and drain are encapsulated by polyimide and PDMS to prevent the contact with electrolyte.

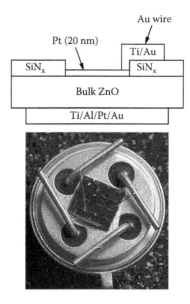

FIGURE 5.24 Schematic of ZnO Schottky rectifier (top) and photograph of packaged device (bottom).

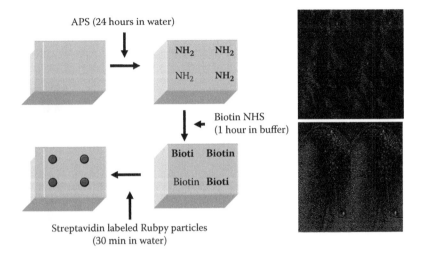

FIGURE 5.55 (Left) Schematics of surface modification of nanowires for biomolecule immobilization for biosensor and biochip applications. (Right) Preliminary results of surface modification: confocal images of a GaN and ZnO surface modified with avidin binding with Rubpy nanoparticles through biotin–avidin linkage (bright image); The same surface modified with BSA and showed no binding to the nanoparticles (dark image).

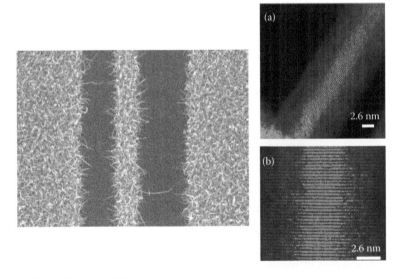

FIGURE 5.57 SEM (left) and Z-STEM (right) of cored ZnMgO nanorods grown on Ag-coated Si substrate.

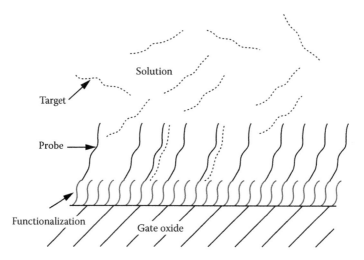

FIGURE 6.5 Molecules on the surface of an oxide in a BioFET. Red: linking layer, black: probe molecules, blue dash: target molecules.

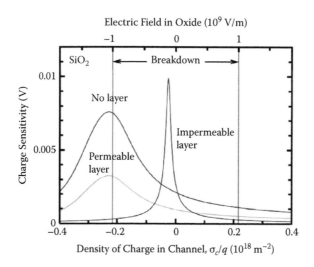

FIGURE 6.10 Charge sensitivity $-S_c$ as a function of charge number density σ_c/q in the semiconductor, for site-binding parameters of SiO_2. The DNA layer is taken to be 10% of close packing, 10 nm thick. The linking layer between the DNA and the oxide is either absent ("no layer"), or has a thickness of 1 nm. The 1 nm layers are either modeled as dielectrics ("impermeable layer") or as filled with solution ("permeable layer"). At the vertical lines, the field in the SiO_2 reaches breakdown.

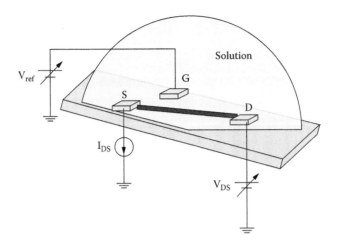

FIGURE 6.18 View of nanowire sensor in solution showing gate electrode bonded to insulating substrate surface.

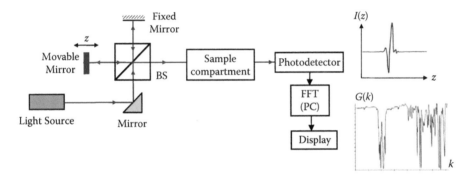

FIGURE 7.8 Simplified block diagram of an FTS system. BS: Beamsplitter.

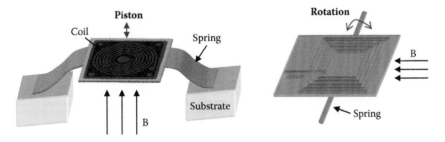

FIGURE 7.11 Magnetic actuators using coils.

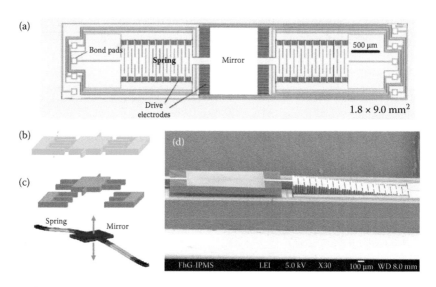

FIGURE 7.18 Translatory electrostatic micromirror.

FIGURE 7.19 Photograph of a prototype FTIR system.

balanced in the alkaline region. Then the CO_2 concentration in the incubator must be increased. In this case, the buffer system reacts as follows:

$$\text{High H}^+ \text{ level: } H_3O^+ + HCO_3^- \Leftrightarrow H_2CO_3 + H_2O \Leftrightarrow CO_2 + 2\ H_2O,$$

$$\text{High OH}^- \text{ level: } OH^- + CO_2 \Leftrightarrow HCO_3^-$$

Because of the high CO_2 concentration and the presence of sodium hydrogen carbonate in the milieu, the pH is kept constant as long as the cell cultures are in the incubator. Once the vessel with cells is removed and exposed to normal atmosphere, which has only 0.3% CO_2 concentration, the milieu becomes alkaline (i.e., a pH of about 8.5) and changes its color to violet, indicated by Phenol red in the media. For this reason, under normal conditions, additional buffer solutions such as HEPES or Buffer All have to be added to the cell media. Cell media that contain the buffer solution are also commercially available (e.g., Leibowitz L15) and are especially designed for work with cell cultures in normal atmosphere.

Prior to any application, such a buffer system has to be tested to prove its biocompatibility with the cell line for which it will be used. These kinds of tests are presented in the following text.

1.5   CELL COUPLING TO AN AlGaN/GaN ISFET

The following sections will report on the cultivation of NG 108-15 nerve on a planar device, the AlGaN/GaN ISFET, and the application of this coupling to monitor the reactions of the cells on inhibitors (drugs). An important feature of the system is its ability to allow repeated measurements over long time periods on the same biological material without damaging the proliferated cell layer. All the concerned parts, for example, cells, cell media, buffer solution, and sensor interface, are complex systems, where a variety of different reactions and processes are possible. Not all of them can be analyzed separately by direct techniques. In order to understand the recorded signal in such a complex system, preliminary experiments in a defined environment were performed in the following order. The basic characterization of the sensor in standard buffer solutions was extended by studying sensitivity to Na^+ and K^+ ions (Section 1.5.3) as well as inhibitors (Section 1.5.4), which are relevant for the cell-coupling experiments. Next, the cell media for the cell growth, which behavior is independently characterized (Section 1.4.6), was brought in contact with the sensor and its response in those media—the buffering and the dosing of neuroinhibitors—was analyzed (Section 1.5). In such media, the behavior of the NG 108-15 cell line (Section 1.5.1) and its cultivation on the AlGaN/GaN sensors (Section 1.5) was studied. Finally, the response of the sensors to the cell activity was monitored, including spontaneous actions such as breathing (Section 1.5.2) as well as the stimulated reaction on inhibitors (Section 1.6). The monitoring in a buffer solution with low ion content (SCZ, Section 1.6.1) supported the identification of relevant ions for the sensor response. The most complex interaction scheme that was studied is the subsequent application of two different neuroinhibitors (Section 1.6.3).

1.5.1   SENSOR PREPARATION

The AlGaN/GaN ISFETs used for the recording of extracellular signal were PIMBE-grown sensor chips with a large active area of 2400×500 μm^2. They were chosen because of the good sensor characteristics shown in the preliminary characterization. The sensor chip was encapsulated on the PBC as described (Figure 1.6b). The corresponding measurement system was the first setup (Figure 1.9a) with the agarose gel-encapsulated reference electrode, and the current I_D versus time was recorded. This system is utilized for sensor calibration in ionic solution and cell media and also for recording of extracellular potential changes as a response to different added neuroinhibitors.

To be able to record an extracellular potential of the nerve cell using the AlGaN/GaN ISFET, it is first necessary to create a stable cell–sensor hybrid. This means that the sensor must be biocompatible, well cleaned, and sterilized. Since the AlGaN/GaN material is highly biocompatible, it enables direct spreading of the cells on the sensor surface without the use of thin films of organic material for improving cellular adhesion and biocompatibility. Prior to their spreading on the sensor, the surface must be cleaned and sterilized in order to avoid the infection of cells with bacteria. This is realized by cleaning the sensor and measurement setup with acetone, isopropanol, and deionized water followed by sterilization in 75% ethanol solution for at least 20 min. Finally, the sensors were dried in N$_2$ flux, and after that mounted in the measurement setup.

After that, the cells can be spread on the sensor surface. Cell spreading is an essential function of cells, which are adherent to a surface and precedes the function of cell proliferation until the cells completely cover the sensor surface. The attachment phase of cell adhesion occurs rapidly and involves physicochemical linkages between cells and the sensor surface, including ionic forces. Diminished cell adhesion is used as a measure of toxicity, if first the initial attachment of cells is investigated [29,89].

The NG 108-15 cells were subcultured in Dulbecco's modified Eagle's medium–high glucose (DMEM) supplemented with 10% FCS and 1% penicillin/streptomycin according to standards set by the ATCC [127]. When the cells showed the best characteristics, they were picked up without enzymatic treatment and were spread on the sensor surface, which was encapsulated in the measurement setup. In the initial step of the attachment, cells seeded on the sensor showed normal morphology. After 6 h, the attached cells had a spherical morphology with rough texture and were more widespread than in the beginning. After one day, some of the attached cells spread radially from the center and developed filopodia (Figure 1.14a). However, not all cells were attached to the surface. The spherical cells were still swimming in the medium. Moreover, the cells were not uniformly distributed; they were proliferating in groups on the sensor surface (Figure 1.14a). After three days in the incubator at 37°C, excellent proliferation was observed, and the sensor surface was completely covered by the cells (Figure 1.14b).

For the measurements, it is very important to have a compact layer of cells covering the whole active area of the sensor. Between cells and the sensor surface, an ion accumulation channel (the cleft) is formed with a thickness of about 30–70 nm [128]. For continuous proliferation, the cells need to exchange nutrients with the media particularly ions. The ion concentration in the cell media is changing depending on the processes, which take place in the cells. For short times, the ion concentration in the channel can increase substantially compared to the media before an equilibrium is reached again by diffusion processes. Because of this short nonequilibrium, it is possible to monitor cell activity by probing the concentration or potential changes in the cleft using different sensors

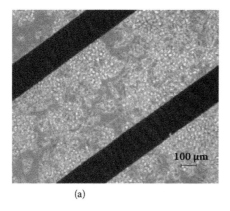

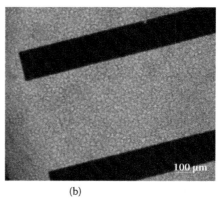

(a)   (b)

FIGURE 1.14 The NG 108-15 cells attached to the active area of the sensor (a) after one day, (b) after three days immediately before starting the experiments for recording of sensor signals.

(MEA, ISFET, electrodes). For reliable measurements, this channel should be confined between the sensor active area and a continuous cell culture. Thus, a complete covering of the sensor surface with cells is beneficial for high signal-to-noise ratios. With the aforementioned procedure, about 10^5 NG 108-15 cells were seeded on the sensor surface. After two days, the NG 108-15 nerve cells showed good proliferation and vitality, and the active area of sensor was covered with a compact layer of cells. Considering a mean cell diameter of 20 µm, about 3500 cells were grown on the active sensor surface. A higher seeding of cells in the beginning can reduce the time for full sensor covering by about a few hours. However, such a fast proliferation procedure is not recommended, because not all the cells have a place to attach to the sensor surface and form a strong oscillating membrane above the adherent surface.

However, even if the cells in the culture appear similar, they might show different functions and characteristics, and two different cell cultures from the same cell type can be quite different, even if the same growth protocol is used [89,126]. This might happen also in the case of the extracellular signal recording with the AlGaN/GaN ISFETs in this work. In biochemical research, this disadvantage of cell culturing can be overcome by highly parallel screening systems. Such measurement methods enable collection of statistical data and determine the error limits; however, they rule out repeating the experiments on the same biological material over a long period. Consequently, for biologists it is important to have an instrument that uses the same cell culture over several days, in order to be certain that cells have the same characteristics. The measurement system with incorporated AlGaN/GaN ISFET permits these types of measurements.

1.5.2 Sensor Response to pH Changes in Cell Media with/without Cells

To be certain that the sensor records changes of ion concentration in the extracellular media, it is necessary to first prove sensor sensitivity to different ions and to ensure a constant pH value in the media during the measurements.

First, the effect of pH changes in the cell media with and without NG 108-15 nerve cells on the AlGaN/GaN sensor sensitivity is evaluated. Buffered and nonbuffered DMEM cell media were brought on the surface of an encapsulated sensor. In the first run, represented by the black curves in Figure 1.15a, the measurement setup filled with cell media was not covered. After that, the measurement setup was covered with a Plexiglas lid, and the measurements are repeated (red curves).

The sensor signal exhibits a drift (continuous decreasing of I_D) corresponding to an increase of the pH value due to alkalization of the cell media. Since the measurements were performed without interruption, the final signal of the measurements without lid is equal to the starting value of the second run with lid, which explains the shift of the curves. Except this, the signal behavior in cell

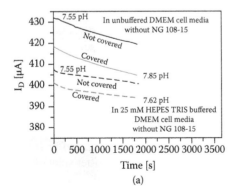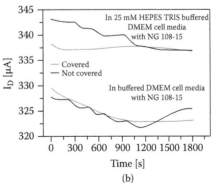

FIGURE 1.15 Influence of the measurement setup (covered: red curves, not covered: black curves) and the buffer solution on the recorded sensor signal for cell media (a) without and (b) with cells (given pH values are measured at the beginning and end of the experiments by MT glass electrode).

media was not influenced by the covered or not-covered measurement setup. In contrast, the presence of 25 mM HEPES-TRIS buffer solution in DMEM cell media resulted in a clearly reduced sensor drift, which demonstrates its function of stabilizing the pH value in the media (Figure 1.15a). The drift did not disappear completely, and the reason could be the continued alkalization of the cell media together with the noise that can appear in the system (see Section 1.4.3.)

In the presence of NG 108-15 nerve cells on the sensor surface, a different phenomenon was observed (Figure 1.15b). Comparing the curves recorded in the covered measurement setup (red curves), the sensor drift within the first 1000 s was similar for the media with (Figure 1.15b) and without cells (Figure 1.15a). In contrast, in buffered cell media and covered setup, the sensor signal was quite stable from the beginning, which again illustrates very well the stabilizing effect of the 25 mM HEPES-TRIS buffer solution in keeping the pH in the cell media with adherent cell relatively constant.

A substantial difference can be seen in the signal shape between covered and not-covered measurement setups. For the covered measurement setup (Figure 1.15b, red curve), the recorded sensor signal was relatively constant in time, while for the not-covered measurement setup (Figure 1.15b, black curves), the signal decreased in steps before approaching a stable value. This phenomenon can be explained as result of an alkalization [89,126] process of cell media due to changes of CO_2 concentration in the cell milieu. Note that the experiments were performed in a normal environment (room temperature and atmosphere with 0.3% CO_2 concentration; see Section 1.7.2) and before the prepared measurement setup with adherent NG 108-15 cell was taken out of the incubator and cell media were refreshed. Thus, the recorded signal oscillations are not related to the sensor drift. It is rather the monitoring of the spontaneous cell activity ("breathing"). The faster stabilization of the sensor signal in the case of buffered cell media is due to the homogenizing function of the 25 mM HEPES-TRIS solution added to the DMEM cell milieu. A similar behavior was observed by Silveira [62], using other AlGaN/GaN sensors and the LTCC-based setup (Figure 1.9b). These measurements clearly demonstrate the ability of the AlGaN/GaN sensors to monitor cell activities.

As a result of these experiments, it can be concluded that in order to ensure stable environmental conditions for NG 108-15 nerve cells cultured on the sensor surface at normal conditions, it is necessary to use a buffered cell media and a covered measurement setup to prevent gas exchange with the environment.

1.5.3 ION SENSITIVITY OF AlGaN/GaN ISFETs

In Section 1.4.7, the inhibition mode of PMSF, DFP, and amiloride, and their effect on nerve cells and membrane potential was presented. These neurotoxins were used in this work for cell transmembrane protein inhibition, since they cause a change of the initial ionic concentration in the extracellular solution. In the case of PMSF, DFP, and amiloride, Na^+, K^+, and Ca^{2+} ions have the most influence on the sensor signal. As discussed later, it can be assumed that Na^+ ions play the most important role in the initiation of the AlGaN/GaN sensor signal. To be certain that the AlGaN/GaN ISFET is sensitive to concentration changes for at least two different ion types, Na^+ and K^+, a special buffer solution is prepared, and new calibration curves are recorded. This buffer solution must ensure a constant pH value and a minimum nutritive factor for culturing NG 108-15 nerve cells for a short period of time (i.e., during the experiments with the cells) without changing the cell characteristics.

The AlGaN/GaN sensitivity to Na^+ and K^+ ions was evaluated in a SCZ buffer solution. The choline is used to maintain a constant ionic strength in the solution. The buffer solution was added in the setup, and after a stabilization time of 10 min, the measurements were started. This stabilization time was introduced to reduce the noise from the device (see Section 1.5.2). The Na^+ and K^+ concentrations were subsequently increased by titration of NaCl and KCl, respectively.

In Figure 1.16, the sensor signal as a response to changes in the ionic concentration is presented. First of all, a drift can be seen as a decreasing sensor signal in time. This drift is not caused by

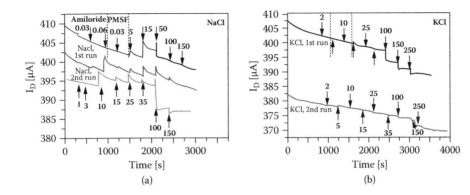

FIGURE 1.16 Sensor response upon dosing of (a) Na^+ and (b) K^+ ions in SCZ solution. In the case of Na^+ ions (a), additionally, amiloride and PMSF are supplied prior to the Na^+ dosing experiments (dosing by titration; numbers correspond to end concentrations in the solution in units of mM).

reactions of the solution, which has a constant pH value of 7.5–7.6. This drift is mainly caused by the used unstable agarose reference electrode. In the case of Na^+ ions (Figure 1.16a), the sensor started to react at a concentration of 5 mM with a signal increase of about 2.5 µA. After every titration, the sensor signal decreases again at a lower rate due to diffusion processes, which equalize the ion concentration in the solution and stabilize the pH value.

The fast increase of the sensor signal after titration requires positive ions on the sensor surface; thus, this proves that the sensor is sensitive to Na^+ ions. At high concentrations of about 100 mM, the positive sensor's response decreased and changed to a negative response, which requires negative ions. Their origin might be recombination processes in the solution, for example, those involving choline. This effect was not further investigated since the Na^+ concentrations are already too high for the cell–transistor coupling experiments. Similar behavior was observed also in the case of K^+ ions (Figure 1.16b). However, the sensor response was weaker, and clear signals could be recorded only at concentrations higher than in the Na^+ case. Thus, the AlGaN/GaN sensor is sensitive at the same time to the ion concentrations as well as to the temporary changes in the solution close to the sensor surface due to chemical reactions.

1.5.4 SENSITIVITY OF AlGaN/GaN ISFETs TO INHIBITORS

This phenomenon is important if the sensor response to changes of inhibitor concentration in different solutions is studied. First, no sensor signal is detected in the case of amiloride and PMSF inhibitors added with different concentrations to a SCZ solution (Figure 1.16, black curve). The same result was obtained in the case of inhibitors added to buffered (25 mM HEPES-TRIS buffer solution) and not-buffered DMEM cell media (Figure 1.17). Also, no influence was observed in the case of mixed inhibitors in the same basic buffer (see Table 1.5). The difference in the slope of recorded curves is a result of cell media alkalization in case of the not-buffered and buffered cell media, as explained in Section 1.5.3.

Furthermore, the sensor response to DFP inhibitor added in SCZ, SCR, SCI, and cell media was studied (Figure 1.18). When the neurotoxin was added to SCZ buffer solution, no sensor signal was recorded. On the other side, no sensor signal was detected for neurotoxins added to cell media, which contain different type of ions in different concentrations (see Figure 1.17). This can be explained either by the fact that the proceeding chemical reactions do not generate a sufficient number of the ions to produce changes in the sensor signal, or by the complexity of the cell media, which neutralize the resulting components before coming close to the sensor surface. It is noteworthy that the presence of Na^+ and K^+ ions in sufficiently high concentrations in the buffer solution generates a sensor signal when the inhibitor DFP is added. Thus, when the cells are adherent on the sensor

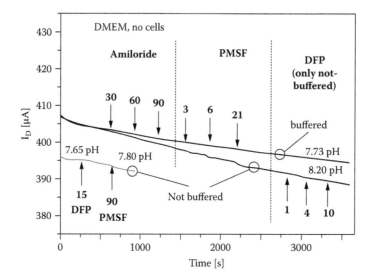

FIGURE 1.17   Sensor response to different inhibitors added to cell medium DMEM, with and without 25 mM HEPES TRIS buffer. DFP was only dosed to not-buffered media (dosing by titration; numbers correspond to end concentrations in the solution in units of μM; given pH values are measured at the beginning and end of the experiments by MT glass electrode).

TABLE 1.5
Concentrations of four different solutions used in this work

Basic buffer	C [mM]	SCR	C [mM]
MgSO$_4$	1	Basic buffer +	
HEPES-TRIS	20	KCl	5
Glycine	5	NaCl	10
l-glutamine	1	CaCl$_2$	0.15
Glucose	10		

SCZ		SC	
Basic buffer +	C [mM]	Basic buffer +	C [mM]
Choline	135	K acetate	150
		NaCl	10
		CaCl$_2$	0.015

Note: SCZ: substituted cell media with zero ions; SCR: substituted cell media
with reduced concentration of ions; SCI: substituted cell media with the
same concentration of ions as in the cell.
All solutions stabilize a pH value of 7.5. The basic buffer solution serves
as the starting basis for the other three solutions.

surface, the same chemical reactions may not take place in the cleft. The response of the sensor on the specific reactions of the NG 108-15 cells is discussed in more detail in Section 1.6.

1.6   RECORDING OF EXTRACELLULAR SIGNAL

As explained in Section 1.4.7, DFP is a very strong inhibitor of AChE. The ligand-gated channel opened after binding of ACh for about 10 ms. In case of differences in the Na$^+$ concentrations inside

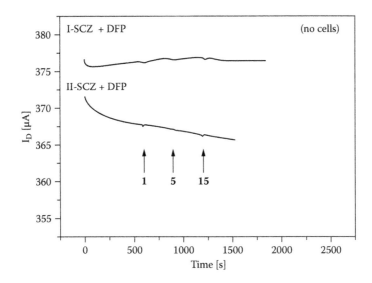

FIGURE 1.18 Sensor response to DFP added to SCZ (dosing by titration, numbers correspond to added concentrations in the solution in units of μM).

and outside the cell, the influx of Na^+ decreases the membrane potential [20,119,129,130]. The cleavage of ACh by AChE proceeds rapidly, and the channel will be closed. Inhibition of AChE leads to a longer holding time of ACh at the receptors and a prolonged opening time of the channel. Repeated use of AChE inhibitors can force the channels to stay open for longer time period, and an equilibration of the ion concentration outside and inside the cell will become possible. If it is considered that a homogenous layer of cells was grown on the surface, then between the cells and the surface a cleft is formed. Due to the exchange processes of the cells and reduced diffusion (long diffusion path) to the major part of the buffered media, on a short timescale the cleft can exhibit nonequilibrium ion concentrations. This nonequilibrium creates a temporary potential that can be measured with the AlGaN/GaN ISFET sensor as will be shown in the next section.

Before starting the recording of extracellular signal, the measurement setup was taken from the incubator to the normal atmosphere, the reference electrode was mounted, the used cell medium was removed, and the cells were washed gently with the new cell medium. $U_{DS} = 0.5$ V and $U_{ref} = -0.35$ V were kept constant, and I_D was recorded versus time. All the experiments were performed in the glove box at a humidity of 45% and a temperature of 22°C. During the recording of the sensor signal, the setup was covered with the lid.

1.6.1 RESPONSE ON SINGLE INHIBITORS IN SCZ BUFFER

First experiments for cell reaction studies were performed in SCZ since the defined composition enables one to determine which ions contribute to a sensor response. The cell media was replaced with a 5 mL SCZ buffer solution, and the cell reaction on the neurotoxin DFP in different concentrations was studied. As described earlier, the SCZ is a buffer solution that contains no Na^+, K^+, and Ca^{2+} ions. Thus, even if neurotoxins block the AChE cleavage and Na^+ channels are kept open for longer time, the changes in the ion concentration will be very small since the cells contain only 7 mM Na^+. Such an ion concentration change, a maximum of 7 mM, by the Na^+ outflow is too small to be detected by the sensor (Figure 1.19a). However, when 50 mM NaCl was added to the SCZ buffer solution, a fast increase of the sensor signal (about 15 μA) was recorded. Then, Na^+ ions flow through the still open channels into the cells. Since the accumulation layer can be assumed to be still Na^+ free, this inflow occurs mainly through channels at the top side of the cells. To establish the membrane equilibrium potential, simultaneously K^+ ions are pumped outward; however, this occurs

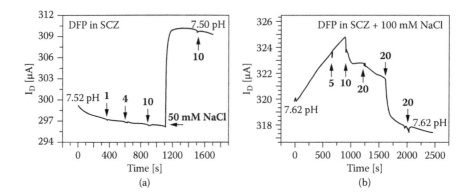

FIGURE 1.19 Extracellular signal from NG 108-15 nerve cells as reaction to the neurotoxin DFP in (a) SCZ buffer solution and (b) SCZ buffer solution with 100 mM NaCl (dosing by titration; numbers correspond to added concentrations in the solution in units of μM; given pH values are measured at beginning and end of the experiments by MT glass electrode).

through all channels of the cells simultaneously, thus increasing the positive ion (K$^+$) concentration in the accumulation layer between the cell and the sensor surface. It is to note that the global pH value of the buffer solution measured using a conventional glass pH electrode was constant during the measurement, about 7.6 units (Figure 1.19a). As a consequence, it is mainly these K$^+$ ions that contributed to the sensor signal.

Using the patch-clamp technique, Hille [129] measured an equilibrium concentration of 140 mM Na$^+$ outside of the mammalian neuron. In the present experiment, only 50 mM Na$^+$ was added to the SCZ buffer solution; thus, full equilibrium could be achieved after the first reactions on DFP, and subsequent addition of the neurotoxin could not generate a new cell reaction. Indeed, no further extracellular response on new inhibitor dosing could be recorded by the AlGaN/GaN ISFET (Figure 1.19a, last dosing of 10 mM).

In a second experiment, Na$^+$ ions were added from the beginning (Figure 1.19b). The same sensor and measurement setup were used; only new nerve cells were cultivated on the sensor. Optical microscopy revealed the same good proliferation and vitality properties of the cell bed as in the previous experiment. The presence of only Na$^+$ ions in SCZ buffer creates nonequilibrium of the cell membrane potential. In this case, according to Bernstein's theory [118], K$^+$ ions flow out from the cell through the selectively permeable cell membrane to offset the potential difference. This mechanism can explain the gradual signal increase within the first 900 s of measurement time. The addition of DHP to SCZ 100 mM NaCl caused opposite sensor response. The very first dosing of 5 μM DFP to the SCZ buffer with 100 mM NaCl had no substantial effect on the sensor signal, and thus, was too small in this experiment to cause a measurable ion flux. Further dosing resulted in a quickly decreased sensor signal due to Na$^+$ flux into the cells, seen basically after the dosing of 10 μM and the second dosing of 20 μM DFP. However, this effect was not fully reproducible during the experiment.

The effect of the DFP can be seen by the stair-like drop of the curve, but the change of the normal cell culture medium to the SCZ buffer generate a more irregular curve, compared to the curve in Figure 1.20a.

The small sensor response on the first dosing of 20 μm DFP might be the result of complex reactions in the medium, as well as the fact that SCZ is not the optimal environment for cells (in contrast to the experiments in cell media, where a staircase-like response was observed; see Figure 1.20). This effect was not further investigated. In any case, taking into account that the pH value of the SCZ buffer solution was constant at 7.62 during the measurements, the sensor signal can only be the result of ion fluxes across the cell membrane from the accumulation channel close to the sensor surface. Further dosing (third 20 μM DFP and beyond) did not cause sensor responses anymore. Thus,

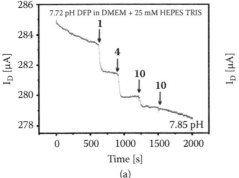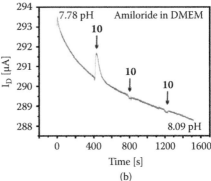

FIGURE 1.20 The recorded NG 108-15 cell reaction with different concentrations of neuroinhibitors: (a) DFP in DMEM with 25 mM HEPES-TRIS buffer solution as cell medium and (b) amiloride (dosing by titration; numbers correspond to added concentrations in the solution in units of µM; given pH values are measured at the beginning and end of the experiments by MT glass electrode).

once the Na^+ concentration is in equilibrium between the NG 108-15 nerve cells and the medium as a result with the reaction to the DFP neuroinhibitor, no sensor signal is recorded.

Previous investigations on ion sensitivity (Section 1.5.1) have shown that the sensor responds both to K^+ and Na^+ ion fluxes. Here, it has been demonstrated that for the monitoring of the cell reactions to DFP neuroinhibitors in SCZ buffer solution, the Na^+ flux is of major importance for sensor signal generation. However, due to the small ion content in SCZ solutions, the cell reaction to neuroinhibitors can denature. As a consequence, further experiments were performed in a more "natural" cell medium, as described in the next section.

1.6.2 Response on Single Inhibitors in DMEM

In the following experiments, instead of the SCZ buffer solution, the complex classical cell medium (DMEM supplemented with 10% FCS and l-glutamine) was used, and 25 mM HEPES-TRIS was added as buffer solution (Figure 1.20a). Also, in this case, the sensor and the measurement setup were not exchanged; only the cell layer was refreshed. Because of the complexity of the cell media, it was difficult to keep its pH value constant in the normal atmosphere, and for this reason, during the measurements an increase of about 0.13 pH was observed.

First, the reaction of the NG 108-15 nerve cells on added neuroinhibitor DFP was recorded (Figure 1.20a). In contrast to the previous experiments, a sensor reaction by decrease of the recorded current appeared already when the inhibitor DFP with concentrations as small as 1 µM DFP was added to a 5 mL cell medium that covered the proliferated nerve cells. A decrease in the recorded sensor signal appeared also if a further 4 and 10 µM of the inhibitor DFP were added. After reaching a concentration of about 15 µM DFP, saturation was observed, where no cell reaction could be recorded anymore even if a new dose of neuroinhibitors was added (Figure 1.20a). Also, in this case, the different ions present in the accumulation channel contribute to the generation of the sensor signal. However, due to the complexity of cell media, it is hard to determine exactly what kind of ions and in which percentage they contribute to these effects. It can be assumed that Na^+ and K^+ ions have the major influence on the sensor signal. To give an assessment of this hypothesis, an analytical estimation of the sensor signal was performed, and is presented in Section 1.7.2.

To see how the AlGaN/GaN ISFET reacts on the response of NG 108-15 cells to other kinds of inhibitors, fresh adherent nerve cells were prepared on the sensor with new buffered cell culture medium. This time, amiloride, a reversible blocker of epithelial Na^+ channels (see Section 1.4.7), was used as neuroinhibitor (Figure 1.20b). In this case, a remarkable instability of the buffered cell medium was observed, which became alkaline in time (pH increased by 0.31).

First, amiloride with a concentration of 10 μM was added in 5 mL cell medium. The sensor signal increased quickly, followed by a slower decrease until the sensor approached the same signal level (taking the alkalization into account). Adding further doses of amiloride generated no sensor reaction anymore. Considering that the neuroinhibitor amiloride closes the Na^+ channels, that is, no Na^+ ions can flow into the cell, and its dosing should lead to increasing concentration of the positive ions (especially K^+ and Na^+) in the accumulation channel, which finally resulted in an increased sensor signal, as shown in Figure 1.20b. Consequently, the equilibrium can be reached via two diffusion processes: (1) laterally out of the cleft into the buffer solution, or (2) by K^+ ion flow through the ion channels into the cells. Thus, the equilibrium state of the membrane potential was established, and the sensor signal approached its equilibrium value. Due to the alkalization, the equilibrium signal is smaller than the initial value as a result of the increased pH. Obviously, the dosing of 10 μM amiloride was already sufficient to block all the channels, since further dosing had no influence on the nerve cells and on the sensor signal (Figure 1.20b).

Further experiments involved the second experimental setup and Amani reference electrode (Figure 1.21) [62], and yielded similar results. The sensor preparation and NG 108-15 nerve cell proliferation were realized as described in Section 1.5.1. In addition, the sensor response upon dosing of the inhibitor PMSF was investigated. As described in Section 1.4.7, PMSF is a weak inhibitor of AChE [130]; that is, ion channels are opened for a longer time upon dosing, which enables equilibrium of the ion concentration outside and inside the cell. Figure 1.22 summarizes the results of PMSF titration experiments in 5 mL cell media using two sensors provided from the same wafer but with different active areas.

Similar to the experiments described earlier, the sensor signal showed a continuous decrease as a result of increased pH due to alkalization. Obviously, upon dosing of PMSF there was a temporary decrease of the sensor signal as an immediate reaction of the nerve cells. Because of the open ion channels, the cleft is affected by a strong local ion flux across the cell membrane, which leads to a different ion concentration than in the remaining cell medium and, thus, a measurable sensor signal. The reaching of equilibrium with the cell medium is responsible for the slower recovery of the sensor signal. The analysis of dependence of sensor signal changes on the dosed PMSF concentration revealed a maximum response at about 50 μM PMSF (Figure 1.22), after which the signal decreases as a result of saturation.

These experiments clearly demonstrate that the AlGaN/GaN sensor is sensitive to cell activities, for example, the "normal" activity (breathing) along with an external stimulation by neurotoxins.

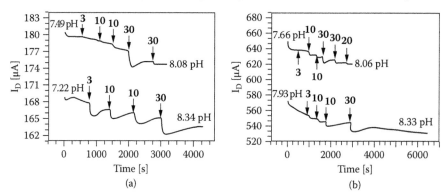

FIGURE 1.21 The recorded NG 108-15 cell reaction to different concentrations of the PMSF neuroinhibitors in DMEM with 25 mM HEPES TRIS, using two different PIMBE-grown sensors (a) 400×100 μm², (b) 2400×100 μm² (dosing by titration; numbers correspond to added concentrations in the solution in units of μM; given pH values are measured at beginning and end of the experiments by MT glass electrode).

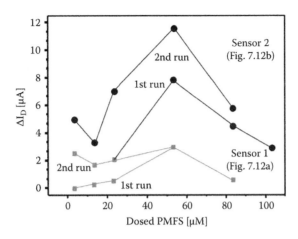

FIGURE 1.22 Summary of the recorded changes of the dependence of the sensor signal on the titrated PMSF concentration.

TABLE 1.6
Summary of the observed sensor response to the cell reaction with different treatments

Experiment	Cell reaction	Sensor response		Underlying effect
Alkalization	Consumption of H_3O^+ and diffusion of CO_2 into the atmosphere	I_D decreases	($\downarrow$)	Increase of OH^- ions
Breathing	Decreasing CO_2 concentration in the cell medium, creating nonequilibrium of membrane potential	I_D oscillates	($\updownarrow$)	Alternating HCO_3^- concentration in the cell medium
Dosing of DFP	Inhibiting of AChE, thus, keeping the Na^+ channels open	I_D decreases	($\downarrow$)	Depletion of positive ions in the cleft (Na^+) due to equilibration of the concentration gradient
Dosing of amiloride	Blocking of Na^+ channels	I_D increases	($\uparrow$)	Accumulation of positive ions in the cleft (Na^+, K^+)
Dosing of PMSF	Inhibition of AChE, thus keeping the Na^+ channels open	I_D decreases	($\downarrow$)	Depletion of positive ions in the cleft (Na^+) due to equilibration of the concentration gradient

The sensor response and the responsible cell reactions are summarized in Table 1.6. In the next section, it will be demonstrated that even complex reactions on different stimuli can be analyzed with those sensors.

1.6.3 Sensor Response on Different Neurotoxins

Finally, the reaction of the NG 108-15 nerve cells to the dosing of more than one inhibitor into the cell media was investigated. The neuroinhibitors used for this study were PMSF and amiloride; that is, first, a neurotoxin that opens ion channels was used, followed by dosing amiloride that blocks specifically Na^+ (and Ca^{2+}) channels (see Section 1.4.7).

The electrical measurements were performed continuously over several days without removing the cells from the sensor. The sensor signal was recorded in cycles, each of them lasting about 1 h. After the measurement cycles, the cells were placed in the incubator with replaced fresh medium for recovery. Figure 1.23 shows the recorded I_D during four measurement cycles (I–IV) as a reaction of

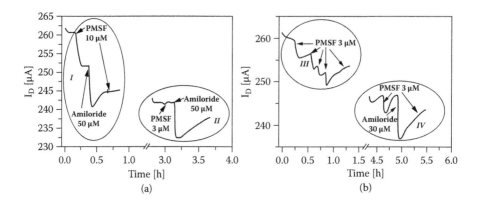

FIGURE 1.23 I_{DS} versus time as reaction of the cells in medium to different inhibitors in DMEM without buffer (I–IV) on (a) the first day with a break of 2 h and (b) on the second day with a break of 3 h.

the dosing of different neuroinhibitors, while the reaction of the pure ISFET without cultivated cells was shown in Figure 1.15. Obviously, a drop of I_D as reaction of the cells on the dosed neuroinhibitors was monitored in contrast to the absence of significant reactions in the case of the uncultivated sensors on the neuroinhibitors added in DMEM cell media (Figure 1.15). The noise in the signals was about 100 nA, which is more than an order of magnitude lower than the evaluated signal.

After a dosing of 10 µM PMSF (I), a strong response of the sensor was observed, while a second dosing after about 30 min and the additional dosing of amiloride did not cause any signal variation. A break of 2 h did not change the situation: the sensor signal was on the same level, and no response on PMSF dosing could be observed. In contrast, a reaction on another neuroinhibitor, on amiloride, is apparent in both cycles (I and II). Consequently, the amount of 10 µM PMSF was sufficient to completely inhibit AChE fission and opened all involved ion channels, or the amiloride action (blocking of Na⁺ channel) limited the response of nerve cell to PMSF inhibitor. The cell media was refreshed, and the measurement setup was placed in incubator. After 18 h, the cells recovered completely (III). The sensor signal was on the initial level and apparently the cells reacted on a new dosing of PMSF. In cycle III, smaller amounts of 3 µM PMSF were fed, resulting in a gradual decrease of the drain current. Similar to the first measurement cycle (I), after a fourth titration, that is, after dosing of 12 µM PMSF, no sensor response was observed (III). A break of 3 h in incubator resulted in partial recovery of the cells (IV): The sensor signal did not reach the initial value; however, the cells responded again on a single titration of PMSF and showed no reaction on the second dosing. In both cases, there was still response on the other used neuroinhibitor, amiloride.

The temporal behavior of the sensor response was fitted by an exponential law

$$I_D = I_{D,e} + \left(I_{D,0} - I_{D,e}\right) * \exp\left(-\frac{t}{\tau}\right),$$ (1.6)

where $I_{D,0}$ and $I_{D,e}$ are the drain currents at times $t = 0$ (titration event) and $t \rightarrow \infty$ (equilibrium value), respectively, and τ is a time constant. Similarly, the recovery (sensor signal increase) was examined. The fitted parameters signal change $\Delta I_D = I_{D,0} - I_{D,e}$ and τ for the PMSF titration experiments are shown in Figure 1.24. The sensor response on every dosing of 3 µM PMSF is approximately constant at about 2.5 µA and drops to zero after saturation (Figure 1.24a, "signal per step"). Accordingly, the sum of the response signals increases. The sensor response to the single dosing of 10 µM PMSF in cycle I fits the sum signal very well after subsequent titration experiments in cycle III, which displays the reproducibility of the sensor.

For an interpretation of the results, it is assumed that the sensor signal drop after titration is caused by a decreased concentration of positive ions. Usually, the Na⁺ concentration in the electrolyte is

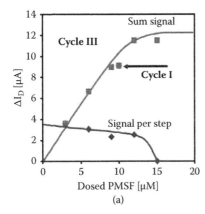
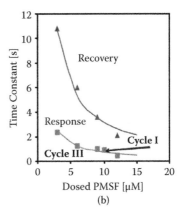

(a)   (b)

FIGURE 1.24 Recorded (a) changes of I_D as reaction on the titration of PMSF during the measurement cycles I and III, and (b) the time constants for the sensor response and recovery after dosing of PMSF.

high while inside the cell the K$^+$ concentration is high. Thus, the possible mechanism could be that after PMSF dosing, ion channels are opened and Na$^+$ ions are transported into the cell. The time constants for the sensor response and the recovery (reproduction of AChE) after every titration event decrease with increased concentration c of PMSF. In Figure 1.24b, the lines represent an inverse proportionality between τ and the dosed amount of PMSF: $\tau \sim 1/V$. Consequently, the ion transport to and from the sensor surface is not purely diffusion controlled. Since a direct reaction of PMSF on the ion cannels is assumed via inhibition of the fission of AChE, a higher dosed volume leads to more opened ion channels and, thus, to a faster sensor response. The recorded data indicate that there is indeed a linear dependence between PMSF concentration and the number of opened ion channels, and saturation sets in when all channels are opened. The good agreement of the time constant for the dosing of 10 μM PMSF in cycle I with the fit for cycle III supports this assumption.

Remarkably, the sensor signal upon dosing amiloride after PMSF was reversed compared to the dosing of amiloride only, as shown in Section 1.6.2. The response on amiloride and, thus, most probably the direction of the ion flux, depends on the ionic concentration in the cleft, which is the result of the previous treatment. It is well known that the interaction of drugs can cause very different effects from the use of the single drug alone. Obviously, a similar effect was observed for the dosing of PMSF and amiloride compared to the combined dosing of both agents. The sensor clearly can distinguish between these different cases; however, a conclusive model to explain the observed behavior is not yet available.

1.7 SENSOR SIGNAL SIMULATION

1.7.1 SELF-CONSISTENT SIMULATION OF THE HETEROSTRUCTURE USING A SITE-BINDING MODEL

For the simulation of the ion-sensitive AlGaN/GaN-based sensors in aqueous solutions, a theoretical model was used that implies self-consistent solution of the Poisson and Poisson–Boltzmann equations in the semiconductor and electrolyte regions, respectively. A more detailed description of the model can be found elsewhere [55,131,132]. The solution of these two equations provides the depth profiles of free carriers across the semiconductor region and of different ions in the electrolyte. In the semiconductor region, where carrier quantization effects play an essential role, the free carriers can be optionally treated quantum-mechanically by self-consistent coupling to the Schrödinger equation. The adsorption/desorption of different ion species onto/from the electrolyte–semiconductor interface is described by the site-binding model illustrated in Figure 1.25. The latter requires knowledge of the ion concentrations in the vicinity of the electrolyte–semiconductor interface and, accordingly, the site-binding model is self-consistently coupled to the Poisson and Poisson–Boltzmann equations.

The AlGaN/GaN heterostructure acts as a sensor, providing a readout of electrical signals depending on the particular ion composition of the electrolyte and the specific chemical bonding of different ion species onto the electrolyte–semiconductor interface. As an example, Figure 1.26 presents the calculated carrier concentrations and potentials for the AlGaN/GaN-based pH sensor exposed to an electrolyte with pH = 4.

These calculations consider only one kind of ion, whereas in the cleft, different ions with changing concentrations can be expected. Each of them can contribute differently to the sensor signal. The

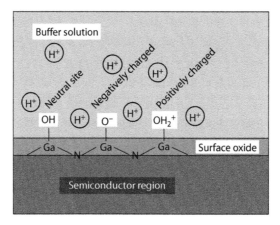

FIGURE 1.25 The site-binding model of the interaction of AlGaN/GaN-transistor with the electrolyte. The presence of a thin, oxidized layer GaO$_x$ (about 2 nm) at top of the AlGaN barrier is assumed. (After Bayer et al., 2005, *J. Appl. Phys.*, 97, 33703.)

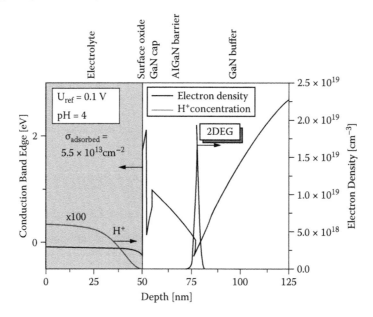

FIGURE 1.26 Self-consistent calculation of electron (black line) and proton H$^+$ (red line) concentrations versus depth. The site-binding model for H$^+$ adsorption onto the electrolyte–semiconductor interface predicts a total positive interface charge of 5.5×10^{13} cm^{-2}. The electrolyte region (representing the cleft) is shown in yellow. The conduction band edge in the semiconductor region and the electrostatic potential in electrolyte are depicted by the blue line.

ion concentrations in the cleft and, as a result, the corresponding electrical signal of the sensor, are strongly affected by nerve cell activity and its reaction to different stimuli. However, the described experiments revealed a maximum current change of about 5% (see the summary in Table 1.7). Based on the simulation results (see following text), this maximum value can be attributed to a possible variation of ion concentrations in the cell media.

The simulation of the electron sheet density versus ion concentration was realized using a program written by the authors. Figure 1.27 demonstrates the small changes in the electron sheet density depending on the concentration of positive ions in the cleft for different pH values and a fixed cell-sensor surface distance as well as for a fixed pH value but different cell-sensor surface distance. The highest simulated concentration of 10^{20} positive ions corresponds to the maximum concentration of Na+, K+, and Ca+ ions of about 170 mM in the cleft or inside the cells [126].

For this highest ion concentration, a variation of the electron sheet density of about 4% is achieved. Neglecting the small changes in 2DEG mobility, this density variation is translated into a maximum current variation of about 4%, which is in reasonably good agreement with the saturation values

TABLE 1.7
Summary of the monitored drain current changes $\Delta I_{D,sum}$ for the different experiments described earlier

Media	Experiment	$I_{D,0}$ (µA)	$\Delta I_{D,sum}$ (µA)	$\Delta I_{D,rel}$ (%)	Reference
DMEM	Stability/breathing	330	4	1.2	Figure 1.15b
SCZ	DFP dosing	296	15	5.1	Figure 1.19a
SCZ + 100 mM NaCl	DFP dosing	325	8	2.5	Figure 1.19b
DMEM	DFP dosing	285	6	2.1	Figure 1.20a
DMEM	Amiloride dosing	290	1.5	0.5	Figure 1.20b
DMEM	PMSF dosing	168	9	5.4	Figure 1.21a
DMEM	PMSF dosing	640	10	1.6	Figure 1.21b
DMEM	PMSF dosing	260	9	3.5	Figure 1.23aI
DMEM	Amiloride dosing	242	13	5.4	Figure 1.23aII
DMEM	PMSF dosing	260	12	4.6	Figure 1.23bIII
DMEM	Amiloride dosing	245	12	4.9	Figure 1.23bIV

Note: $I_{D,0}$ is the starting value at the beginning of the experiments, and $\Delta I_{D,sum}$ and $\Delta I_{D,rel}$ are absolute and relative changes of I_D. For subsequent dosing experiments it corresponds to the sum of changes over all dosings.

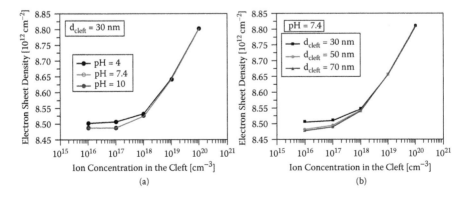

FIGURE 1.27 Calculated electron sheet density depending on the concentration of positive ions for (a) different pH values and fixed cleft dimension and for (b) different cleft dimensions and a constant pH value.

obtained experimentally. This agreement supports the assumption that the sensor signal is indeed generated by the potential changes caused by the positive ion fluxes, and is not a pure pH response.

1.7.2 ESTIMATION OF THE ION FLUX IN CLEFT

To understand the generated sensor signal, especially the ions' contribution to the recorded signal, a calculation of the ions' concentration in the cleft is performed. The total ion flux is calculated from the available data. The distance d_{cleft} of the cell to the sensor surface is assumed to be 50 nm, in accordance with the reported values of 35–70 nm that have been determined by fluorescence interference contrast microscopy [128]. The contact area of a single cell to the transistor was approximated by a circle with mean cell diameter of 20 μm, while the whole active area of the sensor was 2500×500 μm² and a covering of 100% is assumed. The calculated total number of ions in the cleft is summarized in Table 1.8.

For estimation of the flux, which is responsible for the signal generation, the data from the subsequent dosing experiments in Figure 1.23 were used. According to the simulation results in Figure 1.27, the high ion concentration in the cell media causes a nearly linear dependence of the sensor current response on the changing charge concentration, that is, $\Delta I_D \sim dQ/dt$. The saturation current, where no sensor response was measured upon further dosing of PMSF, was $\Delta I_D = 12$ μA. Since I_D changes with time, its temporary behavior needs to be considered, either by a linear approach [$I_D(t) = I_{D,0} \pm \Delta I_D*t$ and thus $Q = \frac{1}{2}\Delta I_D*t$] or by integrating Equation 1.6. The charges calculated in both ways are summarized in Table 1.9. Taking the uncertainties of these calculations into account, a reasonably good agreement could be achieved between the available Na⁺ ions reservoir (see Table 1.8) and the saturation value of the ions that were transported from or into the cleft (see Table 1.9) (about 10^{13} ions for the whole sensor). Here, it has to be considered that an additional supply of Na⁺ ions to the cleft from the surrounding cell media occurs by a diffusion process in a time scale of a few seconds, which is demonstrated by the recovery time below 10 s in Figure 1.24. In contrast, the concentration of K⁺ ions is clearly orders of magnitude too small to contribute significantly to the recorded sensor signals.

In Reference 130, values for a possible ion flux across the ion channels are given. For the three possible cases, the maximum and minimum flux across a passive ion channel as well as the average flux pumped through an active channel, the corresponding number of ion channels that would be necessary to generate the calculated ion fluxes is calculated (Table 1.9). Obviously, the sensor signal is not generated by a flux across active ion channels, which would require an unrealistically high number of channels. A comparison of the calculated passive channel density

TABLE 1.8
Cleft dimensions and total number of Na⁺ and K⁺ ions in the cleft for the cell-transistor coupling experiments

		Contact area of single cells		Sensor area
Area	$A = \dfrac{\pi d^2}{4}$	314 μm²	$A = W*L$	1.25 mm²
Cleft volume ($d_{cleft} = 50$ nm)	$V_{cleft} = A* d_{cleft}$	0.016 pl		125 pl
K⁺ concentration in cell media	$c_{K+} = 5$ mM			
K⁺ total number of ions in cleft		4.8×10^7		3.8×10^{11}
Na⁺ concentration in cell media	$c_{Na+} = 160$ mM			
Na⁺ total number of ions in cleft		1.5×10^9		1.2×10^{13}

of 2–30 channels/μm^2 with values reported in the literature (Table 1.10) shows that the obtained densities seem to be underestimated. However, it needs to be mentioned that the assumption of $\Delta I_D \sim \Delta Q$ necessarily overestimates the number of flowing charges Q by a factor of 3–5, which in turn underestimates the number of ion channels. The estimated number of about 100 ion channels per μm^2, although not from an exact analysis, shows a fairly good agreement with the measured values for axons in Table 1.10.

The major outcome of these calculations is that there is strong evidence that in the monitoring of cell activity upon neurotoxin dosing, the Na^+ flux plays the major role in signal generation. A fairly good agreement between the experimental observations and the model could be found; however, the uncertainties in the employed methods are still too large to perform a reliable quantitative analysis of the sensor behavior. Nevertheless, this chapter has demonstrated that such modeling appears to be possible if the measurement conditions, as well as the employed simulation models, can be improved further.

TABLE 1.9
Measurement conditions for the cell–transistor coupling experiments

		Linear approach	Integration of Equation 7.1
Time period	300 s		
Current change	12 μA		
Flowing charges in the cleft		1.8×10^{-3} C	1.2×10^{-4} C
Average ion flux		3.7×10^{13} ions/s	2.5×10^{12} ions/s
Maximum ion flux for passive ion channels	10^8 ions/s		
Ion channels per sensor		3.7×10^5	2.5×10^4
Ion channels per cell		94	6.2
Ion channels per area [channel/μm^2]		0.3	0.02
Minimum ion flux for passive ion channels	10^6 ions/s		
Ion channels per sensor		3.7×10^7	2.5×10^6
Ion channels per cell		9400	620
Ion channels per area [channel/μm^2]		30	2
Flux for active ion channels	100 ions/s		
Ion channels per sensor		3.7×10^{11}	2.5×10^{10}
Ion channels per cell		9.4×10^7	6.2×10^6
Ion channels per area [channel/μm^2]		3×10^6	2×10^5

TABLE 1.10
Na^+ channel gating charge densities of nerve and muscles

Tissue	Gate charge [charges/μm^2]	Na channel density [channels/μm^2]
Squid giant axon	1,500–1,900	300
Myxicola giant axon	630	105
Crayfish giant axon	2,200	367
Frog node of Ranvier	17,600	3,000
Rat node of Ranvier	12,700	2,100
Frog twitch muscle	3,900	650
Rat ventricle	260	43
Dog Purkinje fiber	1,200	200

Source: After Corry, B., 2006, *Mol. BioSyst.*, 2, 527.

1.8   SUMMARY

In this work, AlGaN/GaN heterostructures, which have been shown to be reliable pH sensors, were characterized and further developed for the in vitro monitoring of cell reactions. This was successfully tested for the reaction of NG108-15 nerve cells on different neuroinhibitors. In particular, the observed difference in the sensor reaction on dosing of amiloride depending on pretreatment demonstrates that biological processes can be distinguished unambiguously, even though they are quite complex and not understood in full detail.

Biocompatibility and stability are two major preconditions for the application of AlGaN/GaN heterostructures as biosensors. Generally, a good proliferation of different cell lines was observed on AlGaN and GaN surfaces. Importantly, it was realized without using any kind of thin film of organic material such as fibroblasts for improving the cellular adhesion and biocompatibility. Furthermore, the impact of several technology- and sensing-relevant treatments on the surface properties of AlGaN/GaN sensors was studied as well as their contaminating effect. With these investigations, technological cleaning procedures were established that maintain good biocompatibility of the GaN surface to living cells and enable study of the effect of drugs or environmental influences on the cell metabolism directly by measuring extracellular acidification or intra- and extracellular potentials.

Primary, the AlGaN/GaN heterostructure is a pH-sensitive device; nevertheless, it reacts to changing ionic concentrations in an electrolyte through the formation of temporary potentials (action potentials in the case of cells). These effects were studied in electrolytes with strongly differing ionic species such as SCZ, SCR, SCI, and cell media.

For the nerve cell–transistor coupling, a reliable seeding procedure was established that could be used for iterative measurements over periods of several days. By calibration experiments, the importance of Na^+ flux in signal generation in the present cell–transistor coupling was verified. The following cell media reactions and cell activities were monitored by the AlGaN/GaN sensors:

- Alkalization due to diffusion of CO_2 to the environment and to consumption of H_3O^+ (I_D decreases)
- Cell breathing, which results in decreased CO_2 concentration in the cell media (I_D oscillates)
- Cell reaction on DFP, which keeps Na^+ channels in an open state and depletes positive ions in the cleft (I_D decreases)
- Cell reaction on PMSF with similar results
- Cell reaction on amiloride, which blocks of Na^+. Here the reaction depends on the preliminary treatment. Dosing of amiloride only results in an accumulation of positive ions in the cleft (I_D increases), while a dosing after previous PMSF dosing has the opposite result (I_D decreases).

These experiments clearly demonstrate the ability of the AlGaN/GaN-ISFETs for quantitative analysis of cell reactions on different neuroinhibitors. However, with the described structure, the kind of participating ions cannot be identified definitively. The calculations using simulation of the heterostructure as well as simplified expressions for the ion flux strongly suggest that the signal in the cell–transistor coupling experiments is primarily generated by the Na^+ flux. For reliable quantitative analysis, however, the models have to be further developed, and various sources of error have to be identified and eliminated. Nevertheless, the AlGaN/GaN-ISFETs show stable operation under physiological conditions, exhibit excellent signal resolution, and are suitable for long-time measurements. They enable easy measurement procedures that can be used in every laboratory using small quantities of biological material, for pharmaceutical screening, drugs detection, or tumor analysis. These sensors can be cleaned and sterilized even with hazardous procedures such as autoclaving or etching in hot alkali-based solutions and are reusable without further conditioning procedures for in vivo measurements.

ACKNOWLEDGMENTS

This work was supported by the Thuringian Ministry of Culture (Pikofluidik), the Federal Ministry of Education and Research (BMBF, Center for Innovation Competence: MacroNano FKZ 03/ZIK062), and the Fraunhofer research grant Attract. The authors would like to thank to M. Klett, L. Silvera, K. Tonisch, F. Niebelschütz, T. Stauden, and K. Friedel for their kind assistance.

REFERENCES

1. Rogers, K. R., 1995, *Biosens. Bioelectron.*, 10, 533.
2. Paddle, B. M., 1996, *Biosens. Bioelectron.*, 11, 1079.
3. Ohlstein, E. H., Ruffolo, R. R. et al., 2000, *Toxicology*, 40, 177.
4. Heck, D. E., Roy, A. et al., 2001, *Biolog. React. Intermed.*, 500, 709.
5. Croston, G. E., 2002, *Trends Biotechnol.*, 20, 110.
6. Fattinger, Ch., 2002, *Carl Zeiss, Innovation,* 12.
7. Jorkasky, D. K., 1998, *Toxicol. Lett.*, 102–103, 539.
8. Kinter, L. B. and Valentin, J. P., 2002, *Fundam. Clin. Pharmacol.*, 16, 175.
9. Bousse, L., 1996, *Sens. Actuat. B: Chem.*, 34, 270.
10. Bentleya, A., Atkinsona, A. et al., 2001, *Toxicol. In Vitro*, 15, 469.
11. Baeumner, A. J., 2003, *Anal. Bioanal. Chem.*, 377, 434.
12. Fromherz, P., Offenhäusser, A., Vetter, T., and Weiss, J., 1991, *Science*, 252, 1290.
13. Gross, G. W., Harsch, A. et al., 1997, *Biosens. Bioelectron.*, 12, 373.
14. Denyer, M. C. T., Riehle, M. et al., 1998, *Med. Biol. Eng. Comput.*, 36, 638.
15. Jung, D. R., Cuttino, D. S. et al., 1998, *J. Vac. Sci. Technol. A*, 16, 1183.
16. Offenhäusser, A. and Knoll, W., 2001, *Trends Biotechnol.*, 19, 62.
17. Krause, M., Ingebrandt, S. et al., 2000, *Sens. Actuat. B: Chem.*, 70, 101.
18. Stett, A., Egert, U. et al., 2003, *Anal. Bioanal. Chem.*, 377, 486.
19. Steinhoff, G., Baur, B., Wrobel, G., Ingebrandt, S., Offenhäusser, A., Dadgar, A., Krost, A., Stutzmann, M., and Eickhoff, M., 2005, *Appl. Phys. Lett.*, 86, 33901.
20. Hodgkin, A. L. and Huxley, A. F., 1952, *J. Physiol.*, 117, 500.
21. Neher, E. and Sakmann, B., 1976, *Nature*, 260, 799.
22. Hamill, O. P., Marty, A., Neher, E., Sackman, B., and Sigworth, F. J., 1981, *Pflügers Archiv.-Eur. J. Physiol.*, 391, 2, 85.
23. Thomas, C. A., Springer, P. A., Loeb, G. E., Berwald-Netter, Y., and Okun, L. M., 1972, *Exp. Cell Res.*, 74, 61.
24. Gross, G. W., Rieske, E., Kreutzberg, G. W., and Meyer, A., 1977, *Neurosci. Lett.*, 6, 101–105.
25. Pine, J., 1980. *J. Neurosci. Meth.*, 2, 19.
26. Novak, J. L. and Wheeler, B. C., 1986, *IEEE Trans. Biomed. Eng., BME*, 33, 196.
27. Drodge, M. H., Gross, G. W., Hightower, M. H., and Czisny, L. E., 1986, *J. Neurosci. Meth.*, 6, 1583.
28. Eggers, M. D., Astolfi, D. K., Liu, S., Zeuli, H. E., Doeleman, S. S., McKay, R., Khuon, T. S., and Ehrlich, D. J., 1990, *J. Vac. Sci. Technol. B*, 8, 1392.
29. Martinoia, S., Bove, M., Carlini, G., Ciccarelli, C., Grattarola, M., Storment, C., and Kovacs, G., 1993, *J. Neurosci. Meth.*, 48, 115.
30. Maeda, E., Robinson, H. P. C., and Kawana, A., 1995, *J. Neurosci.*, 15, 6834.
31. Fromherz, P., Müller, C. O., and Weis, R., 1993, *Phys. Rev. Lett.*, 71, 4079.
32. Fromherz, P. and Stett, A., 1995, *Phys. Rev. Lett.*, 75, 1670.
33. Bergveld, P., 1970, *IEEE Trans. Biomed. Eng.*, 17, 70.
34. Bergveld, P., 2003, *Sens. Actuat. B*, 88, 1.
35. Schöning, M. J. and Poghossian, A., 2002, *Analyst*, 127, 1137.
36. Schöning, M. J. and Poghossian, A., 2006, *Electroanalysis*, 18, 1893.
37. Madou, M. J., 1989, *Biomedical Sensors,* Academic Press, Boston.
38. Neuberger, R., Müller, G., Ambacher, O., and Stutzmann, M., 2001, *Phys. Stat. Sol. (a)*, 183, R10.
39. Eickhoff, M., Schalwig, J., Steinhoff, G., Weidemann, O., Görgens, L., Neuberger, R., Hermann, M., Baur, B., Müller, G., Ambacher, O., and Stutzmann, M., 2003, *Phys. Stat. Sol. (c)*, 0, 1908.
40. Pearton, S. J., Kang, B. S., Kim, S., Ren, F., Gila, B. P., Abernathy, C. R., Lin, J., and Chu, S. N. G., 2004, *J. Phys.: Cond. Matter*, 16, R961.

41. Steinhoff, G., Hermann, M., Schaff, W. J., Eastman, L. F., Stutzmann, M., and Eickhoff, M., 2003, *Appl. Phys. Lett.*, 83, 177.

42. Bindra, A., and Valentine, M., 2007, *RFDESIGN*, 18.

43. Pearton, S. J., Zolper, J. C., Shul, R. J., and Ren, F., 1999, *J. Appl. Phys.*, 86, 1.

44. Cimalla, I., Foerster, Ch., Cimalla, V., Lebedev, V., Cengher, D., and Ambacher, O., 2003, *Phys. Stat. Sol.* (c) 0, 3767.

45. Young, T.-H. and Chen, C.-R., 2006, *Biomaterials*, 27, 3361.

46. Cimalla, I., Will, F., Tonisch, K., Niebelschütz, M., Cimalla, V., Lebedev, V., Kittler, G., Himmerlich, M., Krischok, S., Schaefer, J. A., Gebinoga, M., Schober, A., and Ambacher, O., 2007, *Sens. Actuat. B*, 123, 740.

47. Steinhoff, G., Purrucker, O., Tanaka, M., Stutzmann, M., and Eickhoff, M., 2003, *Adv. Funct. Mater.*, 13, 841.

48. Wong, K.-Y., Tang, W., Lau, K. M., and Chen, K. J., 2007, *Appl. Phys. Lett.*, 90, 213506.

49. Ambacher, O., Majewski, J., Miskys, C., Link, A., Hermann, M., Eickhoff, M., Stutzmann, M., Bernardini, F., Fiorentini, V., Tilak, V., Schaff, B., and Eastman, L. F., 2002, *J. Phys.: Condens. Mat.*, 14, 3399.

50. Ambacher, O., Smart, J., Shealy, J. R., Weimann, N. G., Chu, K., Murphy, M., Schaff, W. J., Eastman, L. F., Dimitrov, R., Wittmer, L., Stutzmann, M., Riegwer, W., and Hilsenbeck, J., 1999, *J. Appl. Phys.*, 85, 3222.

51. Neuberger, R., Müller, G., Ambacher, O., and Stutzmann, M., 2001, *Phys. Stat. Sol.* (a), 185, 85.

52. Bergveld, P., 2003, *IEEE Sens. Conf.*, Toronto.

53. Bousse, L., Rooij, N. F. de, and Bergveld, P., 1983, *IEEE Trans. Electron. Dev.*, 30, 1263.

54. Yates, D. E., Levine, S., and Healy, T. W., 1974, *J. Chem. Soc., Faraday Trans.* 1, 70, 1807.

55. Bayer, M., Uhl, C., and Vogl, P., 2005, *J. Appl. Phys.*, 97, 33703.

56. Kocan, M., Rizzi, A., Lüth, H., Keller, S., and Mishra, U. K., 2002, *Phys. Stat. Sol.* (b), 234, 733.

57. Siu, W. M., and Cobbold, R. S. C., 1979, *IEEE Trans. Electron. Dev.*, 26, 1805.

58. Fung, C. D., Cheung, P. W., and Ko, W. H., 1986, *IEEE Trans. Electron. Dev.*, 33, 8.

59. Ambacher, O., and Cimalla, V., in *Polarization Effects in Semiconductors: From Ab Initio to Device Application* (eds. C. Wood and D. Jena), Springer, New York, 2008, p. 27.

60. Spitznas, A., 2005, *Novel Sensors from Semiconductors,* Diploma thesis, Technical University Ilmenau.

61. Linkohr, S., 2008, *Fabrication of GaN Sensors,* Diploma thesis, Technical University Ilmenau.

62. Costa Silveira, L., 2008, *Cell Kinetics Using Direct Detection*, Diploma thesis, Technical University Ilmenau.

63. Ambacher, O., 1998, *J. Phys. D: Appl. Phys.*, 31, 2653.

64. Jain, S., Willander, M., Narayan, J., and van Overstraeten, R., 2000, *J. Appl. Phys.*, 87, 965.

65. Mehandru, R., Luo, B., Kang, B. S., Kim, J., Ren, F., Pearton, S. J., Pan, C.-C., Chen, G.-T., and Chyi, J.-I., 2004, *Sol. State Electron.*, 48, 351.

66. Alifragis, Y., Georgakilas, A., Konstantinidis, G., Iliopoulos, E., Kostopoulos, A., and Chaniotakis, N. A., 2005, *Appl. Phys. Lett.*, 87, 253507.

67. Kang, B. S., Ren, F., Wang, L., Lofton, C., Tan, W. W., Pearton, S. J., Dabiran, A., Osinsky, A., and Chow, P. P., 2005, *Appl. Phys. Lett.*, 87, 23508.

68. Kang, B. S., Pearton, S. J., Chen, J. J., Ren, F., Johnson, J. W., Therrien, R. J., Rajagopal, P., Roberts, J. C., Piner, E. L., and Linthicum, K. J. et al., 2006, *Appl. Phys. Lett.*, 89, 122102.

69. Baur, B., Howgate, J., Ribbeck, H.-G. von, Gawlina, Y., Bandalo, V., Steinhoff, G., Stutzmann, M., and Eickhoff, M., 2006, *Appl. Phys. Lett.*, 89, 183901.

70. Lebedev, V., Cherkashinin, G., Ecke, G., Cimalla, I., and Ambacher, O., 2007, *J. Appl. Phys.*, 101, 1.

71. Tonisch, K., 2005, *New Sensing Mechanisms for Cells*, Diploma thesis, Technical University Ilmenau.

72. Morkoç, H., 1999, *GaN Electronics,* Springer-Verlag, Berlin, Heidelberg, New York.

73. Lebedev, V., Tonisch, K., Niebelschütz, F., Cimalla, V., Cengher, D., Cimalla, I., Mauder, Ch., Hauguth, S., Morales, F. M., Lozano, J. G., González, D., and Ambacher, O., 2007, *J. Appl. Phys.*, 101, 054906.

74. Motayed, A., Bathe, R., Wood, M. C., Diouf, O. S., Vispute, R. D., and Mohammad, S. N., 2003, *J. Appl. Phys.*, 93, 1087.

75. Lee Kee-Keun, He, J., Singh, A., Massia, St., Ehteshami, Gh., Kim, B., and Raupp, Gr., 2004, *J. Micromech. Microeng.*, 14, 32.

76. Pyralin LX-Series, 1996–2005, Produktinformation DuPont United Kingdom, Ltd.

77. Hintz, M., 2007, Ph.D. thesis, Technical University Ilmenau.

78. Kittler, G., 2008, Ph.D. thesis, Technical University Ilmenau.

79. Geitz, C., 2007, *Processing of Nerve Cell Sensors*, Diploma thesis, Technical University Ilmenau.

80. Stocker, D. A., Schubert, E. F., and Redwing, J. M., 1998, *Appl. Phys. Lett.*, 73, 2654.

81. Zhuang, D. and Edgar, J. H., 2005, *Mater. Sci. Eng.* R48, 1-46.

82. Zhu, K., Kuryatkov, V., Borisov, B., and Kipshidze, G., 2002, *Appl. Phys. Lett.*, 81, 4688.

83. Cho, H., Hahn, Y.-B., Hays, D. C., and Abernathy, C. R., 1999, *J. Vac. Sci. Technol.*, A 17, 2202.
84. Ozasa, K., Nemoto, S., Hara, M., and Maeda, M., 2006, *Phys. Stat. Sol.* (a), 203, 2287.
85. Kim, H., Colavita, P. E., Metz, K. M., Nichols, B. M., Bin Sun, Uhlrich, J., Wang, X., Kuech, Th. F., and Hamers, R. J., 2006, *Langmuir*, 22, 8121.
86. Wang, Yu-Lin., Chu, B. H., Chen, K. H., Chang, C. Y., Lele, T. P., Tseng, Y., Pearton, S. J., Ramage, J., Hooten, D., Dabiran, A., Chow, P. P., and Ren, F., 2008, *Appl. Phys. Lett.*, 93, 262101.
87. Nic, M., Jirat, J., and Kosata, B., IUPAC Compendium of Chemical Terminology, http://goldbook.iupac.org/index.html.
88. Olmsted, J. III, and Williams, G. M., 1997, Wm. C. Brown Publishers, USA.
89. Minuth, W. W., Strehl, R., Schumacher, K., 2002, *Nerve Cell Response,* Pabst Science Publishers.
90. http://www.fermentas.com/catalog/electrophoresis/topvisionagarle.htm.
91. Cimalla, V., Lebedev, V., Linkohr, S., Cimalla, I., Lübbers, B., Tonisch, K., Brückner, K., Niebelschütz, F., and Ambacher, O., 2008, *17th European Workshop on Heterostructure Technology*, Venice, Italy, November 3–5, 2008, p. 33.
92. Lehmann, M., Baumann, W., Brischwein, M., Gahle, H.-J., Freund, I., Ehret, R., Drechsler, S., Palzer, H., Kleintges, M., and Sieben, U. et al., 2001, *Biosens. Bioelectron.*, 16, 195.
93. Schober, A., Kittler, G., Lübbers, B., Buchheim, C., Ali, M., Cimalla, V., Fischer, M., Spitznas, A., Gebinoga, M., and Yanev, V. et al., 2005, *Proc. 7. Dresdener Sensor-Symposium*, TUD Press, Dresden, 143.
94. Borkholder, D. A., 1998, Ph.D. thesis, Stanford University, USA.
95. Imanishi, T., Matsushima, K., Kawaguchi, A., Wada, T., Masuko, T., Yoshida, S., and Ichida, S., 2006, *Biol. Pharm. Bull.*, 29, 701.
96. Krystosek, A., 1985, *J. Cell. Physiol.*, 125, 319.
97. Ma, Wu, Joseph, J. Pancrazio, J. J., Margaret, Coulombe, M., Judith, Dumm, J., RamaSri Sathanoori, R., Jeffery, L. Barker, J. L., Vijay, C. Kowtha, V. C., David, A. Stenger, D. A., James, J., and Hickman, J. J., 1998, *Dev. Brain Res.*, 106, 155.
98. Seidman, K. J. N., Barsuk, J. H., Johnson, R. F., and Weyhenmeyer, J. A., 1996, *J. Neurochem.*, 66, 1011.
99. Goshima, Y., Ohsako, S., and Yamauchi, T., 1993, *J. Neurosci.*, 13, 559.
100. Rouzaire-Dubois, B. and Dubois, J. M., 2004, *Gen. Physiol. Biophys.*, 23, 231.
101. Chen, X.-L., Zhong, Z.-G., Yokoyama, S., Bark, Ch., Meister, B., Berggren, P.-O., Roder, J., Higashida, H., and Jeromin, A., 2001, *J. Physiol.*, 532.3, 649.
102. Schmitt, H. and Meves, H., 1995, *J. Physiol.*, 89, 181.
103. Schmitt, H. and Meves, H., 1995, *J. Membrane Biol.*, 145, 233.
104. Schäfer, S., Béhé, Ph., and Meves, H., 1991, *Pflügers Arch.*, 418, 581.
105. Chin, T.-Y., Hwang, H. M., and Chueh, Sh.-H., 2002, *Mol. Pharmacol.*, 61, 486.
106. Chau, L.-Y., Lin, T. A., Chang, W.-T., Chen, C.-H., Shue, M.-J., Hsu, Y.-S., Hu, C.-Y., Tsai, W.-H., and Sun, G. Y., 1993, *J. Neurochem*, 60, 454.
107. Hsu, L.-S., Chou, W. -Y., and Chueh, Sh.-H., 1995, *J. Biochem.*, 309, 445.
108. Mima, K., Donai, H., and Yamauchi, T., 2002, *Biol. Proced. Online*, 3, 79.
109. Spilker, C., Gundelfinger, E. D., and Braunewell, K.-H., 2002, *Biochim. Biophys. Acta*, 1600, 118.
110. Mohan, D. K., Molnar, P., and Hickman, J. J., 2006, *Biosens. Bioelectron.*, 21, 1804.
111. Yano, K., Higashida, H., Inoue, R., and Nozawa, Y., 1984, *J. Biol. Chem.*, 259, 10201.
112. Zhang, B.-F., Peng, F.-F., Zhang, J.-Z., and Wu, D.-C., 2003, *Acta Pharmacol. Sin.*, 24, 663.
113. Kugawa, F., Arae, K., Ueno, A., and Aoki, M., 1998, *Eur. J. Pharmacol.*, 347, 105.
114. Nelson, Ph., Christian, C., and Nirenberg, M., 1976, *Proc. Natl. Acad. Sci. USA*, 73, 123.
115. Shimohira-Yamasaki, M., Toda, S., Narisawa, Y., and Sugihara, H., 2006, *Cell Structure Function*, 31, 39.
116. Choi, R. C., Pun, S., Dong, T. T., Wan, D. C., and Tsim, K. W. 1997, *Neurosci. Lett.*, 236, 167.
117. Vogel, G. and Angermann, H., 2002, *Nerve Cell Responses to Inhibitors,* Munich: Dt. Taschenbuch-Verl.
118. Becker, W., Deamer, D., 1991, *The World of the Cell*, The Benjamin/Cummings Publishing Company, Menlo Park, CA.
119. Kandel, E. R., Schwartz, J. H., and Jessel, T. M., 2000, *Principles of Neural Science*, McGraw-Hill, New York.
120. Brown, A. G., 2001, *Nerve Cells and Nervous Systems: An Introduction to Neuroscience*, Springer-Verlag, London.
121. Webster, R. A., 2001, *Neurotransmitters, Drugs and Brain Function*, John Wiley & Sons, Baffins Lane, Chichester, West Sussex PO19 1UD, U.K.

122. Han, B. S., Hong, H.-S., Choi Won-Seok, Markelonis, G. J., Oh Tae, H., and Oh Young, J., 2003, *J. Neurosci.*, 23, 5069.
123. Olivera-Bravo, S., Ivorra, I., and Morales, A., 2005, *Brit. J. Pharmacol.*, 144, 88.
124. Du, D., Huang, X., Cai, J., and Zhang, A., 2007, *Biosens. Bioelectron.*, 23, 285.
125. Brenner, G. M., 2000, *Pharmacology*, Philadelphia, PA: W.B. Saunders Company.
126. Freshney, R. I., 2005, *Culture of Animal Cells: A Manual of Basic Technique*, John Wiley & Sons, NY.
127. Braun, D. and Fromherz, P., 1998, *Phys. Rev. Lett.*, 81, 5241.
128. Hille, B., 1991, *Ionic Channels of Excitable Membranes*, Sinauer Associates, Sunderland, MA.
129. Corry, B., 2006, *Mol. BioSyst.*, 2, 527.
130. Polyakov, V. M. and Schwierz, F., 2007, *J. Appl. Phys.*, 101, 033703.
131. Polyakov, V. M., Schwierz, F., Cimalla, I., Kittler, M., Lübbers, B., and Schober, A., 2009, *J. Appl. Phys.*, 106, 023715.

2 Recent Advances in Wide-Bandgap Semiconductor Biological and Gas Sensors

B.H. Chu
Department of Chemical Engineering, University
of Florida, Gainesville, FL 32611 USA

C.Y. Chang
Department of Materials Science and Engineering,
University of Florida, Gainesville, FL 32611 USA

S.J. Pearton
Department of Materials Science and Engineering,
University of Florida, Gainesville, FL 32611 USA

Jenshan Lin
Department of Electrical and Computer Engineering,
University of Florida, Gainesville, FL 32611

F. Ren
Department of Chemical Engineering, University
of Florida, Gainesville, FL 32611 USA

CONTENTS

2.1 INTRODUCTION

Chemical sensors have gained importance in the past decade for applications that include homeland security, medical and environmental monitoring, and food safety. A desirable goal is the ability to simultaneously analyze a wide variety of environmental and biological gases and liquids in the field and to be able to selectively detect a target analyte with high specificity and sensitivity. In the area of detection of medical biomarkers, many different methods have been employed, including enzyme-linked immunosorbent assay (ELISA), particle-based flow cytometric assays, electrochemical measurements based on impedance and capacitance, electrical measurement of microcantilever resonant frequency change, and conductance measurement of semiconductor nanostructures, gas chromatography (GC), ion chromatography, high-density peptide arrays, laser scanning quantitative analysis, chemiluminescence, selected ion flow tube (SIFT), nanomechanical cantilevers, bead-based suspension microarrays, magnetic biosensors, and mass spectrometry (MS) [1–9]. Depending on the sample condition, these methods may show variable results in terms of sensitivity for some applications and may not meet the requirements for a handheld biosensor.

For homeland security applications, reliable detection of biological agents in the field and in real time is challenging. During the anthrax attack on the World Bank in 2002, field tests showed 1200 workers to be positive, and all were sent home. Antibiotics were provided to 100 workers. However, confirmatory testing showed zero positives. False positives and false negatives can result due to very low volumes of samples available for testing and poor device sensitivities. Toxins such as ricin, botulinum toxin, or enterotoxin B are environmentally stable, can be mass-produced, and do not need advanced technologies for production and dispersal. The threat of these toxins is real. This is evident from the ricin detection at White House mail facilities and a U.S. senator's office. Terrorists have already attempted to use botulinum toxin as a bioweapon. Aerosols were dispersed at multiple sites in Tokyo and at U.S. military installations in Japan on at least three occasions between 1990 and 1995 by the Japanese cult Aum Shinrikyo [10]. Four of the countries listed by the U.S. government as "state sponsors of terrorism" (Iran, Iraq, North Korea, and Syria) [10] have developed, or are believed to be developing, botulinum toxin as a weapon [11,12]. After the 1991 Persian Gulf War, Iraq admitted to the United Nations inspection team to having produced 19,000 L of concentrated botulinum toxin, of which approximately 10,000 L were loaded into military weapons. This toxin has not been fully accounted for and constitutes approximately three times the amount needed to kill the entire current human population by inhalation [10]. A significant issue is the absence of a definite diagnostic method and the difficulty in differential diagnosis from other pathogens that would slow the response in case of a terror attack. This is a critical need that has to be met to have an effective response to terrorist attacks. Given the adverse consequences of a lack of reliable biological agent sensing on national security, there is a critical need to develop novel, more sensitive, and reliable technologies for biological detection in the field [13]. Some specific toxins of interest include enterotoxin type B (Category B, NIAID), botulinum toxin (Category A NIAID), and ricin (Category B NIAID).

While the techniques mentioned earlier show excellent performance under lab conditions, there is also a need for small, handheld sensors with wireless connectivity that have the capability for fast responses. The chemical sensor market represents the largest segment for sales of sensors, including chemical detection in gases and liquids, flue gas and fire detection, liquid quality sensors, biosensors, and medical sensors. Some of the major applications in the home include indoor air quality and natural gas detection. Attention is now being paid to more demanding applications where a high degree of chemical specificity and selectivity is required. For biological and medical sensing applications, disease diagnosis by detecting specific biomarkers (functional or structural abnormal enzymes, low-molecular-weight proteins, or antigen) in blood, urine, saliva, or tissue samples has been established. Most of the techniques mentioned earlier such as ELISA possesses a major limitation in that only one analyte is measured at a time. Particle-based assays allow for multiple detection by using multiple beads, but the whole detection process is generally longer than 2 h, which is not practical for in-office or bedside detection. Electrochemical devices have attracted attention due to their low cost and simplicity, but significant improvements in their sensitivities are still needed for use with clinical samples. Microcantilevers are capable of detecting concentrations as low as 10 pg/mL, but suffer from an undesirable resonant frequency change due to the viscosity of the medium and cantilever damping in the solution environment. Nanomaterial devices have provided an excellent option toward fast, label-free, sensitive, selective, and multiple detections for both preclinical and clinical applications. Examples of detection of biomarkers using electrical measurements with semiconductor devices include carbon nanotubes for lupus erythematosus antigen detection [4], compound semiconducting nanowires and In_2O_3 nanowires for prostate-specific antigen detection [5], and silicon nanowire arrays for detecting prostate-specific antigen [9]. In clinical settings, biomarkers for a particular disease state can be used to determine the presence of disease as well as its progress.

Semiconductor-based sensors can be fabricated using mature techniques from the Si chip industry and/or novel nanotechnology approaches. Silicon-based sensors are still the dominant component of the semiconductor segment due to their low cost, reproducibility, and controllable electronic response. However, these sensors are not suited for operation in harsh environments, for instance, high temperature, high pressure, or corrosive ambients. Si will be etched by some of the acidic or basic aqueous solutions encountered in biological sensing. In sharp contrast, GaN is not etched by any acid or base at temperatures below a few hundred degrees. Therefore, wide-bandgap group III nitride compound semiconductors (AlGaInN materials system) are alternative options to supplement silicon in these applications because of their chemical resistance, high temperature/high power capability, high electron saturation velocity, and simple integration with existing GaN-based UV light-emitting diode, UV detectors, and wireless communication chips.

A promising sensing technology utilizes AlGaN/GaN high-electron mobility transistors (HEMTs). HEMT structures have been developed for use in microwave power amplifiers due to their high two-dimensional electron gas (2DEG) mobility and saturation velocity. The conducting 2DEG channel of AlGaN/GaN HEMTs is very close to the surface and extremely sensitive to adsorption of analytes. HEMT sensors can be used for detecting gases, ions, pH values, proteins, and DNA.

The GaN materials system is attracting much interest for commercial applications of green, blue, and UV light emitting diodes (LEDs), laser diodes, as well as high-speed and high-frequency power devices. Due to the wide-bandgap nature of the material, it is very thermally stable, and electronic devices can be operated at temperatures up to 500°C. The GaN-based materials are also chemically stable, and no known wet chemical etchant can etch these materials; this makes them very suitable for operation in chemically harsh environments. Due to the high electron mobility, GaN material-based HEMTs can operate at very high frequency with higher breakdown voltage, better thermal conductivity, and wider transmission bandwidths than Si or GaAs devices [14–16].

An overlooked potential application of the GaN HEMT structure is sensors. The high electron sheet carrier concentration of nitride HEMTs is induced by piezoelectric polarization of the strained AlGaN layer in the heterojunction structure of the AlGaN/GaN HEMT, and the spontaneous polarization is very large in wurtzite III-nitrides. This provides an increased sensitivity relative to simple Schottky diodes fabricated on GaN layers or field-effect transistors (FETs) fabricated on the AlGaN/GaN HEMT structure. The gate region of the HEMT can be used to modulate the drain current in the FET mode or used as the electrode for the Schottky diode. A variety of gas, chemical, and health-related sensors based on HEMT technology have been demonstrated with proper surface functionalization on the gate area of the HEMTs, including the detection of hydrogen, mercury ion, prostate-specific antigen (PSA), DNA, and glucose [17–58].

In this chapter, we discuss recent progress in the functionalization of these semiconductor sensors for applications in detection of gases, pH measurement, biotoxins, and other biologically important chemicals, and the integration of these sensors into wireless packages for remote sensing capability.

2.2 GAS SENSING

2.2.1 O_2 SENSING

The current technology for O_2 measurement, referred to as *oximetry*, is small and convenient to use. However, the O_2 measurement technology does not provide a complete measure of respiratory sufficiency. A patient suffering from hypoventilation (poor gas exchange in the lungs) given 100% oxygen can have excellent blood oxygen levels while still suffering from respiratory acidosis due to excessive CO_2. The O_2 measurement is also not a complete measure of circulatory sufficiency. If there is insufficient blood flow or insufficient hemoglobin in the blood (anemia), tissues can suffer hypoxia despite high oxygen saturation in the blood that does arrive. The current oxide-based O_2 sensors can operate at very high temperatures, such as the commercialized solid electrolyte ZrO_2 (700°C) or the semiconductor metal oxides such as TiO_2, Nb_2O_5, $SrTiO_3$, and CeO_2 (>400°C). However, it remains important to develop a low-operation-temperature and high-sensitivity O_2 sensor to build a small, portable, and low-cost O_2 sensor system for biomedical applications.

Oxide-based materials are widely used and studied for oxygen sensing because of their low cost and good reliability. The commercialized solid electrolyte ZrO_2 [59] has been widely used in automobiles for oxygen sensing in combustion processes. The electrolyte metal oxide oxygen sensor usually uses a reference gas and operates at high temperature (700°C) [60]. Semiconductor metal oxides such as TiO_2, Ga_2O_3, Nb_2O_5, $SrTiO_3$, and CeO_2 do not need the reference gas, but they still need to be operated at a considerably high temperature (>400°C) in order to reach high sensitivity, which means a high power consumption for heating up the sensors [61–66]. For biomedical applications, such as monitoring oxygen in the breath for a lung transplant patient, a portable and low power consumption O_2 sensor system is needed. Therefore, it is crucial to develop a low operating temperature and high sensitivity O_2 sensor for those applications.

The conductivity mechanism of most metal oxides-based semiconductors results from electron hopping from intrinsic defects in the oxide film, and these defects are related to the oxygen vacancies generated during oxide growth. Typically, the higher the concentration of oxygen vacancies in the oxide film, the more conductive is the film. InZnO (IZO) films have been used in fabricating thin-film transistors, and the conductivity of the IZO is also found to depend on the oxygen partial pressure during the oxide growth [67–69]. The IZO is a good candidate for O_2-sensing applications.

The schematic of the oxygen sensor is shown at the top of Figure 2.1. The bottom part of the figure shows the device had a strong response when it was tested at 120°C in pure nitrogen and pure oxygen alternately at Vds = 3 V. When the device was exposed to the oxygen, the drain-source current decreased, whereas when the device was exposed to nitrogen, the current increased. The IZO film provides a high oxygen vacancy concentration, which makes the film readily sense oxygen and create a potential on the gate area of the AlGaN/GaN HEMT. A sharp drain-source current change

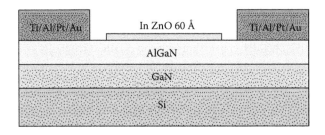

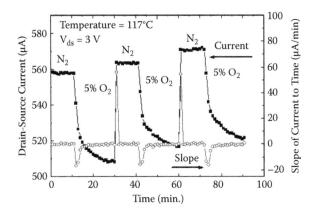

FIGURE 2.1 (See color insert).   (Top) Schematic of AlGaN/GaN HEMT-based O_2 sensor. (Bottom) Drain current of IZO-functionalized HEMT sensor measured at fixed source-drain during the exposure to different O_2 concentration ambients. The drain bias voltage was 0.5 V, and measurements were conducted at 117°C.

demonstrates the combination of the advantage of the high electron mobility of the HEMT and the high oxygen vacancy concentration of the IZO film. Because of these advantages, this oxygen sensor can operate with a high sensitivity at a relatively low temperature compared to many oxide-based oxygen sensors which operate from 400°C to 700°C.

In summary, it is clear that through a combination of IZO films and the AlGaN/GaN HEMT structure, a low operation temperature and low-power consumption oxygen sensor can be achieved. The sensor can be either used in the steady-state or in the annealed mode, which provides flexibility in various applications. This device shows promise for portable, fast-response, and high-sensitivity oxygen detectors.

2.2.2 CO_2 Sensing

The detection of carbon dioxide (CO_2) gas has attracted attention in the context of global warming, biological and health-related applications, such as indoor air quality control, process control in fermentation, and in the measurement of CO_2 concentrations in patients' exhaled breath with lung and stomach diseases [70–73]. In medical applications, it can be critical to monitor the CO_2 and O_2 concentrations in the circulatory systems for patients with lung diseases in the hospital. The current technology for CO_2 measurement typically uses infrared instruments, which can be very expensive and bulky.

The most common approach for CO_2 detection is based on nondispersive infrared (NDIR) sensors, which are the simplest of the spectroscopic sensors [74–77]. The best detection limits for the NDIR sensors are currently in the range of 20–10,000 ppm. The key components of the NDIR approach are an IR source, a light tube, an interference filter, and an infrared (IR) detector. In operation, gas enters the light tube. Radiation from the IR light source passes through the gas in the light tube to impinge on the IR detector. The interference filter is positioned in the optical path in front of the IR detector such that the IR detector receives the radiation of a wavelength that is strongly absorbed by the gas whose concentration is to be determined while filtering out the unwanted wavelengths. The IR detector produces an electrical signal that represents the intensity of the radiation impinging upon it. It is generally considered that the NDIR technology is limited by power consumption and size.

In recent years, monomers or polymers containing amino-groups, such as tetrakis(hydroxyethyl) ethylenediamine, tetraethylene-pentamine, and polyethyleneimine (PEI), have been used for CO_2 sensors to overcome the power consumption and size issues found in the NDIR approach [78–82]. Most of the monomers or polymers are utilized as coatings of surface acoustic wave transducers. The polymers are capable of adsorbing CO_2 and facilitating a carbamate reaction. PEI has also been used as a coating on carbon nanotubes for CO_2 sensing by measuring the conductivity of nanotubes upon exposing to the CO_2 gas. For example, CO_2 adsorbed by a PEI-coated nanotube portion of a NTFET (nanotube FET) sensor lowers the total pH of the polymer layer and alters the charge transfer to the semiconducting nanotube channel, resulting in the change of NTFET electronic characteristics [83–86].

A schematic cross-section of the device is shown in Figure 2.2. The interaction between CO_2 and amino group-containing compounds with the influence of water molecules is based on an acid–base reaction. The purpose of adding starch into the PEI in our experiment was to enhance the absorption of the water molecules into the PEI/starch thin film. Several possible reaction mechanisms have been suggested. The key reaction was that primary amine groups $-NH_2$ on the PEI main chain reacted with CO_2 and water-forming $-NH_3^+$ ions, and the CO_2 molecule became $OCOOH^-$ ions. Thus, the charges, or the polarity, on the PEI main chain were changed. The electrons in the two-dimensional electron gas (2DEG) channel of the AlGaN/GaN HEMT are induced by piezoelectric and spontaneous polarization effects. This 2DEG is located at the interface between the GaN layer and AlGaN layer. There are positive counter charges at the AlGaN surface layer induced by the 2DEG. Any slight changes in the ambient of the AlGaN/GaN HEMT affect the surface charges of

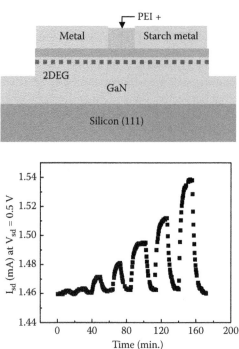

FIGURE 2.2 Schematic of AlGaN/GaN HEMT-based CO_2 sensor (top) and drain current of PEI/starch functionalized HEMT sensor measured at fixed source drain during the exposure to different CO_2 concentration ambients. The drain bias voltage was 0.5 V, and measurements were conducted at 108°C.

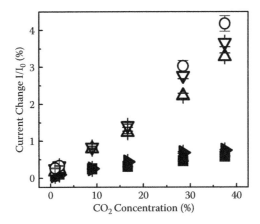

FIGURE 2.3   The drain current changes of HEMT sensor as a function of CO_2 concentration. The inset is the current change of the sensors as function of lower CO_2 concentrations (0.9%–10%).

the AlGaN/GaN HEMT. The PEI/starch was coated on the gate region of the HEMT. The charges of the PE changed through the reactions between -NH_2 and CO_2 as well as water molecules. These are then transduced into a change in the concentration of the 2DEG in the AlGaN/GaN HEMTs.

Figure 2.3 (bottom) shows the drain current of PEI/starch-functionalized HEMT sensors measured exposed to different CO_2 concentration ambients. The measurements were conducted at 108°C and a fixed source-drain bias voltage of 0.5 V. The current increased with the introduction of CO_2 gas. This was due to the net positive charges increased on the gate area, thus inducing electrons in the 2DEG channel. The response to CO_2 gas has a wide dynamic range from 0.9% to 40%, as shown in Figure 2.3. Higher CO_2 concentrations were not tested because there is little interest in these for medical-related applications. The response times were on the order of 100 s. The signal decay time was slower than the rise time and was due to the longer time required to purge CO_2 out from the test chamber.

The effect of ambient temperature on CO_2 detection sensitivity was investigated. The drain current changes were linearly proportional to the CO_2 concentration for all the tested temperatures. However, the HEMT sensors showed higher sensitivity for the higher testing temperatures. There was a noticeable change of the sensitivity from the sensors tested at 61°C to those tested at 108°C. This difference is likely due to higher ambient temperature increasing the reaction rate between amine groups and CO_2 as well as the diffusion of CO_2 molecules into the PEI thin film. The sensors exhibited reversible and reproducible characteristics.

In conclusion, PEI/starch-functionalized HEMT sensors for CO_2 detection with a wide dynamic range from 0.9% to 50%. The sensors were operated at low bias voltage (0.5 V) for low power consumption applications. The sensors showed higher sensitivity at the testing temperature higher than ~100°C. The sensors showed good repeatability. This electronic detection of CO_2 gas is a significant step toward a compact sensor chip that can be integrated with a commercially available handheld wireless transmitter to realize a portable, fast, and high sensitive CO_2 sensor.

2.2.3 CH_4 SENSING

Of particular interest in developing wide-bandgap sensors are methods for detecting ethylene (C_2H_4), which offers problems because of its strong double bonds and hence the difficulty in dissociating it at modest temperatures [87–89]. Ideal sensors have the ability to discriminate between different gases and arrays that contain different metal oxides (e.g., SnO_2, ZnO, CuO, WO_3) on the same chip can be used to obtain this result. Another prime focus should be the thermal stability of the detectors, since they are expected to operate for long periods at elevated temperature [90–97]. MOS diode-based sensors have significantly better thermal stability than a metal-gate structure and also sensitivity than Schottky diodes on GaN. In this work, we show that both AlGaN/GaN MOS diodes and Pt/ZnO bulk Schottky diodes are capable of detecting low concentrations (10%) of ethylene at temperatures between 50°C to 300°C (ZnO) or 25°C to 400°C (GaN).

Figure 2.4 (top) shows a schematic of the completed AlGaN/GaN MOS-HEMT and, at bottom, the difference in forward diode current at 400°C of the $Pt/Sc_2O_3/AlGaN/GaN$ MOS-HEMT diode both in pure N_2 relative to a 10% $C_2H_4/90\%$ N_2 atmosphere. At a given forward bias, the current increases on introduction of the C_2H_4. In analogy with the detection of hydrogen in comparable SiC and Si Schottky diodes, a possible mechanism for the current increases involves atomic hydrogen that is either decomposed from C_2H_4 in the gas phase or chemisorbed on the Pt Schottky contacts then catalytically decomposed to release atomic hydrogen. The hydrogen can then diffuse rapidly through the Pt metallization and the underlying oxide to the interface where it forms a dipole layer and lowers the effective barrier height. We emphasize that other mechanisms could be present; however, the activation energy for the current recovery is ~1 eV, similar to the value for atomic hydrogen diffusion in GaN [98], which suggests that this is at least a plausible mechanism. As the detection temperature is increased, the response of the MOS-HEMT diodes increases due to more efficient cracking of the hydrogen on the metal contact. Note that the changes in both current and voltage are quite large and readily detected. In analogy with results for MOS gas sensors in other materials systems, the effect of the introduction of the atomic hydrogen into the oxide is to create a dipole layer at the oxide–semiconductor interface that will screen some of the piezo-induced charge in the HEMT channel. The time constant for response of the diodes was determined by the mass transport characteristics of the gas into the volume of the test chamber and was not limited by the response of the MOS diode itself.

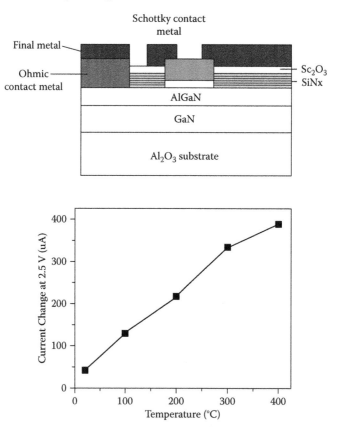

FIGURE 2.4   (Top) Schematic of AlGaN/GaN MOS diode and (bottom) change in current at fixed bias when ethylene is introduced into the test chamber with the sensor held at different temperatures.

2.3 SENSOR FUNCTIONALIZATION

Specific and selective molecular functionalization of the semiconductor surface is necessary to achieve specificity in chemical and biological detection. Devices such as FETs can readily discriminate between adsorption of oxidizing and reducing gas molecules from the changes (increase or decrease) in the channel conductance. However, precise identification of a specific type of molecule requires functionalization of the surface with specific molecules or catalysts. Effective biosensing requires coupling of the unique functional properties of proteins, nucleic acids (DNA, RNA), and other biological molecules with the solid-state "chip" platforms. These devices take advantage of the specific, complementary interactions between biological molecules that are a fundamental aspect of biological function. Specific, complementary interactions are what permit antibodies to recognize antigens in the immune response, enzymes to recognize their target substrates, and the motor proteins of muscle to shorten during muscular contraction. The ability of biological molecules, such as proteins, to bind other molecules in a highly specific manner is the underlying principle of the "sensors" to detect the presence (or absence) of target molecules—just as it is in the biological senses of smell and taste.

One of the key technical challenges in fabricating hybrid biosensors is the junction between biological macromolecules and the inorganic scaffolding material (metals and semiconductors) of the chip. For actual device applications, it is often necessary to selectively modify a surface at micro- and even nanoscale, sometimes with different surface chemistry at different locations. In order to enhance detection speed, especially at very low analyte concentration, the analyte should be delivered directly to the active sensing areas of the sensors. A common theme for bio/chem sensors is that their operation often incorporates moving fluids. For example, sensors must sample a stream of air or water to interact with the specific molecules they are designed to detect.

The general approach to detecting biological species using a semiconductor sensor involves functionalizing the surface (e.g., the gate region of an ungated FET structure) with a layer or substance that will selectively bind the molecules of interest. In applications requiring less specific detection, the adsorption of reactive molecules will directly affect the surface charge and affect the near-surface conductivity. In its simplest form, the sensor consists of a semiconductor film patterned with surface electrodes and often heated to temperatures of a few hundred degrees Celsius to enhance dissociation of molecules on the exposed surface. Changes in resistance between the electrodes signal the adsorption of reactive molecules. It is desirable to be able to use the lowest possible operating temperature to maximize battery life in handheld detection instruments. Electronic oxides such as ZnO as the sensing material are especially sensitive to reaction of target molecules with adsorbed oxygen or the oxygen in the lattice itself.

Biologically modified field-effect transistors (bioFETs) have the potential to directly detect biochemical interactions in aqueous solutions. To enhance their practicality, the device must be sensitive to biochemical interactions on its surface, functionalized to probe-specific biochemical interactions, and must be stable in aqueous solutions for a range of pH and salt concentrations. Typically, the gate region of the device is covered with biological probes, which are used as receptor sites for the molecules of interest. The conductance of the device is changed as reaction occurs between these probes and appropriate species in solution.

Since GaN-based material systems are chemically stable, this should minimize degradation of adsorbed cells. The bond between Ga and N is ionic, and proteins can easily attach to the GaN surface. This is one of the key factors for making a sensitive biosensor with a useful lifetime. HEMT sensors have been used for detecting gases, ions, pH values, proteins, and DNA temperature with good selectivity by the modification of the surface in the gate region of the HEMT. The 2DEG channel is connected to an ohmic-type source and drain contacts. The source-drain current is modulated by a third contact, a Schottky-type gate, on the top of the 2DEG channel. For sensing applications, the third contact is affected by the sensing environment, that is, the sensing targets change the charges on the gate region and behave as a gate. When charged analytes accumulate on the gate area, these charges form a bias and alter the 2DEG resistance. This electrical detection technique is simple, fast, and convenient. The detecting signal from the gate is amplified through the drain-source current and makes this sensor to be very sensitive for sensor applications. The electric signal also can be easily quantified, recorded, and transmit, unlike fluorescence detection methods that need human inspection and are difficult to precisely quantify and transmit the data.

One drawback of HEMT sensors is a lack of selectivity to different analytes due to the chemical inertness of the HEMT surface. This can be solved by surface modification with detecting receptors. Sensor devices of the present disclosure can be used with a variety of fluids having environmental and bodily origins, including saliva, urine, blood, and breath. For use with exhaled breath, the device may include a HEMT bonded on a thermoelectric cooling device, which assists in condensing exhaled breath samples.

In our HEMT devices, the surface is generally functionalized with an antibody or enzyme layer. The success of the functionalization is monitored by a number of methods. Examples are shown in Figures 2.5 and 2.6. The first test is a change in surface tension when the functional layer is in place, and the change in surface bonding can in some cases be seen by x-ray photoelectron spectroscopy. Typically, a layer of Au is deposited on the gate region of the HEMT as a platform to attach a chemical such as thioglycolic acid, whose S-bonds readily attach to the Au. The antibody layer can then be attached to the thioglycolic acid. When the surface is completely covered by these functional layers, the HEMT will not be sensitive to buffer solutions or water that do not contain the antigen of interest, as shown in Figure 2.6. For detecting hydrogen, the gate region is functionalized with a catalyst metal such as Pt or Pd. In other cases, we immobilize an enzyme to catalyze reactions, as is used for the detection of glucose. In the presence of the enzyme glucose oxidase, glucose will react with oxygen to produce gluconic acid and hydrogen peroxide. Table 2.1 shows a summary of the surface functionalization layers we have employed for HEMT sensors to date. There are many

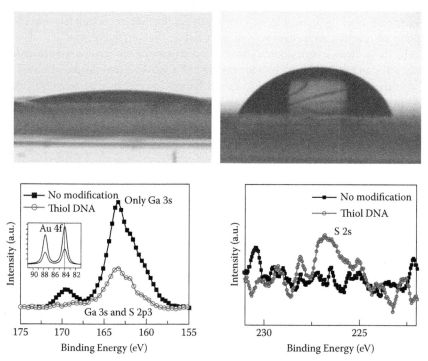

FIGURE 2.5. Examples of a change in surface tension of the HEMT surface functionalization (top) and XPS measurement of formation of Au-S bonds after attachment of thioglycolic acid attachment to Au layer in gate region of HEMT.

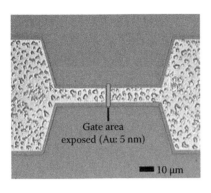
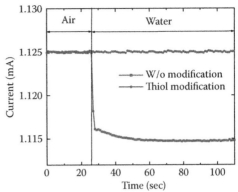

FIGURE 2.6 Example of successful functionalization of HEMT surface; the device is no longer sensitive to water when the surface is completely covered with the functional layer.

TABLE 2.1
Summary of surface functional layers used with HEMT sensors

Detection	Mechanism	Surface functionalization
H_2	Catalytic dissociation	Pd, Pt
Pressure change	Polarization	Polyvinylidene difluoride
Botulinum toxin	Antibody	Thioglycolic acid/antibody
Proteins	Conjugation/hybridization	Aminopropylsilane/biotin
pH	Absorption of polar molecules	Sc_2O_3, ZnO
Hg^{2+}	Chelation	Thioglycolic acid/AU
KIM-1	Antibody	KIM-1 antibody
Glucose	GO_X immobilization	ZnO nanorods
Prostate-specific antigen	PSA antibody	Carboxylate succinimidyl ester/PSA antibody
Lactic acid	LO_X immobilization	ZnO nanorods
Chloride ions	Anodization	Ag/AgCl electrodes; InN
Breast cancer	Antibody	Thioglycolic acid/c-erbB antibody
CO_2	Absorption of water/charge	Polyethylenimine/starch
DNA	Hybridization	3′-thiol-modified oligonucleotides
O_2	Oxidation	InGaZnO

additional options for detection of biotoxins and biological molecules of interest by use of different protein or antibody layers. The advantage of the biofet approach is that large arrays of HEMTs can be produced on a single chip and functionalized with different layers to allow for the detection of a broad range of chemicals or gases.

2.4 pH MEASUREMENT

The measurement of pH is needed in many different applications, including medicine, biology, chemistry, food science, environmental science, and oceanography. Solutions with a pH less than 7 are acidic, and solutions with a pH greater than 7 are basic or alkaline. We have found that ZnO-nanorod surfaces respond electrically to variations of the pH in electrolyte solutions introduced via an integrated microchannel [99]. The ion-induced changes in surface potential are readily measured as a change in conductance of the single ZnO nanorods and suggest that these structures are very promising for a wide variety of sensor applications. An integrated microchannel was made from

SYLGARD@ 184 polymer. Entry and exit holes in the ends of the channel were made with a small puncher (diameter less than 1 mm) and the film immediately applied to the nanorod sensor. The pH solution was applied using a syringe autopipette (2–20 uL).

Prior to the pH measurements, we used pH 4, 7, 10 buffer solutions to calibrate the electrode, and the measurements at 25°C were carried out in the dark or under ultraviolet (UV) illumination from 365 nm light using an Agilent 4156C parameter analyzer to avoid parasitic effects. The pH solution made by the titration method using HNO_3, NaOH, and distilled water. The electrode was a conventional Acumet standard Ag/AgCl electrode. The nanorods showed a very strong photoresponse. The conductivity is greatly increased as a result of the illumination, as evidenced by the higher current. No effect was observed for illumination with below bandgap light. The photoconduction appears predominantly to originate in bulk conduction processes with only a minor surface-trapping component. The adsorption of polar molecules on the surface of ZnO affects the surface potential and device characteristics. The current at a bias of 0.5 V as a function of time from nanorods exposed for 60 s to a series of solutions whose pH was varied from 2 to 12 was reduced upon exposure to these polar liquids as the pH is increased. The experiment was conducted starting at pH = 7 and went to pH = 2 or 12. The current–voltage (I–V) measurement in air was slightly higher that in the pH = 7 (10%–20%). The data shows that the sensor is sensitive to the concentration of the polar liquid and therefore could be used to differentiate between liquids into which a small amount of leakage of another substance has occurred. The conductance of the rods was higher under UV illumination, but the percentage change in conductance is similar with and without illumination. The nanorods exhibited a linear change in conductance between pH 2 and 12 of 8.5 nS/pH in the dark and 20 nS/pH when illuminated with UV (365 nm) light. The nanorods show stable operation with a resolution of ~0.1 pH over the entire pH range, showing the remarkable sensitivity to relatively small changes in concentration of the liquid.

Ungated AlGaN/GaN HEMTs also exhibit large changes in current upon exposing the gate region to polar liquids. The polar nature of the electrolyte introduced led to a change of surface charges, producing a change in surface potential at the semiconductor–liquid interface. The use of Sc_2O_3 gate dielectric produced superior results to either a native oxide or UV ozone-induced oxide in the gate region. The ungated HEMTs with Sc_2O_3 in the gate region exhibited a linear change in current between pH 3 to 10 of 37 µA/pH .The HEMT pH sensors show stable operation with a resolution of <0.1 pH over the entire pH range. 100 Å Sc_2O_3 was deposited as a gate dielectric through a contact window of SiN_x layer. Before oxide deposition, the wafer was exposed to ozone for 25 min. It was then heated in situ at 300°C cleaning for 10 min inside the growth chamber. The Sc_2O_3 was deposited by rf plasma-activated MBE at 100°C using elemental Sc evaporated from a standard effusion at 1130°C and O_2 derived from an Oxford RF plasma source. For comparison, we also fabricated devices with just the native oxide present in the gate region and also with the UV ozone-induced oxide. Figure 2.7 shows a scanning electron microscopy (SEM) image (left) and a cross-sectional

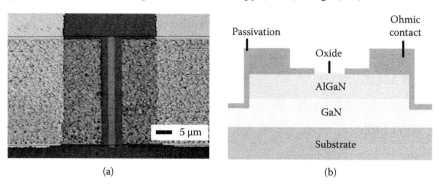

(a)   (b)

FIGURE 2.7 (See color insert).   SEM and schematic of gateless HEMT.

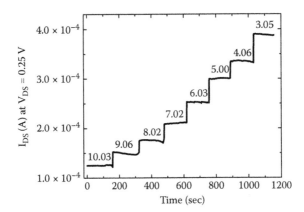

FIGURE 2.8   Change in current in gateless HEMT at fixed source-drain bias of 0.25 V with pH from 3 to 10.

schematic (right) of the completed device. The gate dimension of the device is $2 \times 50\ \mu m^2$. The pH solution was applied using a syringe autopipette (2–20 uL).

Prior to the pH measurements, we used pH 4, 7, 10 buffer solutions from Fisher Scientific to calibrate the electrode, and the measurements at 25°C were carried out in the dark using an Agilent 4156C parameter analyzer to avoid parasitic effects. The pH solution made by the titration method using HCl, NaOH, and distilled water. The electrode was a conventional Acumet standard Ag/AgCl electrode.

The adsorption of polar molecules on the surface of the HEMT affected the surface potential and device characteristics. Figure 2.8 shows the current at a bias of 0.25 V as a function of time from HEMTs with Sc_2O_3 in the gate region exposed for 150 s to a series of solutions whose pH was varied from 3 to 10. The current is significantly increased upon exposure to these polar liquids as the pH is decreased. The change in current was 37 µA/pH.

The HEMTs show stable operation with a resolution of ~0.1 pH over the entire pH range, showing the remarkable sensitivity of the HEMT to relatively small changes in concentration of the liquid. By comparison, devices with the native oxide in the gate region showed a higher sensitivity of ~70 µA/pA but a much poorer resolution of ~0.4 pH and evidence of delays in response of 10 to 15 s. The latter may result from deep traps at the interface between the semiconductor and native oxide, whose density is much higher than at the Sc_2O_3–nitride interface. The devices with UV-ozone oxide in the gate region did not show these incubation times for detection of pH changes and showed similar sensitivities of gate source current as the Sc_2O_3 gate devices (~40 µA/pH) but with poorer resolution (~0.25 pH). Figure 2.9 shows that the HEMT sensor with Sc_2O_3 gate dielectric is sensitive to the concentration of the polar liquid and therefore could be used to differentiate between liquids into which a small amount of leakage of another substance has occurred. As mentioned earlier, the pH range of interest for human blood is 7–8. Figure 2.9 shows the current change in the HEMTs with Sc_2O_3 at a bias of 0.25 V for different pH values in this range. Note that the resolution of the measurement is <0.1 pH. There is still more to understand about the mechanism of the current reduction in relation to the adsorption of the polar liquid molecules on the HEMT surface. These molecules are bonded by van der Waals-type interactions, and they screen surface change that is induced by polarization in the HEMT.

2.5   EXHALED BREATH CONDENSATE

There is significant interest in developing rapid diagnostic approaches and improved sensors for determining early signs of medical problems in humans. Exhaled breath is a unique bodily fluid that can be utilized in this regard [100–112]. Exhaled breath condensate pH is a robust and reproducible assay of airway acidity. For example, the blood pH range that is relevant for humans is 7–8.

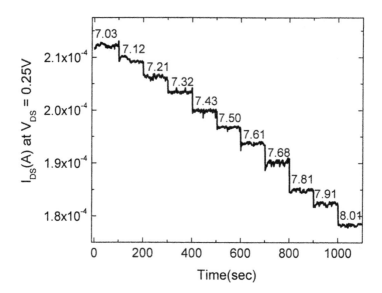

FIGURE 2.9 Change in current in gateless HEMT at fixed source-drain bias of 0.25 V with pH from 7 to 8.

Humans, even when they are extremely ill, will not have a blood (or interstitial = space between cells in tissue) pH below 7. When they do drift below this value, it almost invariably equals mortality.

While most applications will detect substances or diseases in the breath as a gas or aerosol, breath can also be analyzed in the liquid phase as exhaled breath condensate (EBC). Analytes contained in the breath originating from deep within the lungs (alveolar gas) equilibrates with the blood and, therefore, the concentration of molecules present in the breath is closely correlated with those found in the blood at any given time. EBC contains dozens of different biomarkers, such as adenosine, ammonia, hydrogen peroxide, isoprostanes, leukotrienes, peptide, cytokines, and nitrogen oxide [109,111,112]. Analysis of molecules in EBC is noninvasive and can provide a window on the metabolic state of the human body, including certain signs of cancer, respiratory diseases, and liver and kidney functions.

As a diagnostic, exhaled breath offers advantages since samples can be collected and tested with results delivered in real time at the point of testing. Another advantage is that the sample can be collected noninvasively by asking a patient to blow into the disposable portion of a handheld testing device. Therefore, the sample collection method is hygienic for both the patient and the laboratory personnel. Exhaled breath can be used to detect various drugs, medications, their metabolites, and markers, and this can be valuable in measuring both medication adherence and in determining the blood levels of these drugs and medications. Some of today's blood- and urine-based tests might be replaced with simple breath-based testing. In consumer healthcare, diabetics would be able to test their glucose level, replacing painful and inconvenient finger-prick devices. For roadside screening of driving impairment, a point-of-care (POC) device similar in function to a handheld breath alcohol analyzer will detect drugs of abuse such as marijuana and cocaine. In workplace drug testing, a similar desktop device might eliminate the cost, embarrassment, and inconvenience of workplace urine screening. In the setting of chronic oral drug therapy (e.g., treatment of schizophrenia with atypical antipsychotic medications), mortality/morbidity and the cost of healthcare will be markedly reduced by developing breath-based systems that document that drugs were orally ingested and entered the bloodstream.

The glucose oxidase enzyme (GOx) is commonly used in biosensors to detect levels of glucose for diabetics. By keeping track of the number of electrons passed through the enzyme, the concentration of glucose can be measured. Due to the importance and difficulty of glucose immobilization, numerous studies have been focused on the techniques of immobilization of glucose with carbon nanotubes, ZnO nanomaterials, and gold particles [113,114]. ZnO-based nanomaterials are especially interesting due to their nontoxic properties, low cost of fabrication, and favorable electrostatic interaction between ZnO and the GOx lever. However, the activity of GOx is highly dependent on the pH value of the solution [115]. The pH value of a typical healthy person is between 7 and 8. This can vary significantly depending on the health condition of each individual; for example, the pH value for patients with acute asthma was reported as low as 5.23 ± 0.21 ($n = 22$) as compared to 7.65 ± 0.20 ($n = 19$) for the control subjects [116]. To achieve accurate glucose concentration measurement with immobilized GO_x, it is necessary to determine the pH value and glucose concentration with an integrated pH and glucose sensor.

2.6 HEAVY METAL DETECTION

The detection of Superfund contaminants in groundwater, along with testing of their effects on the environment and aquatic wildlife, can greatly improve environmental monitoring and management. While techniques for detection of hazardous environmental chemicals are readily available, they require the transportation of samples to a laboratory for analysis. Data analysis and collection requires skilled expertise, is expensive, and requires prolonged amounts of time. Thus, current detection techniques do not allow real-time monitoring of environmental toxicants.

Conventional methods of detection involve the use of HPLC, mass spectrometry, and colorimetric ELISAs, but these are impractical because such tests can only be carried out at centralized locations, and are too slow to be of practical value in the field. To develop more practical, "field-deployable" sensors, there is a broad range of sensor technologies that are currently being developed. These methods include detectors based on fluorescence, the surface plasmon resonance (SPR) technique, sensors based on semiconductors, and sensors based on measuring mass such as microcantilevers or quartz crystal devices. The demands for detection of environmental toxins in the field cannot be met by most of these methods of detection. In the field, a handheld sensor with real-time sensing capabilities, wireless mode of operation, and robust nature of operation is required. Techniques such as SPR and fluorescence-based sensors are susceptible to artifacts caused by turbidity, particulates, and differences in refractive indices.

Electrochemical devices have attracted attention for the measurement of the target materials due to their low cost and simplicity. Although the electrochemical measurements based on the impedance and capacitance are used as accurate sensor platforms, significant improvements in their sensitivities will be needed for use of environmental samples. Since a reference electrode is required for the electrochemical measurement, the sample size cannot be minimized. Methods based on detection of mass, such as the microcantilever, suffer from an undesirable resonant frequency change due to viscosity of the medium and cantilever damping in the solution environment. Nanowire field-effect sensors coated with antibody have been used for the real-time, highly sensitive detection of biochemicals. However, the low-write speed of electron beam used to control the exact position of nanowire arrays decreases the advantages of the low cost and potential for scale up. Microcantilever functionalized with 1,6-hexane dithiol monolayer and SPR sensor functionalized with dithiothreitol have shown excellent detection limit of arsenic below 10 ppb. However, these types of sensors required laser and detectors, which are quite expensive and not suitable for handheld or portable applications.

Mass-sensitive optical devices tend to lose their reproducibility when used for repeated analysis of samples due to fouling. Besides, most light-based techniques require cumbersome supporting equipment that makes it impossible to have a compact, handheld device. SPR-based handheld sensors have been made, but their response times are in the range of several minutes.

Arsenic and mercury are two of most serious contaminants in the United States. Through different human activities, these contaminants discharge into soil and water resources, which impair the ecological and economic value of those resources. Arsenic contamination of groundwater is a naturally occurring high concentration of arsenic in deeper levels of groundwater, which has became a high-profile problem in recent years and is causing serious arsenic poisoning in large numbers of people around the world. A 2007 study found that over 137 million people in more than 70 countries are probably affected by arsenic poisoning of drinking water [117–119]. Some locations in the United States, such as Fallon and Nevada, have long been known to have groundwater with relatively high arsenic concentrations (in excess of 0.08 mg/L) [120]. Even some surface waters, such as the Verde River in Arizona, sometimes exceed 0.01 mg/L arsenic, especially during low-flow periods when the river flow is dominated by groundwater discharge [121]. A drinking water standard of 0.01 mg/L (10 ppb) was established in January 2006.

Besides the natural occurring of arsenic, the chromated copper arsenate (CCA) wood has posed potential serious treat to the environment [122,123]. Although the EPA banned the sale of CCA-treated wood on December 31, 2003 in residential use, boardwalks, fences, or playground equipment, the industrial use is still allowed, leaving carpenters, linemen, and the environment at risk [124,125]. There has been no action to address the existing CCA structures even though the Consumer Product Safety Commission (CPSC) found a significant increased cancer risk to children from playing on CCA-treated playground equipment.

The disposal of CCA wood is another serious issue. While arsenic never breaks down into a safe form, it does have the ability to change forms. Although CCD-treated wood should not be burned off, millions of board feet entered waste fill sites every year and were often burned at waste disposal sites, releasing very toxic trivalent arsenic and hexavalent chromium into the air. Arsine gas can turn back into a solid form as it did on the outside of our home, which was exposed to the smoke from burning treated wood. Old CCA wood is undistinguishable from old natural wood and is often burned in homes and campsites as scrap firewood, and mulched into wood chips for the garden. Studies have been done that show the metals when burned break down into particles so small that they are easily breathed in. Arsenic can also be drawn up by plants, leached out into soil and water, and be off-gassed from the compost action of the wood breaking down in landfill sites.

Mercury is another serious contaminant in the United States, as of 1998, there were a total of 2,506 fish and wildlife consumption advisories for all substances, of which 1,931 (more than 75%) were for mercury [126,127]. Forty states have issued advisories for mercury, and ten states have statewide advisories for mercury in all freshwater lakes and (or) rivers. The majority (65%) of mercury enters into the environment through stationary combustion, of which coal-fired power plants are the largest aggregate source (40% of U.S. mercury emissions in 1999). There are other industrial resources of mercury contamination, such as gold production and nonferrous metal production, etc. Mercury also enters into the environment through the disposal (e.g., land filling, incineration) of certain products. Products containing mercury include auto parts, batteries, fluorescent bulbs, medical products, thermometers, and thermostats. Mercury is one of the extremely toxic metals, and its compounds produce irreversible neurological damage to human health. Due to its toxic effects, the standard limit of mercury in drinking water is 0.001 mg g^{-1} and in industrial wastewater to mix with surface water is 0.01 mg g^{-1}. Hence, it is important to develop methods to detect their presence in contaminated wastewater to innocuous levels. The sensitive detection of low concentration of mercury (II) (Hg^{2+}) ion is essential because its toxicity has long been recognized as a chronic environmental problem. Traditionally, there are several methods that could be used to detect low Hg^{2+} concentration including spectroscopic (AAS, AES, or ICP-MS) or electrochemical (ISE, or polarography), however, these methods display shortcomings in practical use, being either expensive or too big to be used for detection on-site, where handheld portable devices could be invaluable for metal detections at low concentrations.

A schematic cross section of the device with Hg^{2+} ions bound to thioglycolic acid functionalized on the gold gate region and plan view photomicrograph of a completed device is shown in Figure 2.10. In some cases, we utilized sensors functionalized with thioglycolic acid, HSCH$_2$COOH, which is an organic compound that contains both a thiol (mercaptan) and a carboxylic acid functional group. A self-assembled monolayer of thioglycolic acid molecule was adsorbed onto the gold gate due to strong interaction between gold and the thiol group. The extra thioglycolic acid molecules were rinsed off with DI water. An increase in the hydrophilicity of the treated surface by thioglycolic acid functionalization was confirmed by contact angle measurements that showed a change in contact angle from

58.4° to 16.2° after the surface treatment. As shown in Figure 2.11, the drain current of both sensors (i.e., those with only Au gate and those with the thioglycolic acid functionalized onto the Au layer) further reduced after exposure to different concentrations of Hg^{2+} ion solutions. The drain current was reduced ~60% for the thioglycolic acid-functionalized AlGaN/GaN HEMT sensors, while bare-Au-gate sensors had less than a 3% change in drain current. The mechanism of the drain current reduction for bare Au gate and thioglycolic acid-functionalized AlGaN/GaN HEMT sensors was quite different. For the bare Au-gate devices, Au–mercury amalgam was formed on the surface of the bare Au-gates when the Au-gate electrode was exposed to Hg^{2+} ion solutions. The formation rate of the Au–mercury amalgam depended on the solution temperature and the concentration of the Hg^{2+} ion solution.

Figure 2.11 also shows the time dependence of the drain current for the two types of sensors. For the higher Hg^{2+} ion concentration solution, 10^{-5} M, the bare Au-gate-based sensor took less than

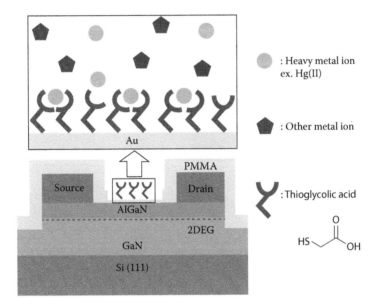

FIGURE 2.10 (See color insert). Schematic of AlGaN/GaN HEMT. The Au-coated gate area was functionalized with thioglycolic acid.

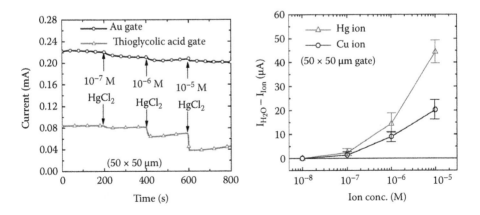

FIGURE 2.11 (Left) Time-dependent response of the drain current for bare Au-gate AlGaN/GaN HEMT sensor and thioglycolic acid functionalized Au-gate HEMT sensor. (Right) Drain current of a thioglycolic acid functionalized Au-gate HEMT sensor as a function of the Hg^{2+} ion concentration.

15 s for the drain to reach steady state. However, the drain current required 30–55 s to reach steady state, when the sensor was exposed to the less concentrated Hg^{2+} ion solutions. A response time of < 5 s was obtained for the thioglycolic acid-functionalized AlGaN/GaN HEMT sensors, when the sensor was exposed to 10^{-5} M of the Hg^{2+} ion solution. This is the shortest response time of Hg^{2+} ion detection ever reported.

For the thioglycolic-acid-functionalized AlGaN/GaN HEMT, the thioglycolic acid molecules on the Au surface align vertically with carboxylic acid functional group toward the solution. The carboxylic acid functional group of the adjacent thioglycolic acid molecules form chelates of R-COO⁻ (Hg^{2+})⁻OOC-R with Hg^{2+} ions when the sensors are exposed to the Hg^{2+} ion solution. The charges of trapped Hg^{2+} ion in the R-COO⁻(Hg^{2+})⁻OOC-R chelates changed the polarity of the thioglycolic acid molecules that were bonded to the Au-gate through -S-Au bonds. This is why the drain current changes in response to mercury ions. Similar surface functionalization was used by Huang and Chang [128], and fluorescence was used for the detection.

The difference in drain current for the device exposed to Hg^{2+} ion concentration and to the DI water is illustrated in Figure 2.12. The Hg^{2+} ion concentration detection limit for the thioglycolic acid-functionalized sensor is 10^{-6} M, which is equivalent to 20 ppb (parts per billion).

The sensors can also be rinsed in DI water after a detection event and reused, as shown in Figure 2.13. This is very convenient when making sure that false positives will not affect the

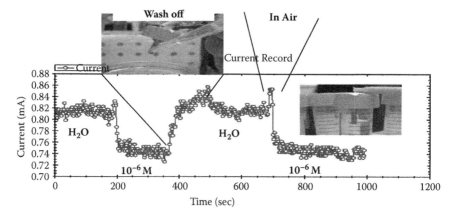

FIGURE 2.12   The difference in drain current for the device exposed to different Hg^{2+} ion concentration to the DI water is illustrated in Figure 2.21.

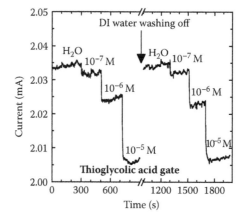

FIGURE 2.13   Time-dependent responses of the drain current as a function of Hg^{2+} ion concentrations after the sensor been washed with DI water.

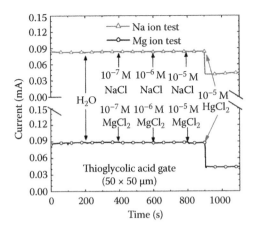

FIGURE 2.14 Time-dependent response of the drain current for detecting Na^+, Mg^{2+}, or Hg^{2+} with a thioglycolic acid-functionalized Au-gate HEMT sensor.

ultimate judgment as to the presence of Hg in the test sample. Since our sensor chip is very compact (1 mm × 5 mm) and operates at extremely low power (8 μW based on 0.3 V of drain voltage and 80 μA of drain current operated at 11 Hz), it can be integrated with a commercial available handheld wireless transmitter to realize a portable, fast-response, and high-sensitivity Hg^{2+} ion detector.

The thioglycolic acid-functionalized sensor also showed excellent sensing selectivity (over 100 times higher selectivity) over Na^+ and Mg^{2+} ions. Time-dependent response of the drain current for detecting Na^+, Mg^{2+}, or Hg^{2+} with a thioglycolic acid-functionalized Au-gate HEMT sensor before and after washing the sensor in water is shown in Figure 2.14.

2.7 BIOTOXIN SENSORS

Reliable detection of biological agents in the field and in real time is challenging. Given the adverse consequences of a lack of reliable biological agent sensing on national security, there is a critical need to develop novel, more sensitive, and reliable technologies for biological detection in the field [129–144]. The objective of this application is to develop and test a wireless sensing technology for detecting toxins. To achieve this objective, we have developed high electron mobility transistors (HEMTs) that have been demonstrated to exhibit some of the highest sensitivities for biological agents. Specific antibodies targeting enterotoxin type B (Category B, NIAID), botulinum toxin (Category A NIAID), and ricin (Category B NIAID), or peptide substrates for testing the toxin's enzymatic activity, have been conjugated to the HEMT surface. While testing still needs to be performed in the presence of cross-contaminants in biologically relevant samples, the initial results are very promising. A significant issue is the absence of a definite diagnostic method and the difficulty in differential diagnosis from other pathogens that would slow the response in case of a terror attack. Our aim is to develop reliable, inexpensive, highly sensitive, handheld sensors with response times on the order of a few seconds, which can be used in the field for detecting biological toxins. This is significant because it would greatly improve our effectiveness in responding to terrorist attacks.

The current methods for toxin sensing in the field are generally not suited for field deployment, and there is a need for new technologies. The current methods involve the use of HPLC, mass spectrometry, and colorimetric ELISAs that are impractical because such tests can only be carried out at centralized locations, and are too slow to be of practical value in the field. These still tend to be the methods of choice in current detection of toxins, for example, the standard test for botulinum toxin detection is the "mouse assay," which relies on the death of mice as an indicator of toxin presence [144]. Clearly, such methods are slow and impractical in the field.

2.7.1 BOTULINUM

Antibody-functionalized Au-gated AlGaN/GaN HEMTs show great sensitivity for detecting botulinum toxin. The botulinum toxin was specifically recognized through botulinum antibody, anchored to the gate area, as shown in Figure 2.15.

We investigated a range of concentrations from 0.1 to 100 ng/mL. The source and drain current from the HEMT were measured before and after the sensor was exposed to 100 ng/mL of botulinum toxin at a constant drain bias voltage of 500 mV, as shown in Figure 2.16 (top). Any slight changes in the ambient of the HEMT affect the surface charges on the AlGaN/GaN. These changes in the surface charge are transduced into a change in the concentration of the 2DEG in the AlGaN/GaN HEMTs, leading to the decrease in the conductance for the device after exposure to botulinum toxin. Figure 2.16 (bottom) shows a real-time botulinum toxin detection in PBS buffer solution using the source and drain current change with constant bias of 500 mV. No current change can be seen with the addition of buffer solution around 100 s, showing the specificity and stability of the device. In clear contrast, the current change showed a rapid response in less than 5 s when target 1 ng/mL botulinum toxin was added to the surface. The abrupt current change due to the exposure of botulinum toxin in a buffer solution was stabilized after the botulinum toxin thoroughly diffused into the buffer solution. Different concentrations (from 0.1 ng/mL to 100 ng/mL) of the exposed target botulinum toxin in a buffer solution were detected. The sensor saturates above 10 ng/mL of the toxin. The experiment at each concentration was repeated four times to calculate the standard deviation of source-drain current response. The limit of detection of this device was below 1 ng/mL of botulinum toxin in PBS buffer solution. The source-drain current change was nonlinearly proportional to botulinum toxin concentration, as shown in Figure 2.16.

Figure 2.17 shows a real-time test of botulinum toxin at different toxin concentrations with intervening washes to break antibody–antigen bonds. This result demonstrates the real-time capabilities and recyclability of the chip. Long-term stability of the botulinum toxin sensor was also investigated with a package sensor. Figure 2.18 shows a photograph of the packaged sensor placed in a petri dish for long-term storage. PBS buffer solution was dropped on the active region of the sensor and the petri dish as well. The petri dish was then covered and sealed in order to keep the antibodies on the sensor in a PBS environment. Sensors were retested for the botulinum detection every 3 months. For those tests, the procedures of toxin detection and sensor surface reactivation were repeated for five times. This experiment demonstrated that after 9-month storage, the sensor still could detect the toxin and could be reactivated right after the test with PBS buffer solution rinse. This indicated

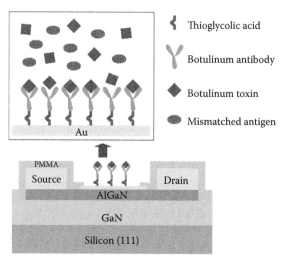

FIGURE 2.15 Schematic of functionalized HEMT for botulinum detection.

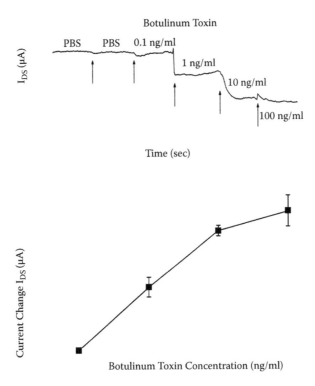

FIGURE 2.16   Drain current of an AlGaN/GaN HEMT versus time for botulinum toxin from 0.1 ng/mL to 100 ng/mL (top) and change of drain current versus different concentrations from 0.1 ng/mL to 100 ng/mL of botulinum toxin (bottom).

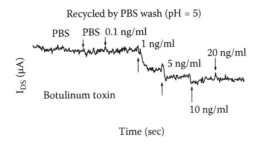

FIGURE 2.17   Real-time test from a used botulinum sensor that was washed with PBS in pH 5 to refresh the sensor.

that the toxin could be completely washed away from the antibodies. However, it was obvious that the detection sensitivity decreased after 9 months of storage. The decrease of the detection sensitivity drop after 9-month storage was not caused by the existence of the unbreakable toxin–antibody binding but was rather due to the decrease of antibody activity. Another important finding was that the response time of the 9-month stored sensor increased from 5 s of the fresh sensor to around 10 s, when target toxins were exposed to the sensor. The longer response time may be also due to the decreased number of highly active sites on the antibodies after long-term storage. This corresponds to the lower sensitivity of the sensor. The detailed mechanism needs further investigation.

Figure 2.19 shows the current changes of the sensors tested after different storage times at a fixed toxin concentration of 10 ng/mL against the first drain current measurement of the sensor. After 3, 6, and 9 months of storage, the current change drops 2%, 12%, and 28%, respectively. Within

FIGURE 2.18   Photograph of a packaged sensor placed in a petri dish for long-term storage.

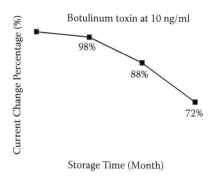

FIGURE 2.19   The drain-source current change percentages of the initial, 3-, 6-, and 9-month stored sensors. The current change in the initial test is defined as 100%. The testing at the subsequent periods was defined relative to the initial test.

3 months of storage, this sensor showed almost the same sensitivity as when it was fresh. Although, after 6 and 9 months of storage, the sensor would need to be recalibrated for toxin concentration determination usage, there is no need for recalibration for the use as the first responder of the detection for the presence or absence of the toxin.

In summary, we have shown that through a chemical modification method, the Au-gated region of an AlGaN/GaN HEMT structure can be functionalized for the detection of botulinum toxin with a limit of detection less than 1 ng/mL. This electronic detection of biomolecules is a significant step toward a field-deployed sensor chip, which can be integrated with a commercial available wireless transmitter to realize a real-time, fast-response, and high-sensitivity botulinum toxin detector.

2.8   BIOMEDICAL APPLICATIONS

2.8.1   pH Sensing in Breath Condensate

AlGaN/GaN HEMTs can be used for measurements of pH in EBC and glucose, through integration of the pH and glucose sensor onto a single chip and with additional integration of the sensors into a portable, wireless package for remote monitoring applications. Figure 2.20 shows an optical microscopy image of an integrated pH and glucose sensor chip and cross-sectional schematics of

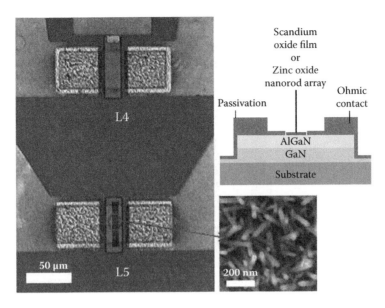

FIGURE 2.20   SEM image of an integrated pH and glucose sensor. The insets show a schematic cross-section of the pH sensor and also an SEM of the ZnO nanorods grown in the gate region of the glucose sensor.

the completed pH and glucose device. The gate dimension of the pH sensor device and glucose sensors was $20 \times 50 \ \mu m^2$.

For the glucose detection, a highly dense array of 20–30 nm diameter and 2 μm tall ZnO nanorods were grown on the $20 \times 50 \ \mu m^2$ gate area. The lower right inset in Figure 2.20 shows closer view of the ZnO nanorod arrays grown on the gate area. The total area of the ZnO was increased significantly with the ZnO nanorods. The ZnO nanorod matrix provides a microenvironment for immobilizing negatively charged GO_x while retaining its bioactivity, and passes charges produced during the GO_x and glucose interaction to the AlGaN/GaN HEMT. The GOx solution was prepared with concentration of 10 mg/mL in 10 mM phosphate buffer saline (pH value of 7.4, Sigma Aldrich). After fabricating the device, 5 μL GO_x (~100 U/mg, Sigma Aldrich) solution was precisely introduced to the surface of the HEMT using a picoliter plotter. The sensor chip was kept at 4°C in the solution for 48 h for GO_x immobilization on the ZnO nanorod arrays followed by an extensively washing to remove the unimmobilized GO_x.

To take the advantage of quick response (less than 1 s) of the HEMT sensor, a real-time EBC collector is needed [145–147]. The amount of the EBC required to cover the HEMT sensing area is very small. Each tidal breath contains around 3 μL of the EBC. The contact angle of EBC on Sc_2O_3 has been measured to be less than 45°, and it is reasonable to assume a perfect half sphere of EBC droplet formed to cover the sensing area $4 \times 50 \ \mu m^2$ gate area. The volume of a half sphere with a diameter of 50 μm is around 3×10^{-11} L. Therefore, 100,000 of 50 μm diameter droplets of EBC can be formed from each tidal breath. To condense the entire 3 μL of water vapor, only ~ 7 J of energy needs to be removed for each tidal breath, which can be easily achieved with a thermal electric module, a Peltier device, as shown in Figure 2.21. The schematic of the system for collecting the EBC is illustrated in Figure 2.22. The AlGaN/GaN HEMT sensor is directly mounted on the top of the Peltier unit (TB-8-0.45-1.3 HT 232, Kryotherm), as also shown in Figure 2.22, which can be cooled to precise temperatures by applying known voltages and currents to the unit. During our measurements, the hotter plate of the Peltier unit was kept at 21°C, and the colder plate was kept at 7°C by applying bias of 0.7 V at 0.2 A. The sensor takes less than 2 s to reach thermal equilibrium with the Peltier unit. This allows the exhaled breath to immediately condense on the gate region of the HEMT sensor.

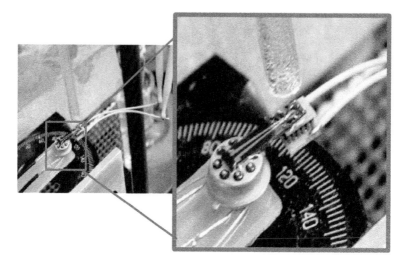

FIGURE 2.21 (See color insert).   Optical image of sensor mounted on Peltier cooler.

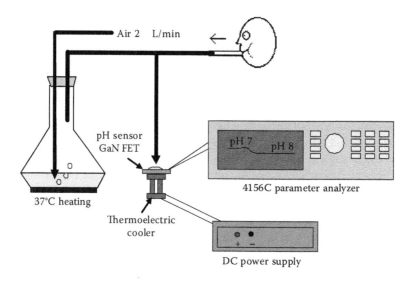

FIGURE 2.22   Schematic of the system for collecting EBC.

Prior to pH measurements of the EBC, a Hewlett-Packard soap film flow meter and a mass flow controller were used to calibrate the flow rate of exhaled breath. The HEMT sensors were also calibrated and exhibited a linear change in current between pH 3 and 10 of 37μA/pH. Due to the difficulty to collect the EBC with different glucose concentration, the samples for glucose concentration detection were prepared from glucose diluted in PBS or DI water.

The HEMT sensors were not sensitive to switching of N_2 gas but responded to applications of exhaled breath pulse inputs from a human test subject, as shown at the top of Figure 2.23, which shows the current of a Sc_2O_3-capped HEMT sensor biased at 0.5 V for exposure to different flow rates of exhaled breath (0.5–3.0 L/min). The flow rates are directly proportional to the intensity exhalation. Deep breath provides a higher flow rate. A similar study was conducted with pure N_2 to eliminate the flow rate effect on sensor sensitivity. The N_2 did not cause any change of drain current, but the increase of exhaled breath flow rate decreased the drain current proportionally from 0.5 L/min to a saturation value of 1 L/min. For every tidal breath, the beginning portion of the

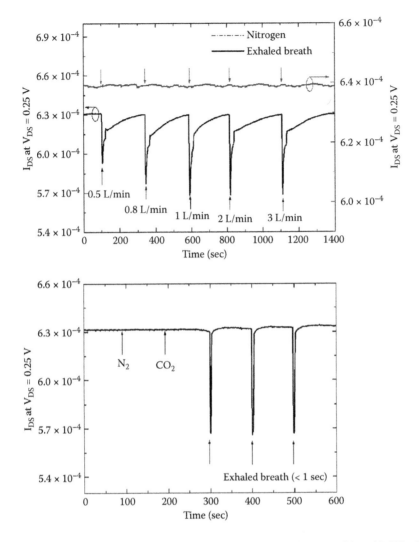

FIGURE 2.23 Changes of drain current for HEMT sensor at fixed drain-source bias of 0.5 V with different flow rates or durations of exhaled breath from tidal breath to hyperventilation. The duration of the breath is 5 s in the bottom figure.

exhalation is from the physiologic dead space, and the gases in this space do not participate in CO_2 and O_2 exchange in the lungs. Therefore, the contents in the tidal breath are diluted by the gases from this dead space. For higher flow rate exhalation, this dilution effect is less effective. Once the exhaled breath flow rate is above 1 L/min, the sensor current change reaches a limit. As a result, the test subject experiences hyperventilation, and the dilution becomes insignificant. Figure 2.23 shows the time response of the sensors to much longer exhaled breaths. The characteristic shape of the response curves is similar and is determined by the evaporation of the condensed EBC from the gate region of the HEMT sensor. The sensor is operated at 50 Hz and 10% duty cycle, which produces heat during operation. It only takes a few seconds for the EBC to vaporize from the sensing area and causes the spike-like response. The principal component of the EBC is water vapor, which represents nearly all of the volume (>99%) of the fluid collected in the EBC. The measured current change of the exhale breath condensate shows that the pH values are within the range between pH 7 and 8. This range is the typical pH range of human blood.

2.8.2 GLUCOSE SENSING

The glucose was sensed by ZnO nanorod-functionalized HEMTs with glucose oxidase enzyme localized on the nanorods, shown in Figure 2.24. This catalyzes the reaction of glucose and oxygen to form gluconic acid and hydrogen peroxide. Figure 2.25 shows the real-time glucose detection in PBS buffer solution using the drain current change in the HEMT sensor with constant bias of 250 mV. No current change can be seen with the addition of buffer solution at around 200 s, showing the specificity and stability of the device. By sharp contrast, the current change showed a rapid response of less than 5 s when target glucose was added to the surface. So far, the glucose detection using Au nano-particle, ZnO nanorod and nanocomb, or carbon nanotube material with GOx immobilization is based on electrochemical measurement [147–151]. Since there is a reference electrode required in the solution, the volume of sample cannot be easily minimized. The current density is measured when a fixed potential is applied between nanomaterials and the reference electrode.

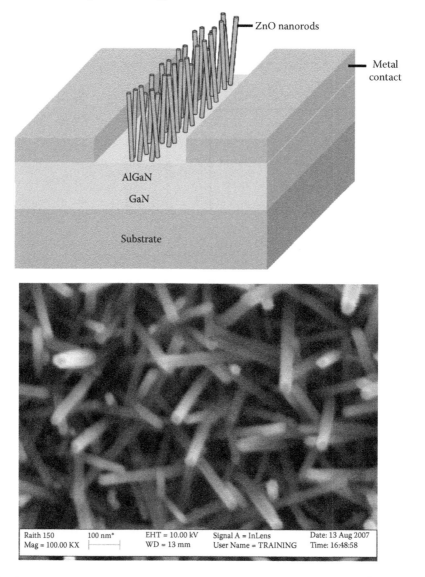

FIGURE 2.24 Schematic of ZnO nanorod-functionalized HEMT (top) and SEM of nanorods on gate area (bottom).

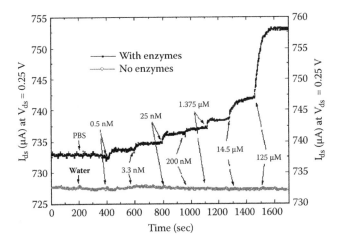

FIGURE 2.25 Plot of drain current versus time with successive exposure of glucose from 500 pM to 125 μM in 10 mM phosphate buffer saline with a pH value of 7.4, both with and without the enzyme located on the nanorods.

This is a first-order detection, and the range of detection limit of these sensors is 0.5–70 μM. Even though the AlGaN/GaN HEMT-based sensor used the same GOx immobilization, the ZnO nanorods were used as the gate of the HEMT. The glucose sensing was measured through the drain current of HEMT with a change of the charges on the ZnO nanorods, and the detection signal was amplified through the HEMT. Although the response of the HEMT-based sensor is similar to that of an electrochemical-based sensor, a much lower detection limit of 0.5 nM was achieved for the HEMT-based sensor due to this amplification effect. Since there is no reference electrode required for the HEMT-based sensor, the amount of sample only depends on the area of gate dimension and can be minimized. The sensors do not respond to glucose unless the enzyme is present, as shown in Figure 2.26.

Although measuring the glucose in the EBC is a noninvasive and convenient method for the diabetic application, the activity of the immobilized GO_x is highly dependent on the pH value of the solution. The GOx activity can be reduced to 80% for pH = 5 to 6. If the pH value of the glucose

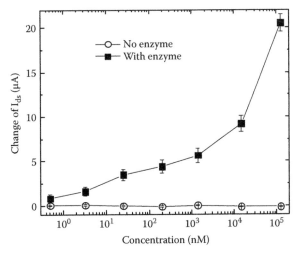

FIGURE 2.26 Change in drain-source current in HEMT glucose sensors both with and without localized enzyme.

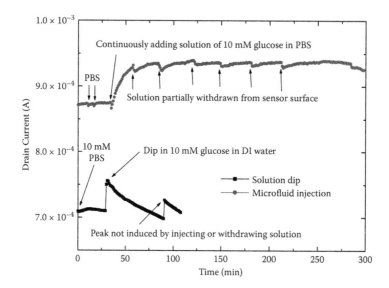

FIGURE 2.27 Plot of drain current versus time by dipping the glucose sensor in 10 mM of glucose dissolved in DI water (black line) and exposing the sensor to continuous flow of 10 mM of glucose dissolved in phosphate buffer saline with a pH value of 7.4 (red line).

solution is larger than 8, the activity drops off very quickly [14]. Figure 2.27 shows the time-dependent source-drain current signals with constant drain bias of 500 mV for glucose detection in DI water and PBS buffer solution. 50 μL of PBS solution was introduced on the glucose sensor, and no current change can be seen with the addition of buffer solution at 20 and 30 min. This stability is important to exclude possible noise from the mechanical change of the buffer solution. By sharp contrast, the current change showed a rapid response in less than 20 s when the sensor was dipped into the 100 mL of 10 mM glucose solution using DI water as the solvent. This sudden drain current increase indicated that GOx immediately reacted with glucose and oxygen was produced as a by-product of this reaction. However, the drain current gradually decreased. This was due to the oxygen produced in the GOx–glucose interaction reacting with water and changing the pH value adjacent the gate area. Since there was not agitation in the glucose solution, the solution around gate area became more basic, and the activity of GOx decreased due to the high pH value environment from 60 to 85 min.

Because the lower activity of GOx in the high pH value condition, the amount of oxygen produced from GOx and glucose decreased as well during the period of 60–85 min. Once the OH− ions produce from reaction between oxygen and water diffused away the gate area, the pH value decreased. Thus, around 85 min, the pH value of the glucose solution around gate area decreased low enough to allow the activity of GOx to resume, and the drain current of the glucose sensor showed another sudden increase. Then, the same process happened again, and drain current of the glucose current gradually decreased for a second time.

On the contrary, when the glucose sensor was used in a pH-controlled environment, the drain current stayed fairly constant, as shown in Figure 2.27. In this experiment, 50 μL of PBS solution was introduced on the glucose sensor to establish the baseline of the sensor as in the previous experiment. Then, glucose of 10 nM concentration prepared in PBS solution was introduced to the gate area of the glucose sensor through a micro-injector. There was no glucose in the 50 μL PBS solution, and the PBS solution was added at 20 and 30 min. It took time for the glucose solution to diffuse to the gate area of the sensor through the blank PBS, and the drain current gradually increased corresponding to the glucose diffusion process. Since the fresh glucose was continuously provided to the sensor surface and the pH value of the glucose was controlled, once the concentration of the glucose reached equilibrium at the gate of the glucose sensor, the drain current of the glucose remained constant except in the presence of glucose solution, which was taken out from time to time using a micropipette. There were small oscillations of the drain current observed, which could be eliminated by using a microfluidic device for this experiment.

The human pH value can vary significantly depending on the health condition. Since we cannot control the pH value of the EBC samples, we needed to measure the pH value while determine the glucose concentration in the EBC. With the fast-response time and low volume of the EBC required for HEMT-based sensor, a handheld and real-time glucose sensing technology can be realized.

2.8.3 PROSTATE CANCER DETECTION

Prostate cancer is the second most common cause of cancer death among men in the United States [152]. The most commonly used serum marker for diagnosis of prostate cancer is prostate-specific antigen (PSA) [153,154]. The market size for prostate cancer testing is enormous. According to the American Cancer Society, prostate cancer is the most common form of cancer among men, other than skin cancer. It is estimated that during 2007, in the United States alone, 218,890 new cases of prostate cancer will be diagnosed and 1 in 6 men will be diagnosed with prostate cancer during his lifetime [155].

The American Cancer Society recommends healthcare professionals to offer PSA blood test and the digital rectal exam (DRE) yearly for men above the age of 50. Those men who have a higher risk, such as African Americans and men who have a first-degree relative diagnosed with prostate cancer, should start testing at 45. Men who have several first-degree relatives diagnosed with prostate cancer should begin testing at 40. Since 1990, a recent awareness of cancers and the benefits of early detection have increased early detection tests for prostate cancer, and they have grown to become fairly common. Prostate cancer can often be found early by testing the amount of PSA in the patient's blood. It can also be detected on a DRE. If a man has routine yearly exams and either one of these test results becomes abnormal, then any cancer present has likely been found at an early, more treatable stage.

The prostate cancer testing market is expected to grow over the upcoming years. As awareness of cancer and early detection increases, so too will the need for testing. Given the high demand for prostate cancer testing, one would think that there are many options for early detection. However, there are only two main ways for preliminary testing for prostate cancer: the prostate cancer antigen blood test and the digital rectal exam. PSA is made by cells in the prostate gland and, although PSA is mostly found in semen, a certain amount is found in the blood as well. Most men have PSA levels under 4 ng/mL of blood. When prostate cancer develops, the PSA level usually goes up above 4 ng/mL; however, about 15% of men with a PSA below 4 will have prostate cancer on biopsy. If the patient's PSA level is between 4 and 10, their chance of having prostate cancer is about 25%. If the patient's PSA level is above 10, there is more than a 50% chance they have prostate cancer, which increases as the PSA level goes up. If the patient's PSA level is high, the doctor may advise a prostate biopsy to find out if they have cancer.

Generally, PSA testing approaches are costly, time-consuming, and need sample transportation. A number of different electrical measurements have been used for the rapid detection of PSA [156–161]. For example, electrochemical measurements based on impedance and capacitance are simple and inexpensive but need improved sensitivities for use with clinical samples [156,157]. Resonant frequency changes of an anti-PSA antibody-coated microcantilever enable a detection sensitivity of ~10 pg/Ml, but this microbalance approach has issues with the effect of the solution on resonant frequency and cantilever damping [157,158]. Antibody-functionalized nanowire FETs coated with antibody provide for low detection levels of PSA [160,161], but the scale-up potential is limited by the expensive e-beam lithography requirements. Antibody-functionalized Au-gated AlGaN/GaN HEMTs shown schematically in Figure 2.28 were found to be effective for detecting PSA at low concentration levels.

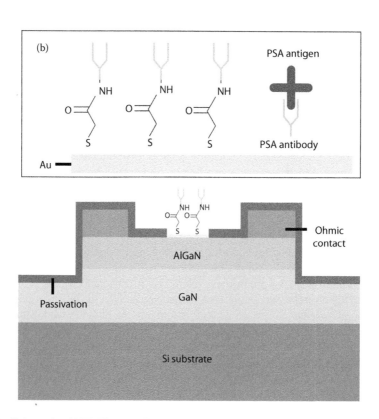

FIGURE 2.28 Schematic of HEMT sensor functionalized for PSA detection.

The PSA antibody was anchored to the gate area through the formation of carboxylate succinimidyl ester bonds with immobilized thioglycolic acid. The HEMT drain-source current showed a response time of less than 5 s when target PSA in a buffer at clinical concentrations was added to the antibody-immobilized surface. The devices could detect a range of concentrations from 1 µg/mL to 10 pg/mL. The lowest detectable concentration was two orders of magnitude lower than the cut-off value of PSA measurements for clinical detection of prostate cancer. Figure 2.29 shows the real-time PSA detection in PBS buffer solution using the source and drain current change with constant bias of 0.5 V [41]. No current change can be seen with the addition of buffer solution or nonspecific bovine serum albumin (BSA), but there was a rapid change when 10 ng/mL PSA was

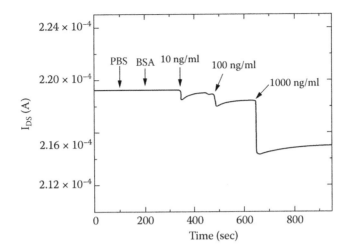

FIGURE 2.29 Drain current versus time for PSA detection when sequentially exposed to PBS, BSA, and PSA.

added to the surface. The abrupt current change due to the exposure of PSA in a buffer solution could be stabilized after it diffused into the buffer solution. The ultimate detection limit appears to be a few picogram per milliliter [41].

2.8.4 KIDNEY INJURY MOLECULE DETECTION

Problems such as acute kidney injury (AKI) or acute renal failure (ARF) are unfortunately still associated with a high mortality rate [162–164]. An important biomarker for early detection of AKI is the urinary antigen known as kidney injury molecule-1 or KIM-1 [165], and this is generally carried out with the ELISA technique discussed earlier [166–168]. The biomarker can also be detected with particle-based flow cytometric assay, but the cycle time is several hours [169]. Electrical measurement approaches based on carbon nanotubes [170], nanowires of In_2O_3 [171] or Si [172–176], or Si or GaN FETs look promising for fast and sensitive detection of antibodies and potentially for molecules such as KIM-1 [162–176].

The functionalization scheme in the gate region began with thioglycolic acid followed by KIM-1 antibody coating [177]. The gate region was deposited with a 5 nm thick Au film. Then the Au was conjugated to specific KIM-1 antibodies with a self-assembled monolayer of thioglycolic acid. The HEMT source-drain current showed a clear dependence on the KIM-1 concentration in phosphate-buffered saline (PBS) buffer solution, as shown in Figure 2.30 where the time-dependent source-drain current at a bias of 0.5 V is plotted for KIM-1 detection in PBS buffer solution. The limit of detection (LOD) was 1 ng/mL using a 20 μm × 50 μm gate-sensing area [177].

2.8.5 BREAST CANCER

The market size for breast cancer testing is vast—nearly 200,000 women and 1,700 men were diagnosed in 2006 alone. Although lucrative, competition in this industry is strong. Growth potential is possible, however, as the most effective and widely used diagnostic exam for breast cancer, the mammogram, is potentially harmful due to radiation exposure. Other, less popular, exams that do not involve radiation tend to be both invasive and expensive. Currently, the overwhelming majority of patients are screened for breast cancer by mammography [178]. This procedure involves a high cost to the patient and is invasive (radiation), which limits the frequency of screening. Breast cancer is currently the most common female malignancy in the world, representing 7% of the more than 7.6 million cancer-related deaths worldwide. Breast cancer accounts for over 30% of all new diagnoses

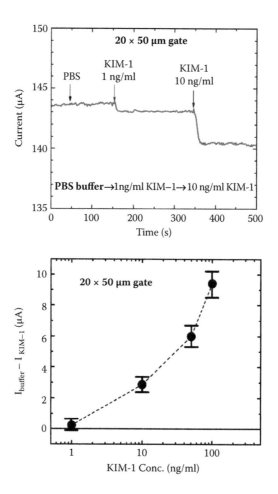

FIGURE 2.30   Time-dependent current signal when exposing the HEMT to 1 and 10 ng/mL KIM-1 in PBS buffer.

in women aged 20–49 and 50–69, and 20% among older women. As a result, more than one million mammograms are performed each year. According to the National Breast Cancer Foundation, it is estimated that nearly 200,000 women and 1,700 men will be diagnosed with breast cancer this year.

When breast cancer is discovered early on, there is a much better chance of successful treatment. Therefore, it is highly recommended that women check their breasts monthly from the age of 20. Clinical breast examinations should be conducted every 3 years from ages 20 to 39 and an annual mammogram for women aged 50 and older. Work by Michaelson et al. [179] indicates a 96% survival rate if patients could be screened every 3 months. Thus, mortality in breast cancer patients could be reduced by increasing the frequency of screening. However, this is not feasible presently due to the lack of cheap and reliable technologies that can screen breast cancer noninvasively.

There is recent evidence to suggest that salivary testing for makers of breast cancer may be used in conjunction with mammography [180–189]. Saliva-based diagnostics for the protein c-erbB-2 have tremendous prognostic potential [187,190]. Soluble fragments of the c-erbB-2 oncoprotein and the cancer antigen 15-3 were found to be significantly higher in the saliva of women who had breast cancer than in those patients with benign tumors [188]. Other studies have shown that epidermal growth factor (EGF) is a promising marker in saliva for breast cancer detection [190,191]. These initial studies indicate that the saliva test is both sensitive and reliable and can be potentially useful in initial detection and follow-up screening for breast cancer. However, to fully realize the potential of salivary biomarkers, technologies are needed that will enable facile, sensitive, specific detection of breast cancer.

Antibody-functionalized Au-gated AlGaN/GaN HEMTs show promise for detecting c-erbB-2 antigen. The c-erbB-2 antigen was specifically recognized through c-erbB antibody, anchored to the gate area. We investigated a range of clinically relevant concentrations from 16.7 μg/mL to 0.25 μg/mL.

The Au surface was functionalized with a specific bifunctional molecule, thioglycolic acid. We anchored a self-assembled monolayer of thioglycolic acid, $HSCH_2COOH$, an organic compound and containing both a thiol (mercaptan) and a carboxylic acid functional group, on the Au surface in the gate area through strong interaction between gold and the thiol group of the thioglycolic acid. The devices were first placed in the ozone/UV chamber and, then, submerged in 1 mM aqueous solution of thioglycolic acid at room temperature. This resulted in binding of the thioglycolic acid to the Au surface in the gate area with the COOH groups available for further chemical linking of other functional groups. The device was incubated in a PBS solution of 500 μg/mL c-erbB-2 monoclonal antibody for 18 h before real-time measurement of c-erbB-2 antigen.

After incubation with a PBS-buffered solution containing c-erbB-2 antibody at a concentration of 1 μg/mL, the device surface was thoroughly rinsed off with deionized water and dried by a nitrogen blower. The source and drain current from the HEMT were measured before and after the sensor was exposed to 0.25 μg/mL of c-erbB-2 antigen at a constant drain bias voltage of 500 mV. Any slight changes in the ambient of the HEMT affect the surface charges on the AlGaN/GaN. These changes in the surface charge are transduced into a change in the concentration of the 2DEG in the AlGaN/GaN HEMTs, leading to the slight decrease in the conductance for the device after exposure to c-erbB-2 antigen.

Figure 2.31 (top) shows real-time c-erbB-2 antigen detection in PBS buffer solution using the source and drain current change with constant bias of 500 mV. No current change can be seen with the addition of buffer solution around 50 s, showing the specificity and stability of the device. In clear contrast, the current change showed a rapid response in less than 5 s when target 0.25 μg/mL c-erbB-2 antigen was added to the surface. The abrupt current change due to the exposure of c-erbB-2 antigen in a buffer solution was stabilized after the c-erbB-2 antigen thoroughly diffused into the buffer solution. Three different concentrations (from 0.25 μg/mL to 16.7 μg/mL) of the exposed target c-erbB-2 antigen in a buffer solution were detected. The experiment at each concentration was repeated five times to calculate the standard deviation of source-drain current response.

The limit of detection of this device was 0.25 μg/mL c-erbB-2 antigen in PBS buffer solution. The source-drain current change was nonlinearly proportional to c-erbB-2 antigen concentration, as

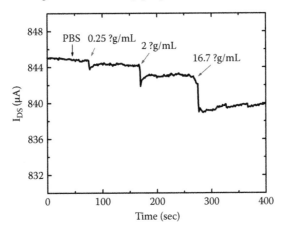

FIGURE 2.31   Drain current of an AlGaN/GaN HEMT over time for c-erbB-2 antigen from 0.25 to 17 μg/mL.

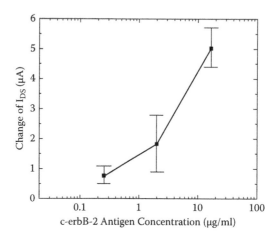

FIGURE 2.32   Drain current as a function of c-erbB-2 antigen concentrations from 0.25 to 17 µg/mL.

shown in Figure 2.32. Between each test, the device was rinsed with a wash buffer of 10 mM, pH 6.0 phosphate buffer solution containing 10 mM KCl to strip the antibody from the antigen.

Clinically relevant concentrations of the c-erbB-2 antigen in the saliva and serum of normal patients are 4–6 µg/mL and 60–90 µg/mL, respectively. For breast cancer patients, the c-erbB-2 antigen concentrations in the saliva and serum are 9–13 µg/mL and 140–210 µg/mL, respectively. Our detection limit suggests that HEMTs can be easily used for the detection of clinically relevant concentrations of biomarkers. Similar methods can be used for detecting other important disease biomarkers, and a compact disease diagnosis array can be realized for multiplex disease analysis.

2.8.6   LACTIC ACID

Lactic acid can also be detected with ZnO nanorod-gated AlGaN/GaN HEMTs. Interest in developing improved methods for detecting lactate acid has been increasing due to its importance in areas such as clinical diagnostics, sports medicine, and food analysis. An accurate measurement of the concentration of lactate acid in blood is critical to patients that are in intensive care or undergoing surgical operations, as abnormal concentrations may lead to shock, metabolic disorder, respiratory insufficiency, and heart failure. Lactate acid concentration can also be used to monitor the physical condition of athletes or of patients with chronic diseases such as diabetes and chronic renal failure. In the food industry, lactate level can serve as an indicator of the freshness, stability, and storage quality. For the preceding reasons, it is desirable to develop a sensor capable of simple and direct measurements, rapid response, high specificity, and low cost. Recent researches on lactate acid detection mainly focus on amperometric sensors with lactate acid-specific enzymes attached to an electrode with a mediator [192–200]. Examples of materials used as mediators include carbon paste, conducting copolymer, nanostructured Si_3N_4, and silica materials. Other methods of detecting lactate acid include utilizing semiconductors [201] and electro-chemiluminescent materials [202].

A ZnO nanorod array, which was used to immobilize lactate oxidase oxidase (LOx), was selectively grown on the gate area using low temperature hydrothermal decomposition as illustrated in Figure 2.33. The array of one-dimensional ZnO nanorods provided a large effective surface area with high surface-to-volume ratio and a favorable environment for the immobilization of LOx.

The AlGaN/GaN HEMT drain-source current showed a rapid response when various concentrations of lactate acid solutions were introduced to the gate area of the HEMT sensor. The HEMT could detect lactate acid concentrations from 167 nM to 139 µM. Figure 2.34 shows a real-time detection of lactate acid by measuring the HEMT drain current at a constant drain-source bias voltage of 500 mV during exposure of HEMT sensor to solutions with different concentrations of lactate

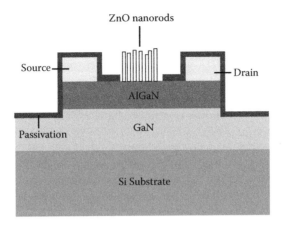

FIGURE 2.33 Schematic cross-sectional view of the ZnO nanorod-gated HEMT for lactic acid detection.

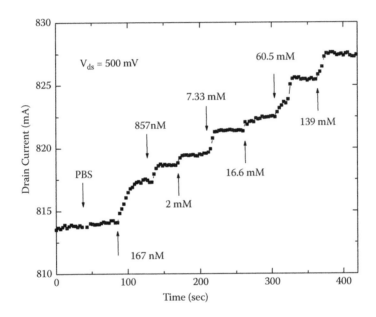

FIGURE 2.34 Drain current as a function of the time with successive exposure to lactate acid from 167 nM to 139 μM (bottom).

acid. The sensor was first exposed to 20 μL of 10 mM PBS, and no current change could be detected with the addition of 10 μL of PBS at approximately 40 s, showing the specificity and stability of the device. By contrast, a rapid increase in the drain current was observed when target lactate acid was introduced to the device surface. The sensor was continuously exposed to lactate acid concentrations from 167 nM to 139 μM.

As compared with the amperometric measurement-based lactate acid sensors, our HEMT sensors do not require a fixed reference electrode in the solution to measure the potential applied between the nano-materials and the reference electrode. The lactate acid sensing with the HEMT sensor was measured through the drain current of HEMT with a change of the charges on the ZnO nanorods, and the detection signal was amplified through the HEMT. Although the time response of the HEMT sensors is similar to that of electrochemical-based sensors, a significant change of drain current was observed for exposing the HEMT to the lactate acid at a low concentration of 167 nM due to this amplification effect. In addition, the amount of sample, which is dependent on the area of gate dimension, can be minimized for the HEMT sensor due to the fact that no reference electrode is required. Thus, measuring lactate acid in the exhaled breath condensate (EBC) can be achieved as a noninvasive method.

2.8.7 CHLORIDE ION DETECTION

Chlorine is widely used in the manufacture of many products and items directly or indirectly, that is, in paper product production, antiseptic, dye-stuffs, food, insecticides, paints, petroleum products, plastics, medicines, textiles, solvents, and many other consumer products. It is used to kill bacteria and other microbes in drinking water supplies and wastewater treatment. Excess chlorine also reacts with organics and forms disinfection by-products such as carcinogenic chloroform, which is harmful to human health. Thus, to ensure the safety of public health, it is very important to accurately and effectively monitor chlorine residues, typically in the form of chloride ion concentration, during the treatment and transport of drinking water [203–211]. In addition, the chloride ion is an essential mineral for humans, and is maintained to a total body chloride balance in body fluids such as serum, blood, urine, exhaled breath condensate, etc., by the kidneys. Variations in the chloride ion concentration in serum may serve as an index of renal diseases, adrenalism, and pneumonia and, thus, the measurement of this parameter is clinically important [212–216].

2.8.7.1 HEMT Functionalized with Ag/AgCl

HEMTs with an Ag/AgCl gate are found to exhibit significant changes in channel conductance upon exposing the gate region to various concentrations of chorine ion solutions, as shown in Figure 2.35. The Ag/AgCl gate electrode, prepared by potentiostatic anodization, changes electrical potential when it encounters chorine ions. This gate potential changes lead to a change of surface charge in the gate region of the HEMT, inducing a higher positive charge on the AlGaN surface, and increasing the piezo-induced charge density in the HEMT channel. These anions create an image positive charge on the Ag gate metal for the required neutrality, thus increasing the drain current of the HEMT. The HEMT source-drain current showed a clear dependence on the chorine concentration [212].

Figure 2.36 shows the time dependence of Ag/AgCl HEMT drain current at a constant drain bias voltage of 500 mV during exposure to solutions with different chlorine ion concentrations. The HEMT sensor was first exposed to DI water, and no change of the drain current was detected with the addition of DI water at 100 s. This stability was important to exclude possible noise from the mechanical change of the NaCl solution. By sharp contrast, there was a rapid response of HEMT drain current observed in less than 30 s when target of 1×10^{-8} M NaCl solution was switched to the surface at 175 s. The abrupt current change due to the exposure of chlorine in NaCl solution stabilized after the chlorine thoroughly diffused into the water to reach a steady state. When Ag/AgCl gate

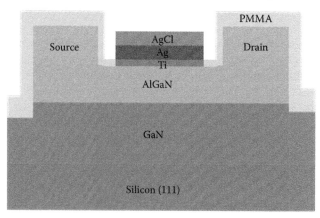

FIGURE 2.35   Schematic cross-sectional view of a Ag/AgCl-gated HEMT.

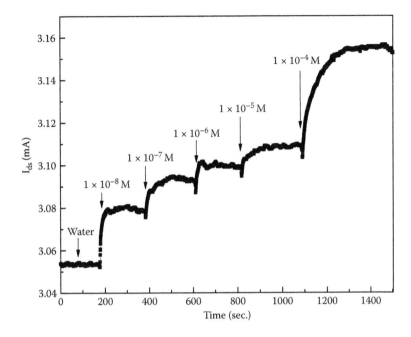

FIGURE 2.36   Time-dependent drain current of an Ag/AgCl-gated AlGaN/GaN HEMT exposed to different concentrations of NaCl solutions (bottom).

metal encountered chorine ions, the electrical potential of the gate was changed, inducing a higher positive charge on the AlGaN surface, and increased the piezo-induced charge density in the HEMT channel. 1×10^{-7} M of NaCl solution was then applied at 382 s, and it was accompanied with a larger signal corresponding to the higher chlorine concentration. Further real-time tests were carried out to explore the detection of higher Cl$^-$ ion concentrations. The sensors was exposed to 10^{-8}, 10^{-7}, 10^{-6}, 10^{-5}, and 10^{-4} M solutions continuously and repeated five times to obtain the standard deviation of source-drain current response for each concentration. The limit of detection of this device was 1×10^{-8} M chlorine in DI-water. Between each test, the device was rinsed with DI water. These results suggest that our HEMT sensors are recyclable with simple DI water rinse. The presence of the Ag/AgCl gate leads to a logarithmic dependence of current on the concentration of NaCl.

2.8.7.2   HEMT Functionalized with InN

Real-time detection of chloride ion detection with AlGaN/GaN HEMTs with an InN thin film in the gate region has also been demonstrated. The sensor, shown schematically in Figure 2.37, exhibited significant changes in channel conductance upon exposure to various concentrations of NaCl solutions. The InN thin film, deposited by molecular beam epitaxy, provided fixed surface sites for reversible anion coordination. The potential change in the gate area induced a change of the piezo-induced charge density in the electron channel in the HEMT. The sensor was tested over the range of 100 nM to 100 μM NaCl solutions.

Figure 2.38 also shows the results of real-time detection of Cl$^-$ ions by measuring the HEMT drain current at a constant drain bias voltage of 500 mV during exposure to solutions of different chloride ion concentrations. The HEMT sensor was first exposed to DI water, and no change of the drain current was detected with the addition of DI water at 100 s. The small spike in the current is due to mechanical disturbance of the HEMT surface when the water was added. By sharp contrast, a rapid response of HEMT drain current was observed in less than 20 s when target of 100 nM NaCl solution was exposed to the surface at 200 s. The abrupt current change stabilized after the sodium chloride solution thoroughly diffused into water and reached a steady state. When the InN gate metal encountered chloride ion, the electrical potential of the gate was changed and resulted in increase of

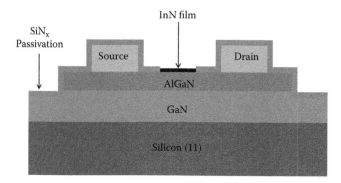

FIGURE 2.37   Cross-section schematic of the InN-gated HEMT.

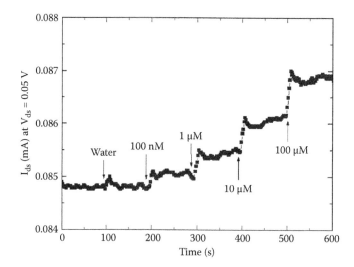

FIGURE 2.38   Real-time source-drain current at a constant bias of 500 mV as different concentrations of Cl⁻ ions were added.

the piezo-induced charge density in the HEMT channel. A larger signal change was observed when 1 μM of NaCl solution was applied at 300 s. The sensor was exposed to higher Cl⁻ ion concentrations of 10 μM and 100 μM sequentially for a further real-time test. The test was repeated with the same sensor for five times to obtain the standard deviation of source-drain current response for each concentration. The sensor can be reusable by washing it with DI water and drying with nitrogen gas. The limit of detection of this device was 100 nM chloride ions in DI water. The presence of the InN gate leads to a logarithmic dependence of current on the concentration for NaCl.

2.8.8   Pressure Sensing

Piezoelectric materials are used widely as sensitive pressure sensors, and piezoelectric gauges are typically fabricated with materials such as PZT, lithium niobate, and quartz [217–221]. In 1969, Kawai found a very high piezoelectric effect in polarized polyvinylidene fluoride (PVDF) [253]. Since then, polarized PVDF has also become an important piezoelectric material due to its flexibility, low density, low mechanical impedance, and easy fabrication as a ferroelectric. Because of its versatility, PVDF has many applications in low cost and disposable pressure sensors [222]. HEMTs with a polarized PVDF film coated on the gate area exhibited significant changes in channel conductance upon exposure to different ambient pressures. The PVDF thin film was deposited on the gate region with an inkjet plotter. Next, the PDVF film was polarized with an electrode located 2 mm above the PVDF film at a bias voltage of 10 kV and 70°C. A schematic of the HEMT is shown in Figure 2.39. Variations in ambient pressure induced changes in the charge in the polarized PVDF, leading to a change of surface charges on the gate region of the HEMT. Changes in the gate charge were amplified through the modulation of the drain current in the HEMT. By reversing the polarity of the polarized PVDF film, the drain current dependence on the pressure could be reversed. Our results indicate that HEMTs have potential for use as pressure sensors. For the pressure-sensing measurement, the HEMTs sensor was mounted on a carrier and put in a pressure chamber. N_2 gas was used for pressurizing the chamber, and a constant drain bias voltage of 500 mV was applied to the drain contact of the sensor.

Figure 2.40 also shows the real-time pressure detection with the polarized PVDF-gated HEMT. The drain current of HEMT sensor showed a rapid decrease in less than 5 s when the ambient pressure was changed to 20 psig [223–228]. A further decrease of the drain current for the HEMT sensor was observed when the chamber pressure increased to 40 psig. These abrupt drain current decreases were due to the change of charges in the PVDF film upon a shift of ambient pressure. An HEMT sensor without the PDVF coating was loaded in the pressure chamber, and there was no change of drain current observed.

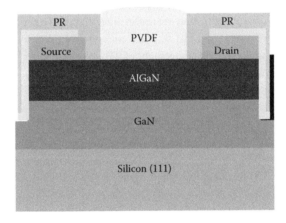

FIGURE 2.39   Schematic of HEMT sensor coated with polarized PVDF.

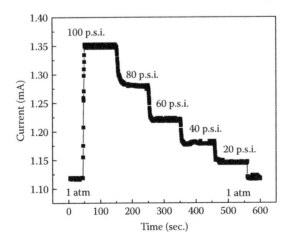

FIGURE 2.40   Drain current of a PVDF-gated AlGaN/GaN HEMT as a function of pressure. The PVDF was polarized by grounding the copper chuck holding the sample, and the copper wire electrode was applied with 10 kV.

2.8.9   TRAUMATIC BRAIN INJURY

Traumatic brain injury (TBI) is one of the most frequent causes of morbidity and mortality on the modern battlefield. U.S. casualties in Iraq are suffering a greater percentage of brain injuries than in previous wars. One of the contributing factors is the proliferation of the use of improvised explosive device (IED) against U.S. war fighters [229,230]. Recent assessments have indicated that about 65% of casualties are correlated with brain injuries. Traumatic brain injury, including concussion, is also a growing medical problem among civilians, with almost 2 million cases in the United States each year [231]. The development of a fast-response and portable TBI sensor can have a tremendous impact in early diagnosis, and proper management of TBI. Accurate and early diagnosis of a soldier's health in acute care environments can significantly simplify decisions about situation management. For example, decisions need to be made about whether to admit or discharge injured soldiers or to transfer them to other facilities with advanced diagonal systems, such as computer tomography (CT) and magnetic resonance imaging (MRI) scans. The capability to detect, in real time, markers in body fluids of soldiers can result in better patient outcomes especially in the battlefield or remote areas, where complicated and expensive CT and MRI scans are not available. For example, TBI is one of the most frequent causes of morbidity and mortality on the modern battlefield [229,230]. U.S. casualties in Iraq are suffering a greater percentage of brain injuries than in previous wars. The development of a fast response and portable TBI sensor can have tremendous impact in early diagnosis, and proper management of TBI.

Preliminary results show that the TBI antibody can be functionalized on the HEMT surface, and fast response of TBI antigen was achieved. The detection of limit of detection (LOD) was in the 10th of μg/mL range; however, this is not low enough for practical use. The typical TBI antigen concentration in the TBI patient's serum is in the range of ng/mL. We have used HEMT sensors to detect kidney injury molecules and prostate-specific antigen, and achieved LOD in the range of 1–10 pg/mL range. The reason for higher LOD for the TBI antigen detection was due to the much smaller size of the TBI antigen. Smaller antigens carry less charges, thus provide less effect on the drain current of the HEMT sensor. Based on the promising biomarker and device data, we have recently used HEMTs for detecting a biomarker UCH-L1 (BA0127) antigen involved in the traumatic injury molecule. The gate region was functionalized with a specific antibody to TBI antigen. The HEMT current showed a decrease as a function of TBI antigen concentration in PBS buffer (Figure 2.41). This shows the time-dependent current change in BA0127 (UCH-L1) antibody-modified HEMTs upon exposure to 2 μg/mL, and then to 16.9, 80, and 188 μg/mL of BA0127 (UCH-L1) in PBS buffer. The response time is around 6 s. The preliminary LOD was found to be 20 μg/mL, demonstrating the potential for TBI detection with accurate, rapid, noninvasive, and high-throughput capabilities.

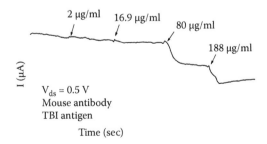

FIGURE 2.41   Time-dependent current signal when exposing the HEMT to 2 to 188 μg/mL BA0127 TBI antigen in PBS buffer.

2.9 ENDOCRINE DISRUPTER EXPOSURE LEVEL MEASUREMENT

There have been many reports evaluating the adverse effects of endocrine disrupters (ED) on repro-duction in wild animals, especially in aquatic environments [232–238]. A wide range of chemicals are considered EDs, including naturally occurring or improperly disposed estrogens and anthro-pogenic chemicals that were heavily used in the past. These chemicals promote feminization in wildlife and also pose a threat to public health. Some reports suggest that ED can influence fetal development [239] or act as a carcinogen [240–242]. Therefore, it is beneficial to develop tools that could accurately monitor the level of ED exposure.

Vitellogenin (Vtg) is a major egg yolk precursor protein that can be used as a biomarker to indicate an organism's exposure to ED [243]. The gene for this protein is expressed in the liver of oviparous animals under the control of estrogen [234,243–250]. Male fish, under natural conditions, should have very low doses of Vtg since they do not produce eggs. However, if male fish is exposed to estrogen or to estrogen mimics in the environment, the Vtg gene is turned on. The dynamic range of this protein in normal male fish is 10–50 ng/mL in plasma and ~20 mg/mL in females produc-ing eggs [248]. There have been reports of finding as much as 100 mg/mL in some fish that were induced with estrogen [249]. While the dynamic range is over 6 orders of magnitude, one normally finds 1~100 µg/mL in plasma in exposed males [248]. Although Vtg from one species is limited in its application as a probe for another, some segments of Vtg are highly conserved among spe-cies, suggesting the possibility of developing antibodies with wide cross-reactivity [248]. Enzyme-linked-immunosorbent-assay (ELISA) [243,246–249], quartz crystal microbalance [244], and yeast cell-based assay [245] are typically used as the Vtg detection methods; however, these methods are not suitable for on-field, real-time measurements.

Figure 2.42 shows the results of real-time detection of Vtg by measuring the HEMT at constant drain bias voltage of 500 mV with an Agilent 4156C parameter analyzer at 25°C. Purified Vtg solu-tions were prepared in 10 mM PBS and introduced to the sensor surface by a syringe autopipette (0.5–2 µL). The drain current measurement began with 10 µL of PBS placing on the HEMT surface. Before introducing the Vtg solutions, an additional 1 µL drop of PBS was added to the sensor. Other than the small disturbance at 50 s, no change of the drain current was detected. The disturbance in the current was due to mechanical disturbance of the HEMT surface, and the level went back to its origi-nal state. In comparison, a rapid response of HEMT drain current was observed in less than 10 s when the sensor was exposed to 5 µg/mL of Vtg at 100 s. The abrupt current change stabilized after the Vtg thoroughly diffused into the solution and reached a steady state. When the antigen encountered the

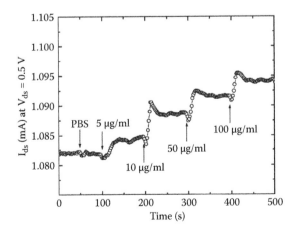

FIGURE 2.42 Real-time source-drain current of sensors when introduced to 5, 10, 50, and 100 µg/mL of vitellogenin.

antibody, the electrical potential of the gate was changed and resulted in the increase of the piezo-induced charge density in the HEMT channel. A larger signal change was observed when 10 μg/mL of Vtg was added at 200 s. There was PBS solution on the sensor already. In order to achieve the 10 μg/mL of Vtg on the sensor, higher concentration than 10 μg/mL was needed to be used to add on the sensor. Thus, an abrupt spike of the drain current change was observed due to the exposure of higher Vtg concentration solution to the sensing area, which was stabilized after the Vtg thoroughly diffused into the solution on the top of the sensing area. The sensor was exposed to higher Vtg concentrations of 50 μg/mL and 100 μg/mL sequentially for a further real-time test. The test was repeated with the same sensor for three times. The sensors were rinsed with 10 mM PBS at Ph = 6 because antibodies have optimal reactivity at pH = 7.4 and will release the antigen at a lower pH.

Figure 2.43 (top) shows Vtg detection results with actual male and female largemouth bass serum samples. The male serum contained no vitellogenin, whereas the female serum contained 8 mg/mL of Vtg. In contrast to PBS solutions, serum has many proteins that can interfere with the correct sensor reading. Therefore, it was necessary to block the unreacted carboxylic groups on the sensor. 1 mg/mL bovine serum albumin (BSA) solution was applied to the sensor for 3 h and washed thoroughly with PBS to remove the excess BSA. An important factor that influences the sensor

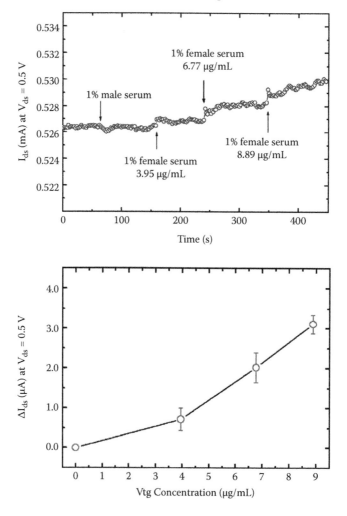

FIGURE 2.43   (Top) Real-time detection of vitellogenin in largemouth bass serum. (Bottom) Drain current change in HEMT as a function of vitellogenin concentration.

performance is the Debye length of the solution [251]. Electrolytes that are present in the solution can screen the field effect from the antigen–antibody interaction. In solutions such as serum, the Debye length is greatly reduced and causes lower sensitivity of the device. Therefore, dilution of largemouth bass serum was needed to detect Vtg in the solution. The serum was diluted to 1% in 10 mM PBS. The measurement started with 1% male serum containing no Vtg on the sensor. At 100 s, additional diluted male serum was added to confirm that there was no current change due to the serum background. By contrast, the current increased when drops of 1% female serum containing Vtg were added every 100 s. Figure 2.43 (bottom) shows the drain current changes as a function of the Vtg concentration. Each concentration was repeated five times to obtain the standard deviation of source-drain current response for each concentration. The source-drain current change is nonlinearly proportional to Vtg concentration. Between each test, the device was rinsed with a wash buffer of 10 μM phosphate buffer solution containing 10 μM KCl with pH 6 to strip the antigen from the antibody. Successful detection in serum samples shows that HEMTs have potential as biological sensors in real-world applications.

2.10 WIRELESS SENSORS

With many of the sensor applications, it is desirable to have the detected signal transmitted wirelessly to a central location. This could be part of an unmanned system for biotoxin detection or part of a personal medical monitoring system in which a patient could breathe into a handheld device that then transmits the encrypted signal to a doctor's office. This would allow for less numerous visits to doctor's offices and less problems with false positive tests because data could be accumulated over an extended period and a more reliable baseline established.

A prototype of a remote hydrogen sensing system was installed in a Ford dealership in Orlando (Greenway Ford), which is a test site for a hydrogen-fueled vehicle program supported by the Florida state government, since August 2006. The hydrogen-fueled buses and cars are stored and have maintenance performed on them in a large work area at the Ford dealership. Our hydrogen sensing system includes four on-site sensors, power management subsystem, wireless transmitter, and receiver connect to a computer. An intelligent monitoring software developed by our team is used to control data logging and tracking of each individual sensor as well as defining and implementing the monitoring states, transitions, and actions of the hydrogen sensor network. It can trigger an alarm and/or send messages to computers, cell phones, or PDAs, when a preset hydrogen threshold level is detected. Also, the software is able to warn the users of potential sensor failure, power outages and network failures through cell phone network and Internet. Currently, the cost of electronic parts including wireless transceiver and detection circuits is less than $20 in small quantity. In large quantity, it can be lower than $10. If the complete wireless transceiver and detection circuit are designed and integrated on a custom IC for mass production, the cost should be in the range of $5–8, similar to Bluetooth wireless chips. The sensors themselves can be mass-produced for 5–10 cents each according to Nitronex, Inc. A schematic block diagram of sensor module and wireless network server is shown in Figure 2.44.

As shown in the Figure 2.45, we have also designed and fabricated a pen-sized portable, reconfigurable wireless transceiver integrated with pH sensor. The wireless transmitter and receiver pair was designed to acquire EBC data and transmit it wirelessly. This system is able to interface multiple different sensors and consists of a transmitter and receiver pair. The transmitter was designed such that it is the size of a marker-pen so that it could be used as an ultra-portable lightweight handheld device. The transmitter is designed to be operated on an ultra-low-power mode. The transmitter is also equipped with an on-board recharging circuit that can be powered by using a standard mini-USB cable. The transmitter consumes on average 80 μA. The transmitter and receiver pair is designed to operate at 2.4 GHz with range of up to 20 ft line-of-sight. The receiver has USB 2.0 connectivity, which relays EBC data from the transmitter to a PC while powering the receiver. The transmitter is designed to integrate with various different sensors through a connector. The

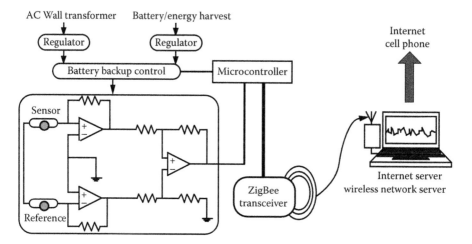

FIGURE 2.44   Schematic of remote data transmission system for sensors.

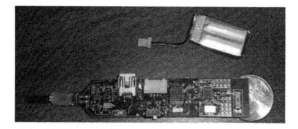

FIGURE 2.45 (See color insert).   Photographs of integrated pH sensor (top) and receiver/transmitter pair (bottom).

transmitter can be reset for the required input signal range to trigger the alarm through the bidirectional wireless communication for a different sensing application. Thus, this system is reconfigurable over the air. The wireless circuits only consume a power level around 1 μW. If the sensor consumes a similar power level, the battery installed on the transmitter package can last more than 1 month. This EBC sensing pair of devices can be mass-produced cost effectively well below $100 each pair. The sensor occupies the tip of the pen-shaped layout in Figure 2.45 and runs off a 75 mA Li ion polymer rechargeable battery. Figure 2.44 illustrated the package sensors mounted on a circuit board containing the detection circuit and microcontroller and the wireless transmitter for data collection. The sensor module is fully integrated on an FR4 PC board and is packaged with battery. The dimension of the sensor module package is 4.5″ × 2.9″ × 2″. The maximum line of sight range between the sensor module and the base station is 150 m. The base station of the wireless sensor network server is also integrated in a single module (3.0" × 2.7" × 1.1") and is ready to be connected to a laptop by a USB cable. The base station draws its power from the laptop's USB interface, thus do not require any battery or wall AC transformer, which reduces its form factor. The PC is used to record the sensing data, send the data to internet, and take actions when the hydrogen is detected.

A client program has also been developed to receive the sensor data remotely. The remote client can get a real-time log of the system for 10 min via the client program. In addition, a full data log will be obtained via accessing the server using an ftp client as the server program incorporates a full data logging functionality. When the current of any of the sensors exceeds a preset level, the server program will automatically execute the phone-dialing program, reporting the emergency to relevant personnel. A Web server was developed using MATLAB® (MathWorks, Inc.) to share the collected sensor data via the Internet. The data display time range can be chosen to show real time, or periods of 85 min, 15 h, and 6 day.

2.11　SUMMARY AND CONCLUSIONS

We have summarized recent progress in AlGaN/GaN HEMT sensors. These devices can take advantage of the microelectronics, including high sensitivity, the possibility of high-density integration, and mass manufacturability. The goal is to realize real-time, portable, and inexpensive chemical and biological sensors and to use these as handheld exhaled breath, saliva, urine, or blood monitors with wireless capability. Frequent screening can catch the early development of diseases, reduce the suffering of the patients due to late diagnoses, and lower the medical cost. For example, a 96% survival rate has been predicted in breast cancer patients if the frequency of screening is every 3 months. This frequency cannot be achieved with current methods of mammography due to high cost to the patient and invasiveness (radiation).

There are many possible applications, including:

- *Diabetes/Glucose Testing*—The population of diabetics is large and growing. There are in excess of 150 million diabetics in the world, and some believe this number will double by 2010. Although the frequent monitoring of glucose levels is strongly encouraged by health professionals, most of the glucose testing products currently on the market are uncomfortable for the user and dissatisfaction is high. Less invasive products are also less effective and have been unable to gain market share. Conditions are favorable for the introduction of an effective, noninvasive product.
- *Hydrogen Sensors*—Because the market for hydrogen sensors is highly dependent upon that of hydrogen fuels, the market for hydrogen sensors is not very large. Because the accurate detection of hydrogen leaks is extremely important, however, a number of still smaller niche markets have developed and competition within them remains high. Further development of the "hydrogen economy" in the near future could increase demand for hydrogen sensors.
- *Breast Cancer Testing*—The market size for breast cancer testing is vast—nearly 200,000 women and 1,700 men were diagnosed in 2006 alone. Although lucrative, competition in this industry is strong. Growth potential is possible, however, as the most effective and widely used diagnostic exam for breast cancer, the mammogram is potentially harmful due to radiation exposure. Other, less popular, exams that do not involve radiation tend to be both invasive and expensive.
- *Asthma Testing*—Asthma testing products are increasingly in demand. 1 in 20 Americans suffers from asthma, and this number stands to increase in the near future. Despite this, the market for preattack testing materials is undersaturated. Only one product has reached the marketplace and, although inexpensive, this product is also relatively inaccurate. Other potential competition may be more accurate but has yet to reach the market.

- *Prostate Testing*—Because one in six men will be diagnosed with prostate cancer during his lifetime, the market for testing products is also quite large. While two tests currently possess the bulk of the market, they are either inaccurate or invasive or both. Because of this relatively weak competitive market new entry possibilities are strong.
- *Narcotics Testing*—Toxicology screens are the most common narcotics testing products currently on the market. Used regularly by law enforcement agencies, medical facilities and corporate businesses tests are inexpensive and effective. More advanced technology has also been developed in an effort to monitor discrete drug levels in an individual's system.

HEMT sensors show promising results for protein, DNA, prostate cancer, kidney injury molecules, pH values of solutions, mercury ions, as well glucose in the exhaled breath condensate. The method relies on an amplification of small changes in antibody structure due to binding to antigens. The characteristics of these sensors include fast response (liquid phase—5 to 10 s, and gas phase—millisecond), digital output signal, small device size (less than $100 \times 100 \ \mu m^2$), and chemical and thermal stability.

Given the ever-increasing incidence of diabetes in both the United States and abroad, the market for diabetes testing and supplies is large and growing. Moreover, prevalent market-wide dissatisfaction with current testing alternatives—due to discomfort, inaccuracy, and/or cost—leads us to believe that the diabetes testing market is by far the most promising one for this technology. Although possible concerns include a difficult government certification process and insurance coverage, a survey of possible competition shows similar products to be either nonexistent or in an embryonic stage of development, meaning possible market entry can be made carefully and strategically.

The gas sensor market, on the other hand, is significantly less attractive. Although the product offers significant advantages over existing technology, these features are limited to what is currently a small niche market within the hydrogen gas and fuel market, which itself has low growth prospects and a high level of competition. Because general gas detection outside of this market is vastly simpler and even more competitive, the product does not offer significant advantages to the consumer, but possible advantages could be introduced to achieve production economies of scale and allow for market entry. In either case, any attempts at mass market entry should be made quickly as still more competition is already in preliminary development.

Similar to the market for diabetes testing, the market for breast cancer testing is also highly promising. Although numerous testing alternatives exist, the market size is huge and growing (in 2005, the testing market in the United States alone was worth well in excess of $1 billion), and the most common diagnostic methods involve some level of discomfort and/or exposure to radiation, giving it the same kind of patient dissatisfaction found among diabetes. Because testing is performed less frequently, however, its larger market may prove less lucrative and market entry is highly dependent upon government regulations and insurance coverage. Barring these difficulties, market entry (if made carefully) should be relatively easy.

There are still some critical issues. First, the sensitivity for certain antigens (such as prostate or breast cancer) needs to be improved further to allow sensing in body fluids other than blood (urine, saliva). Second, a sandwich assay allowing the detection of the same antigen using two different antibodies (similar to ELISA) needs to be tested. Third, integrating multiple sensors on a single chip with automated fluid handling and algorithms to analyze multiple detection signals and, fourth, a package that will result in a cheap final product is needed. Fourth, the stability of surface functionalization layers in some cases is not conducive to long-term storage, and this will limit the applicability of those sensors outside of clinics. There is certainly a need for detection of multiple analytes simultaneously. However, there are many such approaches, and acceptance from the clinical community is generally slow for many reasons, including regulatory concerns.

ACKNOWLEDGMENTS

This work is supported by the ONR-funded Center for Sensor Materials and Technologies, the State of Florida-funded Center for Nano-Bio Sensors, NSF, and ARO. The collaborations with T. Lele, Y. Tseng, K. Wang, D. Dennis, W. Tan, B.P. Gila, W. Johnson, and A. Dabiran are greatly appreciated.

REFERENCES

1. A. L. Burlingame, R. K. Boyd, and S. J. Gaskell, *Mass Spectrometry, Anal. Chem.*, 68 (1996), pp. 599–611.
2. K. W. Jackson and G. Chen, *Anal. Chem.*, 68 (1996), pp. 231–242.
3. J. L. Anderson, E. F. Bowden, and P. G. Pickup, *Anal. Chem.*, 68 (1996), pp. 379–401.
4. R. J. Chen, S. Bangsaruntip, K. A. Drouvalakis, N. W. S. Kam, M. Shim, Y. Li, W. Kim, P. J. Utz, and H. Dai, *Proc. Natl. Acad. Sci. USA*, 100 (2003), pp. 4984–4990.
5. C. Li, M. Curreli, H. Lin, B. Lei, F. N. Ishikawa, R. Datar, R. J. Cote, M. E. Thompson, and C. Zhou, *J. Am. Chem. Soc.*,127 (2005), pp. 12484–12498.
6. J. Zhang, H. P. Lang, F. Huber, A. Bietsch, W. Grange, U. Certa, R. Mckendry, H.-J. Güntherodt, M. Hegner, and Ch. Gerber, *Nature Nanotechnol.*, 1 (2006), pp. 214–220.
7. F. Huber, H. P. Lang, and C. Gerber, *Nature Nanotechnol.*, 3 (2008), pp. 645–646.
8. A. Sandu, *Nature Nanotechnol.*, 2 (2007), pp. 746–748.
9. G. Zheng, F. Patolsky, Y. Cui, W. U. Wang, and C. M. Lieber, *Nature Biotechnol.*, 23 (2005), pp.1294–1296.
10. R. A. Greenfield, B. R. Brown, J. B. Hutchins, J. J. Iandolo, R. Jackson, L. Slater, and M. S. Bronze, *Amer. J. Med. Sci.*, 323, 326 (2002), pp. 326–332.
11. A. H. Cordesman, *Weapons of Mass Destruction in the Gulf and Greater Middle East: Force Trends, Strategy, Tactics and Damage Effects*, Washington, DC: Center for Strategic and International Studies; 9 (1998), pp. 18–31.
12. J. S. Bermudez, *The Armed Forces of North Korea*, London, England: IB Tauris (2001).
13. S. S. Arnon, R. Schechter, and T. V. Inglesby, *JAMA*, 285 (2001), pp. 8–18.
14. T. Makimoto, Y. Yamauchi, and K. Kumakura, *Appl. Phys. Lett.*, 84 (2004), pp. 1964–1966.
15. A. P. Zhang, L. B. Rowland, E. B. Kaminsky, J. W. Kretchmer, R. A. Beaupre, J. L. Garrett, J. B. Tucker, B. J. Edward, J. Foppes, and A. F. Allen, *Solid-State Electron.* 47 (2003), pp. 821–825.
16. W. Saito, T. Domon, I.Omura, M. Kuraguchi, Y.Takada, K. Tsuda, and M. Yamaguchi, *IEEE Electron. Dev. Lett.*, 27 (2006), pp. 326–328.
17. J. Jun, B. Chou, J. Lin, A. Phipps, S. Xu, K. Ngo, D. Johnson, A. Kasyap, T. Nishida, H. T. Wang, B. S. Kang, T. Anderson, F. Ren, L. C. Tien, P. W. Sadik, D. P. Norton, L. F. Voss, and S. J. Pearton, *Solid State Electron.*, 51 (2007), pp. 1018–1022.
18. X. Yu, C. Li, Z. N. Low, J. Lin, T. J. Anderson, H. T. Wang, F. Ren, Y. L. Wang, C. Y. Chang, S. J. Pearton, C. H. Hsu, A. Osinsky, A. Dabiran, P. Chow, C. Balaban, and J. Painter, *Sens. Actuat. B*, 135 (2008), pp. 188–194.
19. H. T. Wang, T. J. Anderson, F. Ren, C. Li, Z. N. Low, J. Lin, B. P. Gila, S. J. Pearton, A. Osinsky, and A. Dabiran, *Appl. Phys. Lett.*, 89 (2006), pp. 242111–242114.
20. H. T. Wang, T. J. Anderson, B. S. Kang, F. Ren, C. Li, Z. N. Low, J. Lin, B. P. Gila, S. J. Pearton, A. Osinsky, and A. Dabiran, *Appl. Phys. Lett.*, 90 (2007), pp. 252109–252111.
21. T. J. Anderson, H. T. Wang, B. S. Kang, F. Ren, S. J. Pearton, A. Osinsky, Amir Dabiran, and P. P. Chow, *Appl. Surf. Sci.*, 255 (2008), pp. 2524–2526.
22. J. Kim, B. P. Gila, G. Y. Chung, C. R. Abernathy, S. J. Pearton, and F. Ren, *Solid-State Electron.*, 47 (2003), pp. 1487–1490.
23. H. T. Wang, T. J. Anderson, F. Ren, C. Li, Z. N. Low, J. Lin, B. P. Gila, S. J. Pearton, A. Osinsky, and A. Dabiran, *Appl. Phys. Lett.*, 89 (2006), pp. 242111–242113.
24. H. T. Wang, B. S. Kang, F. Ren, R. C. Fitch, J. K. Gillespie, N. Moser, G. Jessen, T. Jenkins, R. Dettmer, D. Via, A. Crespo, B. P. Gila, C. R. Abernathy, and S. J. Pearton, *Appl. Phys. Lett.*, 87 (2005), pp. 172105–172107.
25. J. Schalwig, G. Muller, U. Karrer, M. Eickhoff, O. Ambacher, M. Stutzmann, L. Gogens, and G. Dollinger, *Appl. Phys. Lett.*, 80 (2002), pp. 1222–1224.
26. B. P. Luther, S. D. Wolter, and S. E. Mohney, *Sens. Actuat. B*, 56, (1999), pp. 164–168.
27. B. S. Kang, R. Mehandru, S. Kim, F. Ren, R. C. Fitch, J. K. Gillespie, N. Moser, G. Jessen, T. Jenkins, R. Dettmer, D. Via, A. Crespo, K. H. Baik, B. P. Gila, C. R. Abernathy, and S. J. Pearton, *Phys. Status Solidi (c)*, 2 (2005), pp. 2672–2674.

28. H. T. Wang, B. S. Kang, F. Ren, L. C. Tien, P. W. Sadik, D. P. Norton, S. J. Pearton, and J. Lin, *Appl. Phys. A: Mater. Sci. Proc.*, 81 (2005), pp. 1117–1120.
29. J. S. Wright, W. Lim, B. P. Gila, S. J. Pearton, F. Ren, W. Lai, L. C. Chen, M. Hu, and K. H. Chen, *J. Vac. Sci. Technol. B*, 27 (2009), L8–L10.
30. J. L. Johnson, Y. Choi, A. Ural, W. Lim, J. S. Wright, B. P. Gila, F. Ren, and S. J. Pearton, *J. Electron. Mater.*, 38 (2009), pp. 490–494.
31. W. Lim, J. S. Wright, B. P. Gila, J. L. Johnson, A. Ural, T. Anderson, F. Ren, and S. J. Pearton, *Appl. Phys. Lett.*, 93 (2008), pp. 072110–072112.
32. L. Tien, P. Sadik, D. P. Norton, L. Voss, S. J. Pearton, H. T. Wang, B. S. Kang, F. Ren, J. Jun, and J. Lin, *Appl. Phys. Lett.*, 87 (2005), pp. 222106–222108.
33. O. Kryliouk, H. J. Park, H. T. Wang, B. S. Kang, T. J. Anderson, F. Ren, and S. J. Pearton, *J. Vac. Sci. Technol. B*, 23, 1891 (2005), pp.1891–1894.
34. L. Tien, H. T. Wang, B. S. Kang, F. Ren, P. W. Sadik, D. P. Norton, S. J. Pearton, and J. Lin, *Electrochem. Solid-State Lett.*, 8 (2005), pp. G239–G241.
35. H. T. Wang, B. S. Kang, F. Ren, L. C. Tien, P. W. Sadik, D. P. Norton, S. J. Pearton, and J. Lin, *Appl. Phys. Lett.*, 86 (2005), pp. 243503–243505.
36. M. Eickhoff, J. Schalwig, G. Steinhoff, O. Weidemann, L. Görgens, R. Neuberger, M. Hermann, B. Baur, G. Müller, O. Ambacher, and M. Stutzmann, *Phys. Stat. Sol. (c)* 0, No. 6 (2003), pp. 1908–1918.
37. R. Mehandru, B. Luo, B. S. Kang, J. Kim, F. Ren, S. J. Pearton, C.-C. Pan, G.-T. Chen, and J.-I. Chyi, *Solid-State Electron.*, 48 (2004), pp. 351–353.
38. R. Neuberger, G. Muller, O. Ambacher, and M. Stutzmann, *Phys. Stat. Sol. (A)*, 183, 2 (2001), pp. R10–R12.
39. P. Gangwani, S. Pandey, S. Haldar, M. Gupta, and R. S. Gupta, *Solid-State Electron.*, 51 (2007), pp. 130–135.
40. L. Shen, R. Coffie, D. Buttari, S. Heikman, A. Chakraborty, A. Chini, S. Keller, S. P. DenBaars, and U. K. Mishra, *IEEE Electron. Dev. Lett.*, 25 (2004), pp. 7–9.
41. B. S. Kang, H. T. Wang, T. P. Lele, F. Ren, S. J. Pearton, J. W. Johnson, P. Rajagopal, J. C. Roberts, E. L. Piner, and K. J. Linthicum, *Appl. Phys. Lett.*, 91 (2007), pp. 112106–112108.
42. A. El. Kouche, J. Lin, M. E. Law, S. Kim, B. S. Kim, F. Ren, and S. J. Pearton, *Sens. Actuat. B: Chem.*, 105 (2005), pp. 329–333.
43. H. T. Wang, B. S. Kang, F. Ren, S. J. Pearton, J. W. Johnson, P. Rajagopal, J. C. Roberts, E. L. Piner, and K. J. Linthicum, *Appl. Phys. Lett.*, 91 (2007), pp. 222101–222103.
44. B. S. Kang, S. Kim, F. Ren, B. P. Gila, C. R. Abernathy, and S. J. Pearton, *Sens. Actuat. B: Chem.*, 104 (2005), 232–236.
45. H. T. Wang, B. S. Kang, T. F. Chancellor, Jr., T. P. Lele, Y. Tseng, F. Ren, S. J. Pearton, A. Dabiran, A. Osinsky, and P. P. Chow, *Electrochem. Solid-State Lett.*, 10 (2007), pp. J150–152.
46. K. H. Chen, H. W. Wang, B. S. Kang, C. Y. Chang, Y. L. Wang, T. P. Lele, F. Ren, S. J. Pearton, A. Dabiran, A. Osinsky, and P. P. Chow, *Sens. Actuat. B: Chem.*, 134 (2008), pp. 386–389.
47. S. J. Pearton, T. Lele, Y. Tseng, and F. Ren, *Trends Biotechnol.*, 25 (2007), pp. 481–482.
48. H. T. Wang, B. S. Kang, T. F. Chancellor, Jr., T. P. Lele, Y. Tseng, F. Ren, S. J. Pearton, J. W. Johnson, P. Rajagopal, J. C. Roberts, E. L. Piner, and K. J. Linthicum, *Appl. Phys. Lett.*, 91, (2007), pp. 042114–042116.
49. B. S. Kang, H. T. Wang, F. Ren, B. P. Gila, C. R. Abernathy, S. J. Pearton, D. M. Dennis, J. W. Johnson, P. Rajagopal, J. C. Roberts, E. L. Piner, and K. J. Linthicum, *Electrochem. Solid-State Lett.*, 11, (2008), pp. J19–J21.
50. B. S. Kang, H. T. Wang, F. Ren, B. P. Gila, C. R. Abernathy, S. J. Pearton, J. W. Johnson, P. Rajagopal, J. C. Roberts, E. L. Pine, and K. J. Linthicum, *Appl. Phys. Lett.*, 91 (2007), pp. 012110–012112.
51. B. S. Kang, G. Louche, R. S. Duran, Y. Gnanou, S. J. Pearton, and F. Ren, *Solid-State Electron.*, 48 (2004), pp. 851–854.
52. J. R. Lothian, J. M. Kuo, F. Ren, and S. J. Pearton, *J. Electron. Mater.*, 21 (1992), pp. 441–445.
53. J. W. Johnson, B. Luo, F. Ren, B. P. Gila, W. Krishnamoorthy, C. R. Abernathy, S. J. Pearton, J. I. Chyi, T. E. Nee, C. M. Lee, and C. C. Chuo, *Appl. Phys. Lett.*, 77, 3230 (2000).
54. B. S. Kang, S. J. Pearton, J. J. Chen, F. Ren, J. W. Johnson, R. J. Therrien, P. Rajagopal, J. C. Roberts, E. L. Piner, and K. J. Linthicum, *Appl. Phys. Lett.*, 89 (2006), pp. 122102–122104.
55. B. S. Kang, F. Ren, L. Wang, C. Lofton, W. Tan, S. J. Pearton, A. Dabiran, A. Osinsky, and P. P. Chow, *Appl. Phys. Lett.*, 87 (2005), pp. 023508–023510.
56. B. S. Kang, H. Wang, F. Ren, S. J. Pearton, T. Morey, D. Dennis, J. Johnson, P. Rajagopal, J. C. Roberts, E. L. Piner, and K.J. Linthicum, *Appl. Phys. Lett.*, 91 (2007), pp. 252103–252105.

57. B. S. Kang, S. Kim, F. Ren, J. W. Johnson, R. Therrien, P. Rajagopal, J. Roberts, E. Piner, K. J. Linthicum, S. N. G. Chu, K. Baik, B. P. Gila, C. R. Abernathy, and S. J. Pearton, *Appl. Phys. Lett.*, 85 (2004), pp. 2962–2964.

58. S. J. Pearton, B. S. Kang, S. Kim, F. Ren, B. P. Gila, C. R. Abernathy, J. Lin, and S. N. G. Chu, *J. Phys: Condensed Matter*, 16 (2004), R961–985.

59. E. M. Logothetis, Automotive oxygen sensors, in N. Yamazoe (Ed.), *Chemical Sensor Technology*, vol. 3, Elsevier, Amsterdam, 1991.

60. Y. Xu, X. Zhou, and O. T. Sorensen, *Sens. Actuat. B*, 65 (2000), pp. 2–9.

61. L. Castañeda, *Mater. Sci. Eng. B*, 139 (2007), pp. 149–157.

62. J. Gerblinger, W. Lohwasser, U. Lampe, and H. Meixner, *Sens. Actuat. B*, 26, 93 (1995), pp. 93–98.

63. R. Yakimova, G. Steinhoff, R. M. Petoral Jr., C. Vahlberg, V. Khranovskyy, G. R. Yazdi, K. Uvdal, and A. Lloyd Spetz, *Biosens. Bioelectron.*, 22 (2007), pp. 2780–2785.

64. A. Trinchi, Y. X. Li, W. Wlodarski, S. Kaciulis, L. Pandolfi, S. P. Russo, J. Duplessis, and S. Viticoli, *Sens. Actuat. A*, 108 (2003), pp. 263–270.

65. M. R. Mohammadi and D. J. Fray, *Acta Mater.*, 55 (2007), pp. 4455–4461.

66. E. Sotter, X. Vilanova, E. Llobet, A. Vasiliev, and X. Correig, *Sens. Actuat. B*, 127 (2007), pp. 567–572.

67. Y.-L. Wang, L. N. Covert, T. J. Anderson, W. Lim, J. Lin, S. J. Pearton, D. P. Norton, J. M. Zavada, and F. Ren, *Electrochem. Solid-State Lett.* 11 (3) (2007), pp. H60–H62.

68. Y.-L. Wang, F. Ren, W. Lim, D. P. Norton, S. J. Pearton, I. I. Kravchenko, and J. M. Zavada, *Appl. Phys. Lett.* 90 (2007), pp. 232103–232105.

69. W. Lim, Y.-L. Wang, F. Ren, D. P. Norton, I. I. Kravchenko, J. M. Zavada, and S. J. Pearton, *Electrochem. Solid-State Lett.* 10 (9) (2007), pp. H267–H269.

70. M. J. Thorpe, K. D. Moll, R. J. Jones, B. Safdi, and J. Ye, *Science* 311 (2006), pp. 1595–1598.

71. M. J. Thorpe, D. Balslev-Clausen, M. S. Kirchner, and J. Ye, *Opt. Express* 16 (2008), pp. 2387–2393.

72. K. Namjou, C. B. Roller, and P. J. McCann, *IEEE Circuits Dev. Mag.*, September/October (2006), pp. 22–27.

73. R. F. Machado, D. Laskowski, O. Deffenderfer, T. Burch, S. Zheng, P. J. Mazzone, T. Mekhail, C. Jennings, J. K. Stoller, and J. Pyle., *Am. J. Respir. Crit. Care. Med.* 171 (2005), pp. 1286–1292.

74. Wormhoudt, J., Ed., *Infrared Methods for Gaseous Measurements*, Marcel Dekker, New York (1985).

75. T. J. Manuccia and J. G. Eden, Infrared Optical Measurement of Blood Gas Concentrations and Fiber Optical Catheter, U.S. Patent 4,509,522 (1985).

76. C. S. Chu and Y. L. Lo, *Sens. Actuat. B: Chem.* 129 (2008), pp. 120–126.

77. L. Kimmig, P. Krause, M. Ludwig, and K. Schmidt, Non-Dispersive Infrared Gas Analyzer, U.S. Patent 6,166,383 (2000).

78. R. Zhou, A. Hierlemann, U. Weimar, D. Schmeiber, and W. Gopel, The 8th International Conference on Solid-State Sensors and Actuators, and Eurosensors IX. Stockholm, Sweden, June 25–29, 225-PD6 (1995).

79. M. Shim, A. Javey, N. Wong, S. Kam, and H. Dai, *J. Am. Chem. Soc.*, 123 (2001), pp. 11512–11515.

80. J. Kong and H. Dai, *J. Phys. Chem. B*, 105 (2001), pp. 2890–2895.

81. S. Satyapal, T. Filburn, J. Trela, and J. Strange, *Energy Fuel.*, 15, 250 (2001), pp. 250–254.

82. D. B. Dell'Amico, F. Calderazzo, L. Labella, F. Marchetti, and G. Pampaloni, *Chem. Rev.*, 103 (2003), pp. 3857–3897.

83. K. G. Ong and C. A. Grimes, *Sensors* 1 (2001), pp. 193–200.

84. O. K. Varghese, P. D. Kichambre, D. Gong, K. G. Ong, E. C. Dickey, and C. A. Grimes, *Sens. Actuat. B: Chem.*, 81 (2001), pp. 32–38.

85. A. Star, T. R. Han, V. Joshi, J. P. Gabriel, and G. Gruner, *Adv. Mater.*, 16 (2004), pp. 2049–2056.

86. O. Kuzmych, B. L. Allen, and A. Star, *Nanotechnology*, 18 (2007), pp. 375502–375502.

87. A. Vasiliev, W. Moritz, V. Fillipov, L. Bartholomäus, A. Terentjev, and T. Gabusjan, *Sens. Actuat. B*, 49 (1998), pp. 133–138.

88. S. M. Savage, A. Konstantinov, A. M. Saroukan, and C. Harris, *Proc. ICSCRM '99* (2000), pp. 511–515.

89. K. D. Mitzner, J. Sternhagen, and D. W. Galipeau, *Sens. Actuat. B*, 9 (2003), pp. 92–97.

90. J. Wollenstein, J. A. Plaza, C. Cane, Y. Min, H. Botttner, and H. L. Tuller, *Sens. Actuat. B*, 93 (2003), pp. 350–356.

91. Y. Hu, X. Zhou, Q. Han, Q. Cao, and Y. Huang, *Mat. Sci. Eng. B*, 99 (2003), pp. 41–46.

92. Z. Ling, C. Leach, and R. Freer, *J. Eur. Ceramic Soc.*, 21 (2001), pp. 1977–1981.

93. B. B. Rao, *Mater. Chem. Phys.*, 64 (2000), pp. 62–67.

94. P. Mitra, A. P. Chatterjee, and H. S. Maiti, *Mater. Lett.*, 35 (1998), pp. 33–38.

95. J. F. Chang, H. H. Kuo, I. C. Leu, and M. H. Hon, *Sens. Actuat. B*, 84 (2003), pp. 258–264.

96. B. P. Gila, J. W. Johnson, R. Mehandru, B. Luo, A. H. Onstine, V. Krishnamoorthy, S. Bates, C. R. Abernathy, F. Ren, and S. J. Pearton, *Phys. Stat. Solid A*, 188 (2001), pp. 239–243.
97. J. Kim, R. Mehandru, B. Luo, F. Ren, B. P. Gila, A. H. Onstine, C. R. Abernathy, S. J. Pearton, and Y. Irokawa, *Appl. Phys. Lett.*, 80 (2000), pp. 4555–4557.
98. N. H. Nickel and K. Fleischer, *Phys. Rev. Lett.*, 90 (2003), pp. 197402-1–197402-4.
99. B. S. Kang, F. Ren, Y. W. Heo, L. C. Tien, D. P. Norton, and S. J. Pearton, *Appl. Phys. Lett.*, 86 (2005), pp. 112105–112107.
100. I. Horvath, J. Hunt, and P. J. Barnes, *Eur. Respir.*, 26 (2005), pp. 523–529.
101. K. Namjou, C. B. Roller, and P. J. McCann, *IEEE Circuits Dev. Mag.*, September–October 22 (2006), pp. 22–28.
102. R. F. Machado, D. Laskowski, O. Deffenderfer, T. Burch, S. Zheng, P. J. Mazzone, T. Mekhail, C. Jennings, J. K. Stoller, J. Pyle, J. Duncan, R. A. Dweik, and S. C. Erzurum, *Am. J. Respir. Crit. Care Med.*, 171 (2005), pp. 1286–1295.
103. T. Kullmann, I. Barta, Z. Lazar, B. Szili, E. Barat, M. Valyon, M. Kollai, and I. Horvath, *Eur. Respir.*, 29 (2007), pp. 496–502.
104. J. Vaughan, L. Ngamtrakulparit, T. N. Pajewski, R. Turner, T. A. Nguyen, A. Smith, P. Urban, S. Hom, B. Gaston, and J. Hunt., *Eur. Respir. J.*, 22 (2003), pp. 889–895.
105. J. F. Hunt, K. Fang, R. Malik et al., *Am. J. Respir. Crit. Care Med.*, 171 (2005), pp. 1286–1292.
106. K. Kostikas, G. Papatheodorou, K. Ganas, K. Psathakis, P. Panagou, and S. Loukides, *Am. J. Respir. Crit. Care Med.*, 165 (2002), pp. 1364–1369.
107. G. E. Carpagnano, M. P. Foschino Barbaro, and O. Resta, *Eur. J. Pharmacol.*, 519 (2005), pp. 175–181.
108. C. Gessner, S. Hammerschmidt, H. Kuhn et al., *Resp. Med.*, 97 (2003), pp. 1188–1194.
109. T. Kullmann, I. Barta, B. Antus, M. Valyon, and I. Horvath, *Eur. Respir. J.*, 31 (2), Feb. (2008), pp. 474–475.
110. I. Horvath, J. Hunt, and P. J. Barnes, *Eur. Respir. J.*, 26(9), Sept. 2005, pp. 523–548.
111. R. Accordino, A. Visentin, A. Bordin, S. Ferrazzoni, E. Marian, F. Rizzato, C. Canova, R. Venturini, and P. Maestrelli, *Resp. Med.*, 102 (March 3, 2008), pp. 377–381.
112. K. Czebe, I. Barta, B. Antus, M. Valyon, I. Horváth, and T. Kullmann, *Resp. Med.*, 102 (5), May 2008, pp. 720–725.
113. K. Bloemen, G. Lissens, K. Desager, and G. Schoeters, *Resp. Med.*, 101 (6), June 2007, pp. 1331–1337.
114. S. Park, H. Boo, and T. D. Chung, *Anal. Chi. Act.*, Vol. 46, June 2006, pp. 556–560.
115. P. Pandey, S. P. Singh, S. K. Arya, V. Gupta, M. Datta, S. Singh, and B. D. Malhotra, *Langmuir*, 23, April 2007, pp. 3333–3339.
116. G. K. Kouassi, J. Irudayaraj, and G. McCarty, *BioMag. Res. Tech.*, May 3, 2005, pp. 1–8.
117. A. L. Burlingame, R. K. Boyd, and S. J. Gaskell, *Anal. Chem.*, 68 (1996), pp. 599–604.
118. K. W. Jackson and G. Chen, *Anal. Chem.*, 68 (1996), pp. 231–235.
119. J. L. Anderson, E. F. Bowden, and P. G. Pickup, *Anal. Chem.*, 68 (1996), pp. 379–384.
120. Z. X. Cai, H. Yang, Y. Zhang, and X. P. Yan, *Anal. Chim. Acta*, 559 (2006), pp. 234–243.
121. J. L. Chen, Y. C. Gao, Z. B. Xu, G. H. Wu, Y. C. Chen, C. Q. Zhu, *Anal. Chim. Acta*, 577 (2006), pp. 77–83.
122. T. Balaji, M. Sasidharan, and H. Matsunaga, *Analyst*, 130 (2005), pp. 1162–1167.
123. G. Q. Shi and G. Jiang, *Anal. Sci.*, November 18 (2002), pp. 1215–1219.
124. A. Caballero, R. Martínez, V. Lloveras, I. Ratera, J. Vidal-Gancedo, K. Wurst, A. Tárraga, P. Molina, and J. Veciana, *J. Am. Chem. Soc.*, 127 (2005), pp. 15666–15672.
125. E. Coronado, J. R. Galán-Mascarós, C. Martí-Gastaldo, E. Palomares, J. R. Durrant, R. Vilar, M. Gratzel, and Md. K. Nazeeruddin, *J. Am. Chem. Soc.*, 127 (2005), pp. 12351–12356.
126. Y. K. Yang, K. J. Yook, and J. Tae, *J. Am. Chem. Soc.*, 127 (2005), pp. 16760–16765.
127. M. Matsushita, M. M. Meijler, P. Wirsching, R. A. Lerner, and K. D. Janda, *Org. Lett.*, 7 (2005), pp. 4943–4948.
128. C. C. Huang and H. T. Chang, *Anal. Chem.*, 78 (2006), pp. 8332–8843.
129. S. S. Arnon, R. Schechter, and T. V. Inglesby, *JAMA*, 285 (8), (2001), pp. 256–265.
130. R. A. Greenfield, B. R. Brown, J. B. Hutchins, J. J. Iandolo, R. Jackson, L. N. Slater, and M. S. Bronze, *Amer. J. Med. Sci.*, 323 (2002), pp. 326–334.
131. J. S. Michaelson, E. Halpern, and D. B. Kopans, *Radiology*, 212 (2), (1999), pp. 551–558.
132. T. Harrison, L. Bigler, M. Tucci, L. Pratt, F. Malamud, J. T. Thigpen, C. Streckfus, and H. Younger, *Spec. Care Dentist*, 18 (3), (1998), pp. 109–115.
133. R. McIntyre, L. Bigler, T. Dellinger, M. Pfeifer, T. Mannery, and C. Streckfus, *Oral Surg. Oral Med. Oral Pathol. Oral Radiol. Endod.*, 88 (6), (1999), pp. 687–693.

134. C. Streckfus, L. Bigelr, T. Dellinger, M. Pfeifer, A. Rose, and J. T. Thigpen, *Clin. Oral Investig.*, 3 (3), (1999), pp. 138–144.
135. C. Streckfus, L. Bigler, T. Dellinger, X. Dai, A. Kingman, and J.T. Thigpen, *Clin. Cancer. Res.*, 6 (6), (2000), pp. 2363–2365.
136. C. Streckfus, L. Bigler, M. Tucci, and J. T. Thigpen, *Cancer Invest.*, 18 (2), (2000), pp. 101–108.
137. C. Streckfus, L. Bigler, T. Dellinger, X. Dai, W. J. Cox, A. McArthur, A. Kingman and J. T. Thigpen, *Oral Surg Oral Med Oral Pathol. Oral Radiol. Endod.*, 91 (2), (2001), pp. 174–178.
138. L. R. Bigler, C. F. Streckfus, L. Copeland, R. Burns, X. Dai, M. Kuhn, P. Martin, and S. Bigler, *J. Oral Pathol. Med.*, 31 (7), (2002), pp. 421–434.
139. C. Streckfus and L. Bigler, *Adv. Dent. Res.*, 18 (1), (2005), pp. 17–22.
140. C. F. Streckfus, L. R. Bigler, and M. Zwick, *J. Oral Pathol. Med.*, 35 (5), (2006), pp. 292–299.
141. W. R. Chase, *J. Mich. Dent. Assoc.*, 82 (2), (2000), pp. 12–18.
142. S.Z. Paige and C.F. Streckfus, *Gen Dent*, 55 (2), (2007), pp. 156–166.
143. M. A. Navarro, R. Mesia, O. Diez-Giber, A. Rueda, B. Ojeda, and M. C. Alonso, *Breast Cancer Res. Treat.* 42(1), (1997), pp. 83–88.
144. K. Bagramyan, J. R. Barash, S. S. Arnon, and M. Kalkum, and M. Matrices, *PLoS ONE*, 3 (2008), pp. 2041–2045.
145. P. Montuschi and P. J. Barnes, *Trends Pharmacol. Sci.*, vol. 23 (May 5, 2002), pp. 232–237.
146. T. Dam, V. Anh, W. Olthuis, and P. Bergveld, *Sens. Actuat. B*, Vol.111/112(11), Nov. (2005), pp. 494–499.
147. G. M. Multu, *Am. J. Res. Crit. Care Med.* 164(11), Nov. (2001), pp. 731–737.
148. J. X. Wang, X. W. Sun, A. Wei, Y. Lei, X. P. Cai, C. M. Li, and Z. L. Dong, *Appl. Phys. Lett.*, 88 (2006), pp. 233106–233108.
149. A. Wei, X. W. Sun, J. X. Wang, Y. Lei, X. P. Cai, C. M. Li, Z. L. Dong, and W. Huang, *Appl. Phys. Lett.*, 89 (2006), pp. 123902–123904.
150. Y. H. Yang, H. F. Yang, M. H. Yang, Y. L. Liu, G. L. Shen, and R. Q. Yu, *Anal. Chim. Acta*, 525 (2004), pp. 213–220.
151. S. Hrapovic, Y. L. Liu, K. B. Male, and J. H. T. Luong, *Anal. Chem.*, 76 (2004), pp. 1083–1088.
152. G. J. Kelloff, D. S. Coffey, B. A. Chabner, A. P. Dicker, K. Z. Guyton, P. D. Nisen, H. R. Soule, and A. V. D'Amico, *Clin. Cancer Res.*, 10 (2007), pp. 3927–3934.
153. I. M. Thomson and D. P. Ankerst, *CMAJ*, 176 (2007), pp. 1853–1857.
154. D. C. Healy, C. J. Hayes, P. Leonard, L. McKenna, and R. O'Kennedy, *Trends Biotechnol.*, 25 (3), (2007), pp. 125–132.
155. Detailed Guide: Prostate Cancer. What Are the Key Statistics about Prostate Cancer, American Cancer Society. 2007. 08 NOV 2007 <http://www.cancer.org/docroot/CRI/content/CRI_2_4_1X_What_are_the_key_-statistics_for_prostate_cancer_36.asp?rnav=cri>.
156. J. Wang, *Biosens. Bioelectron.* 21 (2006), pp. 1887–1994.
157. C. F. Sanchez, C. J. McNeil, K. Rawson, and O. Nilsson, *Anal. Chem.*, 76 (2004), pp. 5649–5654.
158. K. S. Hwang, J. H. Lee, J. Park, D. S. Yoon, J. H. Park, and T. S. Kim, *Lab Chip.* 4 (2004), pp. 547–554.
159. K. W. Wee, G. Y. Kang, J. Park, J. Y. Kang, D. S. Yoon, J. H. Park, and T. S. Kim, *Biosens. Bioelectron.*, 20 (2005), pp. 1932–1936.
160. Y.-L. Wang, B. H. Chu, K. H. Chen, C. Y. Chang, T. P. Lele, G. Papadi, J. K. Coleman, B. J. Sheppard, C. F. Dungen, S. J. Pearton, J. W. Johnson, P. Rajagopal, J. C. Roberts, E. L. Piner, and K. J. Linthicum, *Appl. Phys. Lett.*, 94 (2009), pp. 243901–243903.
161. T. Anderson, F. Ren, S. J.Pearton, B.S. Kang, H.-T.Wang, C.-Y. Chang, and J. Lin, *Sensors*, 9(6), (2009), pp. 4669–4702.
162. R. Thadhani, M. Pascual, and J. V. Bonventre, *N. Engl. J. Med.*, 334 (1996), pp. 1448–1452.
163. G. M. Chertow, E. M. Levy, K. E. Hammermeister, F. Grover, and J. Daley, *Amer. J. Med.*, 104 (1998), pp. 343–347.
164. J. V. Bonventre and J. M. Weinberg, *J. Am. Soc. Nephrol.*, 14 (2003), pp. 2199–2203.
165. T. Ichimura, J. V. Bonventre, V. Bailly, H. Wei, C. A. Hession, R. L. Cate, and M. Sanicola, *J. Biol. Chem.*, 273 (1998), pp. 4135–4140.
166. V. S. Vaidya, V. Ramirez, T. Ichimura, N. A. Bobadilla, and J. V. Bonventre, *Am. J. Physiol. Renal. Physiol.*, 290 (2006), pp. F517–F522.
167. V. S. Vaidya and J. V. Bonventre, *Expert Opin. Drug Metab. Toxicol.*, 2 (2006), pp. 697–704.
168. R. Lequin, *Clin. Chem.*, 51 (2005), pp. 2415–2420.
169. Dario A.A. Vignali, *J. Immunol. Methods* 243 (2000), pp. 243–248.

170. R. J. Chen, S. Bangsaruntip, K. A. Drouvalakis, N. W. S. Kam, M. Shim, Y. Li, W. Kim, P. J. Utz, and H. Dai, *Proc. Natl. Acad. Sci. USA*, 100 (2003), pp. 4984–4989.
171. C. Li, M. Curreli, H. Lin, B. Lei, F. N. Ishikawa, R. Datar, R. J. Cote, M. E. Thompson, and C. Zhou, *J. Am. Chem. Soc.*, 127 (2005), pp. 12484–12489.
172. G. Zheng, F. Patolsky, Y. Cui, W. U. Wang, and C. M. Lieber, *Nature Biotechnol.*, 23 (2005), pp. 1294–1296.
173. F. Patolsky, G. Zheng, and C. M. Lieber, *Nanomedicine*, 1 (2006), pp. 51–56.
174. F. Patolsky, G. Zheng, and C. M. Lieber, *Nature Protocols*, 1 (2006), pp. 1711–1715.
175. F. Patolsky, B. P. Timko, G. Zheng, and C. M. Lieber, *MRS Bull.*, 32 (2007), pp. 142–148.
176. D. I. Han, D. S. Kim, J. E. Park, J. K. Shin, S. H. Kong, P. Choi, J. H. Lee, and G. Lim, *Jpn. J. Appl. Phys.*, 44 (2005), pp. 5496–5499.
177. G. Shekhawat, S. H. Tark, and V. P. Dravid, *Science*, 311 (2006), pp. 1592–1597.
178. H. T. Wang, B. S. Kang, F. Ren, S. J. Pearton, J. W. Johnson, P. Rajagopal, J. C. Roberts, E. L. Piner, and K. J. Linthicum, *Appl. Phys. Lett.*, 91 (2007), pp. 222101–222103.
179. What is Breast Cancer? United States Department of Health and Human Services, Nov. 3, 2007. <http://www.hhs.gov/breastcancer/whatis.html.>.
180. J. S. Michaelson, E. Halpern, and D. B. Kopans, *Radiology*, 212 (2), (1999), pp. 551–558.
181. T. Harrison, L. Bigler, M. Tucci, L. Pratt, F. Malamud, J. T. Thigpen, C. Streckfus, and H. Younger, *Spec. Care Dentist*, 18 (3), (1998), pp. 109–115.
182. R. McIntyre, L. Bigler, T. Dellinger, M. Pfeifer, T. Mannery, and C. Streckfus, *Oral Surg. Oral Med. Oral Pathol. Oral Radiol. Endod.*, 88 (6), (1999), pp. 687–695.
183. C. Streckfus, L. Bigler, T. Dellinger, M. Pfeifer, A. Rose, and J. T. Thigpen, *Clin. Oral Investig.*, 3 (3), (1999), pp. 138–144.
184. C. Streckfus, L. Bigler, T. Dellinger, X.Dai, A. Kingman, and J.T. Thigpen, *Clin. Cancer Res.*, 6 (6), (2000), pp. 2363–2368.
185. C. Streckfus, L. Bigler, M. Tucci, and J. T. Thigpen, *Cancer Invest.*, 18 (2), (2000), pp. 101–109.
186. C. Streckfus, L. Bigler, T. Dellinger, X. Dai, W. J. Cox, A. McArthur, A. Kingman, and J. T. Thigpen, *Oral Surg. Oral Med. Oral Pathol. Oral Radiol. Endod.*, 91 (2), (2001), pp. 174–178.
187. L. R. Bigler, C.F. Streckfus, L. Copeland, R. Burns, X. Dai, M. Kuhn, P. Martin, and S. Bigler, *J Oral Pathol. Med.*, 31 (7), (2002), pp. 421–426.
188. C. Streckfus and L. Bigler, *Adv. Dent. Res.*, 18 (1), (2005), pp. 17–23.
189. C. F. Streckfus, L. R. Bigler and M. Zwick, *J Oral Pathol. Med.*, 35 (5), (2006), pp. 292–297.
190. W.R. Chase, *J Mich. Dent. Assoc.*, 82 (2), 12 (2000), pp. 12–18.
191. S.Z. Paige, C.F. Streckfus, *Gen Dent.*, 55 (2), (2007), pp. 156–162.
192. M.A. Navarro, R. Mesia, O. Diez-Giber, A. Rueda, B. Ojeda, and M.C. Alonso, *Breast Cancer Res. Trea.*, 42(1), (1997), pp. 83–88.
193. Parra, E. Casero, L. Vazquez, F. Pariente, and E. Lorenzo, *Anal. Chim. Acta*, 555 (2006), pp. 308–312.
194. B. Phypers and T. Pierce, *Crit. Care Pain*, 6 (3), (2006), pp. 128–134.
195. C. Lin, C. Shih, and L. Chau, *Anal. Chem.*, 79 (2007), pp. 3757–3767.
196. U. Spohn, D. Narasaiah, L. Gorton, and D. Pfeiffer, *Anal. Chim. Acta*, 319 (1996), pp. 79–88.
197. J. Tong, J. Hu, Z. Huang, M. Pan, and Y. Chen, *Proceedings of the 2005 IEEE Engineering in Medicine and Biology 27th Annual Conference (Shanghai, China)* (2005), pp. 252–255.
198. M. Pohanka and P. Zbořil, *Food Technol. Biotechnol.*, 46 (1), (2008), pp. 107–114.
199. S. Suman, R. Singhal, A. Sharma, B.D. Malthotra, and C.S. Pundir, *Sens. Actuat. B*, 107 (2005), pp. 768–778.
200. J. Haccoun, B. Piro, V. Noël, and M.C. Pham, *Bioelectrochemistry*, 68 (2006), pp. 218–223.
201. J. Di, J. Cheng, Q. Xu, H. Zheng, J. Zhuang, Y. Sun, K. Wang, X. Mo, and S. Bi, *Biosens. Bioelectron.* 23, (2007), pp. 682–689.
202. A. Lupu, A. Valsesia, F. Bretagnol, P. Colpo, and F. Rossi, *Sens. Actuat. B*, 127 (2007), pp. 606–611.
203. C. A. Marquette, A. Degiuli, and L. Blum, *Biosens. Bioelectron.*, 19 (2003), pp. 433–438.
204. J. Taylor and S. Hong, *J. Lab. Med.*, 31-10 (2000), pp. 563–567.
205. H. Shekhar, V. Chathapuram, S. H. Hyun, S. Hong, and H. J. Cho, *IEEE Sensor J.*, 1 (2003), pp. 67–73.
206. H. K. Walker, W. D. Hall and J. W. Hurst, *Clinical Methods; The History, Physical, and Laboratory Examinations*, 3rd ed., Butterworth, London, 1990, pp. 189–197.
207. J. M. Cook and D. L. Miles, *Inst. Geol. Sci. Rep.*, 80 (1980), pp. 5–11.
208. H. N. Elsheimer, *Geostand Newsl.*, 11 (1987), pp. 115–122.
209. R. Verma and R. Parthasarthy, *J. Radioanal. Nucl. Chem. Lett.*, 214 (1996), pp. 391–398.

210. T. Graule, A. von Bohlen, J. A. C. Broekaert, E. Grallath, R. Klockenkamper, P. Tschopel, and G. Telg, *Fresenius Z. Anal. Chem.*, 335 (1989), pp. 637–645.
211. S. D. Kumar, K. Venkatesh, and B. Maiti, *Chromatograpia*, 59 (2004), pp. 243–249.
212. P. A. Blackwell, M. R. Cave, A. E. Davis, and S. A. Malik, *J. Chromatogr. A*, 770 (1997), pp. 93–99.
213. H. K. Walker, W. D. Hall, and J. W. Hurst, *Clinical Methods; The History, Physical, and Laboratory Examinations*, 3rd ed., Butterworth, London, 1990, pp. 34–38.
214. A. Davidsson, M. Söderström, K. Naidu Sjöswärd, and B. Schmekel, *Respiration*, 74 (2007), pp. 184–191.
215. O. Niimi, L. T. Nguyen, O. Usmani, B. Mann, and K. F. Chung, *Thorax*, 59 (2004), pp. 608–612.
216. R. M. Effros, K. W. Hoagland, M. Bosbous, D. Castillo, B. Foss, M. Dunning, M. Gare, W. Lin, and F. Sun, *Amer. J. Resp. Crit. Care Med.*, 165 (2002), pp. 663–669.
217. B. Davidsson, K. Naidu Sjöswärd, L. Lundman, and B. Schmekel, *Respiration*, 72 (2005), pp. 529–536.
218. V. Mortet, R. Petersen, K. Haenen, and M. D'Olieslaeger, *IEEE Ultrasonics Symp.*, (2005), pp. 1456–1460.
219. S. C. Ko, Y. C. Kim, S. S. Lee, S. H. Choi, and S. R. Kim, *Sens. Actuat. B*, 103 (2003), pp. 130–134.
220. R. Greaves and G. Sawyer, *Phys. Technol.*, 14 (1983), pp. 15–21.
221. E. S. Kim and R. S. Muller, *IEEE, IEDM 86*, (1986), pp. 8–10.
222. A. Odon, *Measurement Sci. Rev.*, 3 (2003), pp. 111–116.
223. A. V. Shirinov and W. K. Schomburg, *Sens. Actuat. A*, 142 (2008), pp. 48–53.
224. S. C. Hung, B. H. Chou, C. Y. Chang, K. H. Chen, Y. L. Wang, S. J. Pearton, A. Dabiran, P. P. Chow, G. C. Chi, and F. Ren, *Appl. Phys. Lett.*, 94 (2009), pp. 043903–043905.
225. Y. L. Wang, B. H. Chu, K. H. Chen, C. Y. Chang, T. P. Lele, Y. Tseng, S. J. Pearton, J. Ramage, D. Hooten, A. Dabiran, P. P. Chow, and F. Ren, *Appl. Phys. Lett.*, 93 (2008), pp. 262101–262103.
226. B. H. Chu, B. S. Kang, F. Ren, C. Y. Chang, Y. L. Wang, S. J. Pearton, A. V. Glushakov, D. M. Dennis, J. W. Johnson, P. Rajagopal, J. C. Roberts, E. L. Piner, and K. J. Linthicum, *Appl. Phys. Lett.*, 93 (2008), pp. 042114–042116.
227. S. C. Hung, Y. L. Wang, B. Hicks, S. J. Pearton, F. Ren, J. W. Johnson, P. Rajagopal, J. C. Roberts, E. Piner, K. Linthicum, and G. C. Chi, *Electrochem. Solid. State Lett.*, 11 (2008), pp. H241–H243.
228. C. Y. Chang, B. S. Kang, H. T. Wang, F. Ren, Y. L. Wang, S. J. Pearton, D. M. Dennis, J. W. Johnson, P. Rajagopal, J. C. Roberts, E. L. Piner, and K. J. Linthicum, *Appl. Phys. Lett.*, 92 (2008), pp. 232102–232104.
229. S. Okie, *N. Engl. J. Med.*, 352 (2005), pp. 2043–2047.
230. D. Warden, Defense and Veterans Brain Injury center—Blast Injury. http://www.dvbic.org/blastinjury.htm (Accessed April 7, 2007).
231. J. A. Langlois, D. A. Rutland-Brown, K. E. Thomas, (2004). Traumatic brain injury in the United States: Emergency department visits, hospitalizations, and deaths. Atlanta (GA): Centers for Disease Control and Prevention, National Center for Injury Prevention and Control.
232. C. Porte, G. Janer, L. C. Lorusso, M. Ortiz-Zarragoitia, M. P. Cajaraville, M. C. Fossi, and L. Canesi, *Comp. Biochem. Physiol.*, Part C, 143, 303 (2006).
233. G. Mosconi, O. Carnevali, M. F. Franzoni, E. Cottone, I. Lutz, W. Kloas, K. Yamamoto, S. Kikuyama, and A. M. Polzonetti-Magni, *Gen. Comp. Endocrinol.* 126, 125 (2002).
234. J. P. Sumpter and S. Jobling, *Environ. Health Perspect.*, Vol. 103, Supplement 7: Estrogens in the Environment (Oct. 1995), pp. 173–178.
235. V. Matozzo, F. Gagné, M. Gabriella Marin, F. Ricciardi, and C. Blaise, *Environ. Int.*, 34, 531 (2008).
236. C. S. Watson, N. N. Bulayeva, A. L. Wozniak, and R. A. Alyea, *Steroids,* 72, 124 (2007).
237. N. Garcia-Reyero, D. S. Barber, T. S. Gross, K. G. Johnson, M. S. Sep´ulveda, N. J. Szabo, and N. D. Denslow, *Aquatic Toxicol.*, 78, 358 (2006).
238. J. E. Hinck, V. S. Blazer, N. D. Denslow, K. R. Echols, R. W. Gale, C. Wieser, T. W. May, M. Ellersieck, J. J. Coyle, and D. E. Tillitt, *Sci. Total Environ.*, 390, 538 (2008).
239. H.A. Bern, *J. Clean Tech. Environ. Toxicol. Occup. Med.*, 7, 25 (1998).
240. J. G. Liehr, *Hum. Reprod.* Update 7, 273 (2001).
241. J. D. Yager and J. G. Liehr, *Annu. Rev. Pharmacol. Toxicol.*, 36, 203 (1996).
242. D. L. Davis, H. L. Bradlow, M. Wolff, T. Woodruff, D. G. Hoel, and H. Anton-Culver, *Environ. Health Perspect.*, 101, 372 (1993).
243. S. A. Heppell, N. D. Denslow, L. C. Folmar, and C. V. Sullivan, *Environ. Health Perspect.*, 103, 9 (1995).
244. K. S. Carmon, R. E. Baltus, and L. A. Luck, *Anal. Biochem.*, 345, 277 (2005).
245. T. Hahn, K. Tag, Klaus Riedel, S. Uhlig, K. Baronian, G. Gellissen, and G. Kunze, *Biosens. Bioelectron.*, 21, 2078 (2006).

246. J. K. Eidem, H. Kleivdal, K. Kroll, N. Denslow, R. van Aerle, C. Tyler, G. Panter, T. Hutchinson, and A. Goksøyr, *Aquatic Toxicol.*, 78, 202 (2006).
247. T. Sabo-Attwood, J. L. Blum, K. J. Kroll, V. Patel, D. Birkholz, N. J. Szabo, S. Z. Fisher, R. McKenna, M. Campbell-Thompson, and N. D. Denslow, *J. Mol. Endo.*, 39, 22 (2007).
248. N. D. Denslow, *Ecotoxicology*, 8, 385 (1999).
249. N. Garcia-Reyero, K. J. Kroll, L. Liu, E. F. Orlando, K. H. Watanabe, M. S. Sepúlveda, D. L. Villeneuve, E. J. Perkins, G. T. Ankley, and N. D. Denslow, *BMC Genomics*, 10, 308 (2009).
250. G. M. Chertow, E. M. Levy, K. E. Hammermeister, F. Grover, and J. Daley, *Amer. J. Med.*, 104, 343 (1998).
251. M. Curreli, R. Zhang, F. N. Ishikawa, H. K. Chang, R. J. Cote, C. Zhou, and M. E. Thompson, *IEEE Trans. Nanotechnol.*, 7, 651 (2008).

3 Advances in Hydrogen Gas Sensor Technology and Implementation in Wireless Sensor Networks

Travis J. Anderson
Naval Research Laboratory, Washington DC 20375

Byung Hwan Chu
Department of Chemical Engineering, University
of Florida, Gainesville, FL 32611

Yu-Lin Wang
National Tsing Hua University, Hsinchu, Taiwan 30013

Stephen J. Pearton
Department of Electrical and Computer Engineering,
University of Florida, Gainesville, FL 32611

Jenshan Lin
Department of Materials Science Engineering,
University of Florida, Gainesville, FL 32611

Fan Ren
Department of Chemical Engineering, University
of Florida, Gainesville, FL 32611

CONTENTS

3.1   INTRODUCTION

Hydrogen gas (H_2) is a colorless, odorless, combustible gas. It has many industrial applications, mostly in the processing of fossil fuels, metal refining, and the production of hydrochloric acid, methanol, and ammonia. It has been used as the combustion material on the space shuttle solid rocket booster, and was used as a lifting gas until the now-infamous Hindenburg disaster occurred. Hydrogen poses several safety concerns. The upper and lower flammability limits of hydrogen in air are 4% and 75% by volume. Furthermore, pure hydrogen flame emits ultraviolet light; thus, it is nearly invisible to the naked eye. Finally, when stored in liquid form, it is a cryogen. Therefore, in an industrial process where hydrogen is used, a robust sensor is necessary.

An emerging application for hydrogen is in fuel cell electric vehicles. Such vehicles operate on the principle of electricity generation from the reaction of hydrogen and oxygen to form H_2O. This electricity is used to directly drive a motor on each wheel. The only emission is water vapor. This emerging technology has been proposed for next-generation zero carbon emission vehicle technology. A concern with this technology, though, is the wide-scale use of hydrogen gas. All fueling stations will have to store and dispense gas and, in the event of a crash, vehicles would leak a hydrogen plume if the tank ruptures. Therefore, a sensor will have to be very robust, detect low hydrogen concentrations, have a large detection range, and have the ability to be deployed in city-scale networks.

Hydrogen sensor technology has been traditionally based upon cracking of the molecule in the presence of a catalytic metal such as Pt or Pd. By integrating these metals on the gate electrode of the high electron mobility transistors (HEMTs), the change of the sensing material's conductivity can be amplified through the Schottky diode or field-effect transistor (FET) operation. It is generally accepted that H_2 is dissociated when adsorbed on Pt and Pd at room temperature. The reaction is as follows:

$$H_{2(ads)} \rightarrow 2\,H^+ + e^-$$

Dissociated hydrogen causes a change in the channel and conductance change, creating a measurable signal. This makes the integrated semiconductor device-based sensors extremely sensitive to a broad dynamic range of hydrogen concentrations.

Gallium nitride (GaN)- and silicon carbide (SiC)-based wide-bandgap semiconductor sensors can be operated at lower current levels than conventional Si-based devices and offer the capability of detection at elevated temperatures due to their low intrinsic carrier concentrations. The ability of electronic devices fabricated in these materials to function at high temperature, high power, and under high flux/energy radiation conditions enable performance enhancements in a wide variety of spacecraft, satellite, homeland defense, mining, automobile, nuclear power, and radar applications. This chapter will discuss advances in hydrogen sensor technology using wide-bandgap semiconductor devices and the implementation of sensors in wireless networks.

3.2 AlGaN/GaN HEMT SCHOTTKY-DIODE-BASED HYDROGEN SENSOR

GaN materials system is attracting significant interest for commercial and military applications, ranging from radar and unmanned vehicles to cable television amplifiers and wireless base stations. Due to the wide-bandgap nature of the material, it is very thermally stable, and electronic devices can be operated to temperatures up to 500°C. The material is also chemically stable, with the only known wet etchant being molten NaOH or KOH, making it very suitable for operation in chemically harsh environments or in radiation fluxes. Due to the high electron mobility, nitride-based HEMTs can operate from very high frequency (VHF) through X-band frequencies with higher breakdown voltage, better thermal conductivity, and wider transmission bandwidths than Si or GaAs devices. GaN-based HEMTs can also operate at significantly higher power densities and higher impedance than currently used GaAs devices [1–16].

3.2.1 Basic Schottky Diode Hydrogen Sensor

AlGaN/GaN HEMT shows promising performance for use in broadband power amplifiers in base station applications due to the high sheet carrier concentration, electron mobility in the two-dimensional electron gas (2DEG) channel, and high saturation velocity. The high electron sheet carrier concentration of nitride HEMT is induced by piezoelectric polarization of the strained AlGaN layer, and spontaneous polarization is very large in wurtzite III-nitrides. An overlooked potential application of the AlGaN/GaN HEMT structure is in sensors. The high electron sheet carrier concentration of nitride HEMTs is induced by piezoelectric polarization of the strained AlGaN layer, and the spontaneous polarization is very large in wurtzite III-nitrides. This provides an increased sensitivity relative to simple Schottky diodes fabricated on GaN layers [17–35]. The gate region can be functionalized so that current changes can be detected for a variety of gases, liquids, and biomolecules. Hydrogen sensors are particularly interesting for the emerging fuel cell vehicle market. There are also applications for the detection of combustion gases for fuel leak detection in spacecraft, automobiles, aircraft, fire detectors, exhaust diagnosis, and emissions from industrial processes [36–45]. A variety of gas, chemical, and health-related sensors based on HEMT technology have been demonstrated with proper surface functionalization on the gate area of the HEMTs, including the detection of hydrogen, mercury ion, prostate-specific antigen (PSA), DNA, and glucose [46–49].

In GaN HEMT, the sensing mechanism is ascribed to the dissociation of the molecular hydrogen on a catalytic metal gate contact, followed by diffusion of the atomic hydrogen, thus changing the effective barrier height on Schottky diode structures. This effect has been used in Si-, SiC-, ZnO-, and GaN-based Schottky diode combustion gas sensors [50–59]. Furthermore, under forward bias conditions, the positively charged atomic hydrogen will be screened by the applied bias, whereas under reverse bias conditions the positive charge will be attracted to the surface. Therefore, reverse bias conditions are expected to yield improved sensitivity.

3.2.1.1 Device Structure and Fabrication

Simple two-terminal Schottky diodes are effective hydrogen sensors. The use of an AlGaN/GaN HEMT substrate improves the sensitivity, as the 2DEG will serve to amplify the effect of hydrogen absorption, discussed in the following text. The HEMT structure is grown via metal organic chemical vapor deposition (MOCVD) or molecular beam epitaxy (MBE) and consists of an AlN buffer layer on a sapphire (Al_2O_3) substrate, followed by a GaN buffer layer (0.5–2 um), followed by an $Al_{0.3}$GaN barrier layer (25–35 nm). The basic device fabrication begins with a mesa etch for isolation in a Cl$_2$/Ar inductively coupled plasma (ICP) etch system. The next step is the deposition of ohmic metal via e-beam evaporation and liftoff. The metal scheme is typically Ti/Al/Pt/Au, although Ti/Al/TiB2/Ti/Au has been shown to improve reliability. The metal stack is annealed in a rapid thermal anneal (RTA) system at 850°C for 30 s in flowing N$_2$ to form the contact. The next step is the deposition of the sensing metal, which is also the Schottky contact. In an FET configuration, this would be the gate metal step, and would consist of a thin Ni or Pt layer to form the contact, followed by a thick Au layer for stability and probing. In this configuration, however, a thin Ni or Pt layer is desirable, as it will serve as the catalytic surface. Finally, an overlay metal is deposited, typically Ti/Au by e-beam evaporation and liftoff, to enable probing. This is deposited as a ring around the catalytic metal with a large metal trace for wire bonding and probing. A device cross-section schematic and optical image of a packaged device are shown in Figure 3.1.

3.2.1.2 Testing Procedure

The system built for gas sensor testing consists of an environmental chamber with an electrical feedthrough to a semiconductor parameter analyzer for monitoring the device current–voltage (I–V) characteristics. The equipment schematic is illustrated in Figure 3.2. Mass flow controllers are used

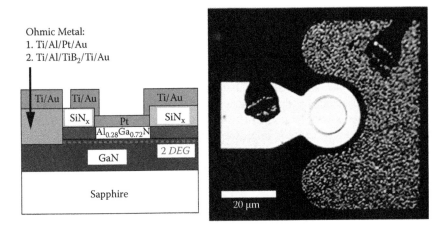

FIGURE 3.1 (See color insert). Cross-sectional schematic of completed Schottky diode on AlGaN/GaN HEMT layer structure (left) and plan-view photograph.

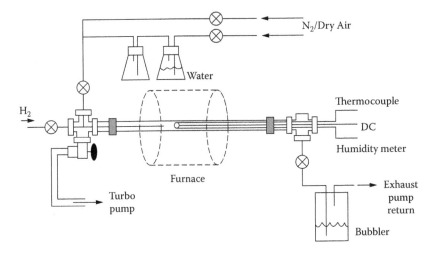

FIGURE 3.2 Schematic of the hydrogen-sensing system.

to introduce test gases and nitrogen to the chamber and vary the concentration, and the chamber passes through a furnace for testing at elevated temperature. The hydrogen concentration can be controlled down to 1 ppm, or up to percent levels, and the temperature range is from 25°C to 500°C.

3.2.1.3   Experimental Results

A basic demonstration of the hydrogen sensitivity of an AlGaN/GaN Schottky diode is shown in the following text. Figure 3.3 shows the linear (top) and log scale (bottom) forward I–V characteristics at 25°C of the HEMT diode, both in air and in a 1% H_2 in air atmosphere. For these diodes, there is a clear increase in current upon introduction of the H_2, as a result of a lowering of the effective barrier height through the mechanism previously discussed. The H_2 catalytically decomposes on the Pt metallization and diffuses rapidly to the interface where it forms a dipole layer. A more detailed calculation of the barrier height reduction is discussed in the following text. The differential change in forward current upon introduction of the hydrogen into the ambient is ~1 mA.

To test the time response of the sensors, a 10% H_2/90% N_2 ambient was switched into the chamber through a mass flow controller for periods of 10, 20, or 30 s and then switched back to pure N_2. Figure 3.6 shows the time dependence of forward current at a fixed bias of 2 V under these conditions. The response of the sensor is rapid (<1 s), with saturation taking close to 30 s. Upon switching off the hydrogen-containing ambient, the forward current decays exponentially back to its initial value. This time constant is determined by the transport properties of the test chamber and is not limited by the response of the diode itself.

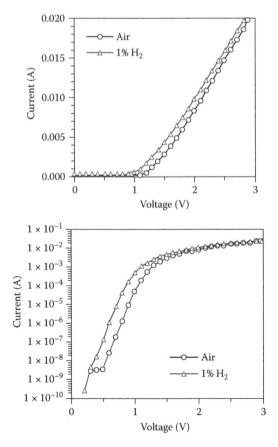

FIGURE 3.3   I–V characteristics in linear (top) or log (bottom) form of Pt-gated diode measured in air or 1% hydrogen ambient at 25°C.

For further study, devices were tested under both forward and reverse bias conditions at room temperature (25°C) in a nitrogen atmosphere at hydrogen concentrations ranging from 0 to 500 ppm, controlled by diluting the gas mix with nitrogen using mass flow controllers. There was again an increase in current under both forward and reverse bias conditions upon exposure to hydrogen, as shown in Figure 3.5. This result is consistent with previously discussed mechanisms [60,61]. The hydrogen atoms form a dipole layer, lowering the Schottky barrier height, and increasing net positive charges on the AlGaN surface as well as negative charges in the 2DEG channel. The calculated barrier height decrease for 500 ppm and 100 ppm hydrogen is 5 meV and 1 meV, respectively. The ideality factors were calculated to be 1.25 and 1.23 in 500 and 100 ppm hydrogen, respectively, compared to 1.26 in 100% nitrogen.

However, a plot of hydrogen sensitivity (defined as the drain current change over the initial drain current) versus bias voltage shows different characteristics for forward and reverse bias polarity conditions at 500 ppm of H_2, as shown in Figure 3.6. For the forward bias condition, there is a maximum sensitivity obtained around 1 V, and further increase of bias voltage reduces the sensitivity.

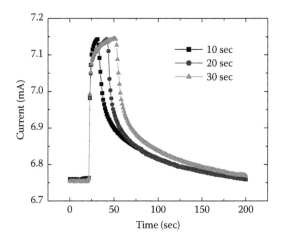

FIGURE 3.4   Time response at 25°C of MOS-HEMT-based diode forward current at a fixed bias of 2 V when switching the ambient from N_2 to 10% H_2/90% N_2 for periods of 10, 20, or 30 s and then back to pure N_2.

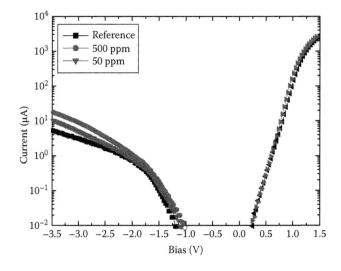

FIGURE 3.5   Forward and reverse bias plot of diode current in varying atmospheres.

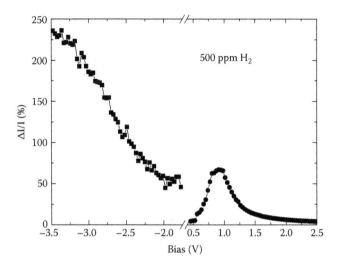

FIGURE 3.6 Percentage change in current as a function of bias at 500 ppm H_2.

The sensitivity for the reverse bias condition is quite different and increases proportionally to the bias voltage. We have proposed the following mechanism for the change in sensitivity under forward and reverse bias conditions: (1) The initial increase in the sensitivity is due to the Schottky barrier height reduction. (2) Further increase in forward bias allows electrons to flow across the Schottky barrier. These excess electrons bind with H^+ to form atomic hydrogen, and gradually destroy the dipole layer at the interface, thereby losing the hydrogen detection sensitivity. (3) For the reverse bias condition, electrons given away by the hydrogen atom may be swept across the depletion region. At higher reverse bias voltage, a higher driving force is applied to the electrons to move across the depletion region. Thus, the dipole layer is amplified at the Pt/AlGaN interface for higher reverse bias voltage. Due to this dipole layer amplification, the detection sensitivity is enhanced at higher reverse bias voltage.

The detection sensitivity as a function of hydrogen concentration is quantified in Figure 3.7. It is clear that the diodes are much more sensitive under reverse bias conditions, as predicted from the proposed mechanism. A detection limit of 100 ppm is achieved under forward bias, but the reverse bias detection limit is an order of magnitude lower, 10 ppm. The change in current at 10 ppm is 14% and over 200% at 500 ppm under reverse bias conditions, where forward bias operation results in changes of 25%–75% over the 100–500 ppm range. This is consistent with published reports indicating improved sensitivity under reverse bias [62]. The reliability of the hydrogen sensor may be quite different under the two bias voltage polarities, since different degradation mechanisms in GaN devices are accelerated by either the presence of high-voltage depletion regions (reverse bias) or current injection (forward bias) [63].

3.2.2 TiB$_2$ Ohmic Contacts

Boride-based ohmic contacts on HEMTs show lower contact resistance than Ti/Al/Pt/Au after extended aging at 350°C [64]. TiB$_2$-based ohmic contacts show improved stability for long-term operation [65]. The structure of the ohmic contacts is Ti(200 Å)/Al(1000 Å)/TiB$_2$(400 Å)/Ti(200 Å)/(Au(800 Å). All of the metals were deposited by Ar plasma-assisted rf sputtering at pressures of 15–40 mTorr and rf (13.56 MHz) powers of 200–250 W. The contacts were annealed at 850°C for 45 s under a flowing N_2 ambient in a Heatpulse 610T system. Figure 3.8 shows the time dependence of forward current at 1.5 V gate bias for devices with both types of ohmic contacts. These tests were carried out in the field, where temperature and humidity were not controlled. There are several features of note. First, the current is much higher in the diodes with TiB$_2$-based contacts because of

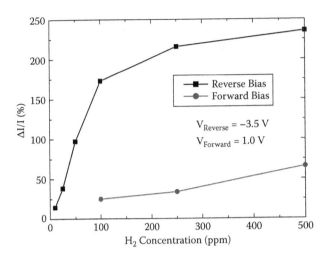

FIGURE 3.7   Percentage change in current as a function of hydrogen concentration under both forward and reverse bias.

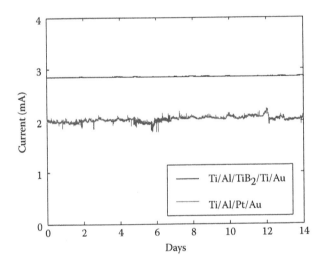

FIGURE 3.8   Variation in forward current at fixed bias for diodes with boride-based ohmic contacts (top) or conventional ohmic contacts (bottom) as a function of time under field conditions where the temperature increases during the day and decreases at night.

their lower contact resistance (1.6×10^{-6} Ω.cm^2 versus 7.5×10^{-6} Ω.cm^2 for the conventional Ti/Al/ Pt/Au). Second, there is much better stability of the devices with TiB$_2$-based contacts. There is much less temperature dependence to the contact resistance of the boride contacts, and this translates to less variation in gate current as the temperature cycles from day to night.

3.2.3   EFFECT OF HUMIDITY ON HYDROGEN SENSORS

Most of the hydrogen-sensing studies to date have been conducted with hydrogen balanced with dry nitrogen, and a few experiments were performed in dry air conditions. However, in the real applications for detecting hydrogen leaks, humidity may play a significant role on the hydrogen sensing, and the humidity in the major cities in the United States are often quite high, >50% [66]. Therefore, it is important to study the effect of humidity on the AlGaN/GaN HEMT-based hydrogen sensors.

Figure 3.9 shows the time dependence of the diode current for a HEMT sensor biased at a −1.5 V and exposed 1% H_2 balanced with air with different relative humidity. The 1% H_2 mixture was switched into the chamber through a mass flow controller for 180 s and then switched back to the humid air. Both sensing response and recovery times of the sensor were less than 1 min, and the sensor also demonstrated good recyclability.

However, the sensitivity for 1% H_2 decreases linearly as the relative humidity increased, as illustrated in Figure 3.10. Both relative humidity and oxygen partial pressure have been reported to degrade the electrocatalytic activity of platinum [67–71]. The absorbed water vapor and oxygen blocked available surface adsorption sites of platinum for H_2, and lowered the concentration of hydrogen at the metal–semiconductor interface.

The sensor exhibited a fast response for the recovery time, when the gas ambient switched from hydrogen-containing humid air to non-hydrogen-containing humid air. However, a slow recovery time was observed, when the gas ambient switched from hydrogen-containing humid air to non-hydrogen-containing dry air, as illustrated in Figure 3.11. This slow recovery behavior was similar to the results obtained in our previous study conducted using dry nitrogen as the background ambient, which showed a long and slow recovery time [65,72]. The slow recovery time in that work was, we now believe, mistakenly attributed to longer times required to purge the residue hydrogen out of the gas chamber. In this work we note that the presence of water molecules in the ambient shortened

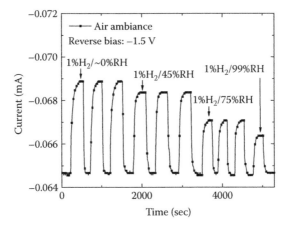

FIGURE 3.9 Time dependence of absolute value of the AlGaN/GaN HEMT diode current biased at −1.5 V as the gas ambient switched back and forth between 1% hydrogen balance and air with different humidity.

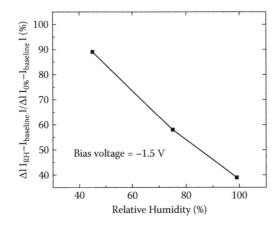

FIGURE 3.10 Hydrogen sensitivity as a function of the humidity.

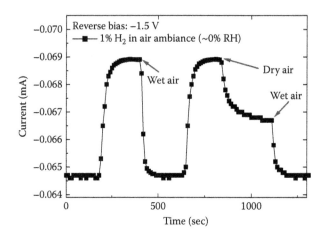

FIGURE 3.11   Time dependence of recovery time characteristics of the AlGaN/GaN HEMT diode current biased at −1.5 V as the gas ambient switched back and forth between 1% hydrogen balance and air with 100% humidity as well as dry air.

the recovery time. It has been reported that Pd or Pt adsorbs H_2O and catalytically dissociates the adsorbed H_2O to form OH molecules with assistance of surface-chemisorbed atomic oxygen [73,74]. The atomic H would readily react with OH to form H_2O, thus the presence of humidity consumed the atomic hydrogen and reduced the sensor recovery time.

3.2.4   DIFFERENTIAL SENSOR

A differential pair of AlGaN/GaN HEMT diodes could be used for hydrogen sensing near room temperature. This configuration provides a built-in control diode to reduce false alarms due to temperature swings, humidity changes, or voltage transients. Figure 3.12 shows an optical microscope image of the completed devices. The active device has 10 nm Pt exposed to the atmosphere, while the reference diode has Ti/Au covering the Pt layer.

Figure 3.13 shows the absolute and differential forward current–voltage (I–V) characteristics at 25°C of the HEMT active (top) and reference (bottom) diodes, both in air and in a 1% H_2 in air atmosphere. For the active diode, the current increases upon introduction of the H_2, through a lowering of the effective barrier height. The H_2 catalytically decomposes on the Pt metallization and diffuses rapidly to the interface where it forms a dipole layer. The differential change in forward current upon introduction of the hydrogen into the ambient is ~1–4 mA over the voltage range examined and peaks at low bias. This is roughly double the detection sensitivity of comparable GaN Schottky gas sensors tested under the same conditions, confirming that the HEMT-based diode has advantages for applications requiring the ability to detect hydrogen even at room temperature.

As the detection temperature is increased to 50°C, the differential current response of the HEMT diode pair was almost constant over a wide range of voltages due to more efficient cracking of the hydrogen on the metal contact, as shown in Figure 3.14. The maximum differential current is similar to that at 25°C, but the voltage control to achieve maximum detection sensitivity for hydrogen is not as important at 50°C.

To test the time response of the HEMT diode sensors, the 1% H_2 ambient was switched into the chamber through a mass flow controller for 300 s and then switched back to air. Figure 3.15 shows the time dependence of forward current for the active and reference diodes at a fixed bias of 2.5 V under these conditions. The response of the sensor is rapid (<1 s based on a series of switching tests). Upon switching out of the hydrogen-containing ambient, the forward current decays exponentially

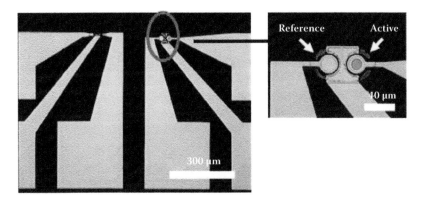

FIGURE 3.12 Microscopic images of differential sensing diodes. The opening of the active diode was deposited with 10 nm Pt, and the reference diode was deposited with Ti/Au.

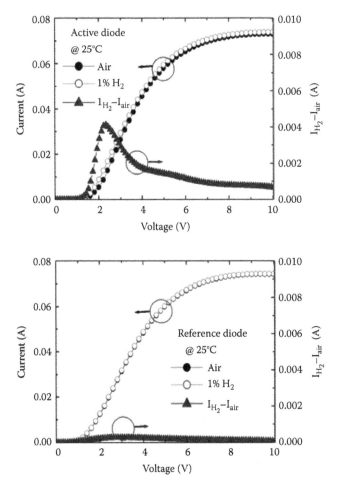

FIGURE 3.13 Absolute and differential current of HEMT diodes measured at 25°C.

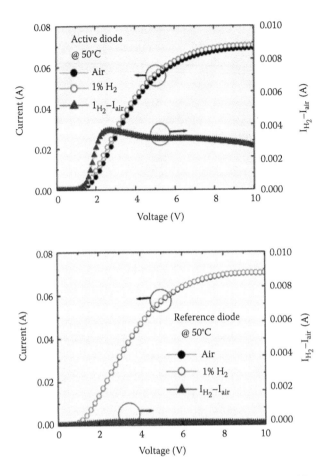

FIGURE 3.14   Absolute and differential current of HEMT diode measured at 50°C.

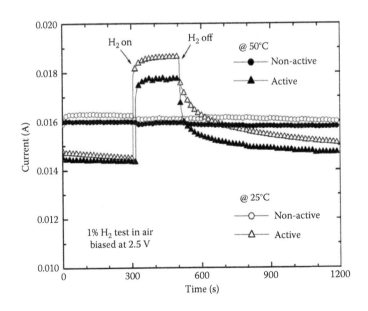

FIGURE 3.15   Differential diode test results using 1% H_2 at room temperature and elevated temperature.

back to its initial value. This time constant is determined by the volume of the test chamber and the flow rate of the input gases, and is not limited by the response of the HEMT diode itself. Note that the use of the differential pair geometry removes false alarms due to changes in ambient temperature or voltage drifts.

3.3   GaN SCHOTTKY DIODE SENSOR

3.3.1   N-FACE AND GA-FACE COMPARISON

Wurtzite GaN is a polar material. Therefore, along the c-axis, there is N-face (N-polar) or Ga-face (Ga-polar) orientations on the GaN surface. Since there are different reactivities for the N-face and Ga-faces [75–77], it is interesting how these different surfaces react with different chemicals. The polarity determined by the GaN surface has been used to improve performance of AlGaN/GaN HEMTs [78]. The Ga-polar surface is intentionally grown to make the spontaneous polarization compatible with the piezoelectric polarization to enhance the 2DEG in AlGaN/GaN HEMTs [78]. However, the density functional theory suggests that hydrogen has a much higher affinity for the N-face surface of GaN than the Ga-face [76]. High-resolution electron energy loss spectroscopy (HREELS) showed a strong preference of N sites for the adsorption of hydrogen gas or atomic H [75]. It was also shown that below 820°C, N-polar GaN has much faster reaction rate than that of Ga-polar GaN surface [77]. Thus, one might expect the polarity of the surface to play an important role in hydrogen-sensing characteristics. Recent results showed that Schottky diodes fabricated on N-face GaN provided much higher sensitivity than Ga-face GaN and Ga-face AlGaN/GaN HEMTs [79,80].

In N-polar GaN Schottky diodes, because of the higher surface reactivity with hydrogen-inducing higher polarity, the Schottky barrier heights are reduced much more than those of Ga-polar Schottky diodes. Therefore, the N-polar GaN Schottky diodes have much higher sensitivity than Ga-polar GaN.

3.3.1.1   Device Structure and Fabrication

The GaN layer structures were grown on C-plane Al_2O_3 substrates with a low-temperature GaN buffer by metal-organic chemical vapor deposition (MOCVD) as described previously [81,82]. A key aspect for the N-polar growth is the control of heteroepitaxial evolution with the use of two dissimilar conditions, one to enhance the vertical, island formation and the other to promote the lateral coalescence process. The degree of surface roughening through island nucleation can be controlled as seen by the low reflectance region, followed by designed rapid lateral growth (coalescence) to smooth out the surface. All epilayers at the end of coalescence are optically smooth and free of pits [81,82]. Both Ga- and N-polar layers were grown to a thickness of ~1.3 μm on sapphire substrates. The layers exhibited n-type carrier densities of 1.5×10^{18} cm^{-3} (mobility of 245 cm²/V.s) for the undoped N-face samples and 1×10^{18} cm^{-3} (mobility of 410 cm²/V·s) for the Si-doped Ga-face samples. For comparison, we also used standard Ga-polar AlGaN/GaN heterostructure samples with sheet carrier density of 9×10^{12} cm^{-2} for comparison as sensors, since we have found these to exhibit higher hydrogen sensitivities than GaN diodes [25,44].

For all samples, ohmic contacts were formed by lift-off of e-beam deposited Ti (200 Å)/Al (1000 Å)/Ni (400 Å)/Au (1200 Å) subsequently annealed at 850°C for 45 s under a flowing N_2 ambient. The surface was encapsulated with 2000 Å of plasma-enhanced chemical vapor deposited SiN$_x$ at 300°C. Windows in the SiN$_x$ were opened by dry etching and 100 Å of Pt deposited by e-beam evaporation for Schottky contacts. The final metal was e-beam-deposited Ti/Au (200 Å/1200 Å) interconnection contacts. The schematics of device structure and top view are depicted in Figure 3.16.

3.3.1.2   Experimental Results

Schottky diode I–V characteristics made from N-polar GaN, Ga-polar GaN, and Ga-polar HEMT before and after exposure to 4% H_2 in N_2 are illustrated in Figure 3.17 (top). The absolute and

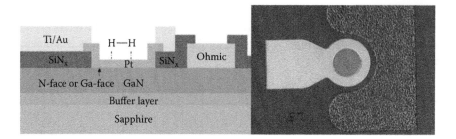

FIGURE 3.16 Cross-sectional schematic of Schottky diode made of Ga-polar or N-polar (left) and plan-view photograph of device.

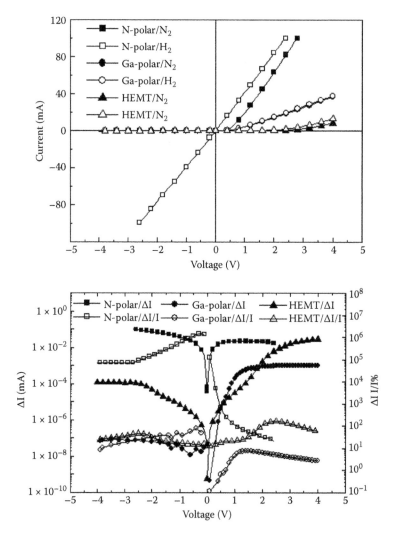

FIGURE 3.17 (Top) I–V characteristics from the three types of diodes (Schottky diodes made of N-polar GaN, Ga-polar GaN, and Ga-polar HEMT) before and after exposure to 4% H_2 in N_2 (bottom). The absolute and percentage change in current as a function of voltage as a result of hydrogen exposure.

percentage change in current as a function of voltage as a result of hydrogen exposure are shown in Figure 3.17 (bottom). Note the remarkably larger effect of hydrogen exposure on the N-face diodes. This diode actually reverts from rectifying to near-ohmic behavior after exposure to the 4% H_2 in N_2. The maximum percentage current changes were correspondingly much larger than for the Ga face GaN or HEMT diodes, namely, 10^6 compared to 10 for the Ga-polar and 170 for the HEMT diode.

Figure 3.18 shows the time dependence of current change in the three types of sensors as a function of cycling the ambient from N_2 to 4% H_2 in N_2 and then back to N_2. The devices were operated at slightly different biases that were found to maximize their response. The N-polar diodes show much larger relative responses, but do not recover their original current within the time frame of the measurement, in contrast to the Ga face and HEMT diodes. It is possible that some hydrogen was strongly bonded with nitrogen, with the thermal energy at room temperature not being high enough to break the bonding.

As illustrated in Figure 3.19 (top), a shift from rectifying to near-ohmic behavior in the top frame upon initial exposure to the hydrogen-containing ambient was observed. This sudden current change in the reversed bias region demonstrated the effectiveness of using N-polar GaN Schottky diode as the hydrogen sensor. However, the recovery time of the N-polar GaN Schottky diode was significantly longer as compared to that of the GaN-polar Schottky diode as previously discussed. Figure 3.19 (bottom) shows the recovery of the I–V characteristics in N-polar Schottky diode in 5 min intervals after switching back to an N_2 ambient at room temperature. Even cycling the diodes to 150°C in N_2 after hydrogen sensing was not sufficient to restore the initial current, in sharp contrast to the other types of diodes where full recovery of the current was achieved at room temperature after approximately 15 min. This is consistent with the predicted stability of the H-covered N-polar GaN surface [76,77].

3.3.2 W/Pt-Contacted GaN Schottky Diodes

W/Pt-contacted GaN Schottky diodes also show forward current changes of >1 mA at low bias (3 V) in the temperature range 350°C–600°C when the measurement ambient is changed from pure N_2 to 10% H_2/90% N_2. We have found that the use of a metal-oxide-semiconductor (MOS) diode structure with Sc_2O_3 gate dielectric and the same W/Pt metallization shows these same reversible changes in forward current upon exposure to H_2-containing ambient over a much broader temperature range (90°C to >625°C). The increase in current in both cases is the result of a decrease in effective barrier

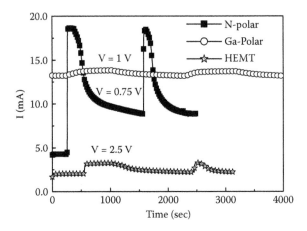

FIGURE 3.18 Time dependence of current change in the three types of sensors as a function of cycling the ambient from N_2 to 4% H_2 in N_2 at room temperature.

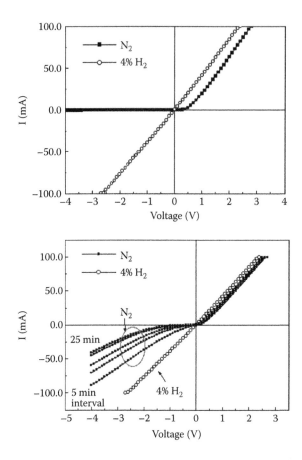

FIGURE 3.19   (Top) I–V characteristics of the N-polar diodes before and after exposure to the H_2-containing ambient. (Bottom) Recovery of I–V characteristics in N-face diode as a function of time in N_2 ambient after initial sensing of hydrogen.

height of the MOS and Schottky gates of 30–50 mV 10% H_2/90% N_2 ambient relative to pure N_2 and is due to catalytic dissociation of the H_2 on the Pt contact, followed by diffusion to the W–GaN or Sc_2O_3–GaN interface. The presence of the oxide lowers the temperature at which the hydrogen can be detected and, in conjunction with the use of the high temperature, stable W metallization enhances the potential applications of these wide-bandgap sensors. Figure 3.20 shows that the relative change in current is larger with the MOS structure.

3.4   NANOSTRUCTURED WIDE-BANDGAP MATERIALS

Nanostructured wide-bandgap materials functionalized with Pd or Pt are even sensitive than their thin-film counterparts because of the large surface-to-volume ratio [83,84]. 1-D semiconductor nanomaterials, such as carbon nanotubes (CNTs), Si nanowires, GaN nanowires, and ZnO nanowires, are good candidates to replace 2-D semiconductors due to several advantages. First, 1-D structure has large surface-to-volume ratio, which means that a significant fraction of the atoms can participate in surface reactions. Second, the Debye length (λ_D) for 1-D nanomaterial is comparable to their radius over a wide temperature and doping range, which causes them more sensitive than 2-D thin film. Third, 1-D nanostructure is usually stoichiometrically controlled better than 2-D thin film, and has a greater level of crystallinity than the 2-D thin film. With 1-D structures, common defect problems in 2-D semiconductors could be easily solved. Fourth,

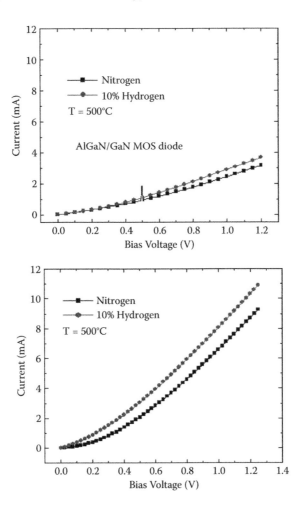

FIGURE 3.20 Change in current in W/Pt GaN and AlGaN/GaN diodes at 500°C when the ambient is switched from N_2 to 10% H_2 in N_2.

further decreasing the diameter, onset of quantum effects is expected to be appeared. Finally, low cost, low power consumption, and high compatibility with microelectronic processing make 1-D nanostructures practical materials for sensors. Impressive results have been demonstrated with GaN, InN, and ZnO nanowires or nanobelts that are sensitive to hydrogen down to approximately 20 ppm at room temperature.

3.4.1 HYDROGEN SENSORS BASED ON ZnO NANORODS

One of the main requirements of such sensors is the ability to selectively detect hydrogen at room temperature in the presence of air. In addition, for most of these applications, the sensors should have very low power requirements and minimal weight. Nanostructures are natural candidates for this type of sensing. One important aspect is to increase their sensitivity for detecting gases such as hydrogen at low concentrations or temperatures, since typically an on-chip heater is used to increase the dissociation efficiency of molecular hydrogen to the atomic form, and this adds complexity and power requirements. Previous work has shown that Pd-coated or doped carbon nanotubes (CNTs) become more sensitive to the detection of hydrogen through catalytic dissociation of H_2 to atomic hydrogen [85–87]. ZnO is also an attractive material for sensing applications, and nanowires and nanorods in this system have been reported for pH, gas, humidity, and chemical sensing [88–91]. These nanostructures could also have novel applications in biomedical science because ZnO is biosafe [92]. ZnO nanorods are relatively straightforward to synthesize by a number of different methods [92–105].

In ZnO devices, gas-sensing mechanisms suggested include the exchange of charges between adsorbed gas species and the ZnO surface leading to changes in depletion depth [106] and changes in surface or grain boundary conduction by gas adsorption or desorption [107,108]. It should be noted that hydrogen introduces a shallow donor state in ZnO, and this change in near-surface conductivity may also play a role. In the presence of a catalytic metal, the resulting charged atomic hydrogen will further change the conductivity, thus improving sensitivity.

3.4.1.1 Device Structure and Fabrication

The first step in this case is the growth of ZnO nanorods. The structures are grown by MBE using Zn metal and O_2 plasma discharge as the source chemicals. To establish nucleation sites for the nanorods, a thin Au layer (2 nm) is deposited and annealed on a sapphire substrate. This process will form islands rather than a continuous thin film. After 2 h MBE growth time at 600°C, single-crystal nanorods with a typical length around 2~10 μm and diameter in the range of 30–150 nm have been grown. Figure 3.21 shows a scanning electron micrograph of the as-grown rods.

The device structure is a simple two-terminal resistor, so Al/Ti/Au electrodes are deposited using e-beam evaporation with a shadow mask with a spacing of ~30 um. The nanowires alone are sensitive to hydrogen, but studies have shown that a blanket deposition of catalytic metal clusters will improve the sensitivity. The metal is deposited using e-beam evaporation, and the film is only 10 nm thick, so that nucleation clusters form but no continuous thin film. Current–voltage measurements confirmed that there was no conduction through the metal clusters, and there was no thin ZnO film formed using the nanorod growth conditions. A cross-section schematic and packaged device optical image are shown in Figure 3.22.

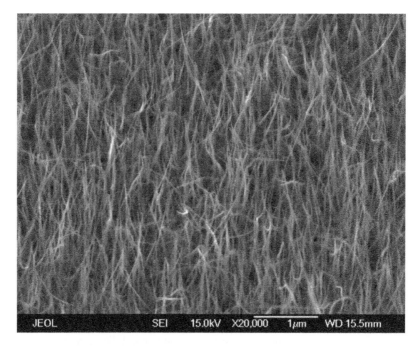

FIGURE 3.21   SEM image of ZnO nanorods.

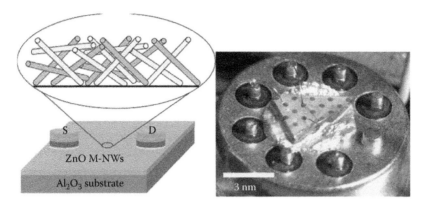

FIGURE 3.22 Schematic of contact geometry for multiple nanorod gas sensor (left) and packaged, wire-bonded device for testing (right).

3.4.1.2 Experimental Results

A preliminary investigation established the effect of catalytic metal coatings on ZnO nanorods. The time-dependent response of uncoated and Pd-coated nanorods to varying hydrogen concentrations from 10 to 500 ppm, with recovery in N_2, is shown in Figure 3.23. There is clearly a strong increase (approximately a factor of 5) in the response of the Pd-coated nanorods to hydrogen relative to the uncoated devices. The addition of the Pd appears to be effective in catalytic dissociation of the H_2 to atomic hydrogen. In addition, there was no response to the presence of O_2 in the ambient at room temperature, and the relative response of Pt-coated nanorods is a function of H_2 concentration in N_2. The Pd-coated CNTs detected hydrogen down to <10 ppm, with relative responses of >2.6% at 10 ppm and >4.2% at 500 ppm H_2 in N_2 after 10 min exposure. By comparison, the uncoated devices showed relative resistance changes of ~0.25% for 500 ppm H_2 in N_2 after 10 min exposure, and the results were not consistent for lower concentrations.

To study the transient sensor response, the resistance change of the Pt-coated multiple ZnO nanorods was monitored as the gas ambient was switched from vacuum to N_2, oxygen, or various concentrations of H_2 in air (10–500 ppm), and then back to air, as shown in Figure 3.24. This data confirms

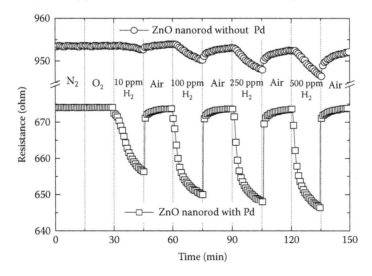

FIGURE 3.23 Time dependence of resistance of either Pd-coated or uncoated multiple ZnO nanorods as the gas ambient is switched from N_2 to various concentrations of H_2 in air (10–500 ppm).

the absence of sensitivity to O_2. The resistance change during the exposure to hydrogen was slower in the beginning, and the rate resistance change reached a maximum at 1.5 min of the exposure time. This could be due to some of the Pd becoming covered with native oxide and then removed by exposure to hydrogen. Since the available surface of Pd for catalytic chemical absorption of hydrogen increased after the removal of oxides, the rate of resistance change increased. However, the Pd surface gradually saturated with the hydrogen and resistance change rate decreased. When the gas ambient switched from hydrogen to air, the oxygen reacted with hydrogen right away and the resistance of the nanorods changed back to the original value instantly.

Figure 3.25 shows the Arrhenius plot of the rate of nanorod resistance change. The rate of resistance change for the nanorods exposed to 500 ppm H_2 in N_2 was measured at different temperatures. An activation energy of 12 kJ/mol was extracted from the slope of the Arrhenius plot. This value is larger than that of the typical diffusion process. Therefore, the dominant mechanism for this sensing process should be the chemisorption of hydrogen to the Pd surface.

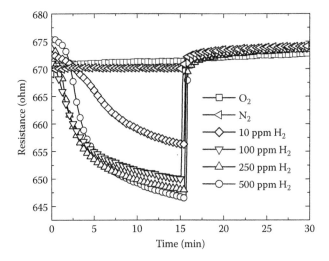

FIGURE 3.24 Relative response of Pd-coated nanorods as a function of H_2 concentration in N_2.

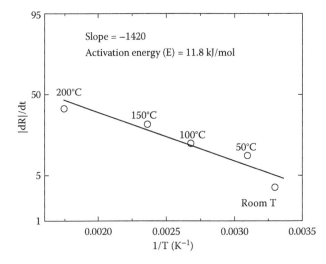

FIGURE 3.25 Time dependence of resistance change of Pd-coated multiple ZnO nanorods as the gas ambient is switched from N_2 to oxygen or various concentrations of H_2 in air (10–500 ppm) and then back to N_2.

Having established improved sensitivity by coating the nanorods with catalytic metal, we then investigated the effects of different metals to further improve sensitivity and response. Using the same coating procedure described earlier, we investigated the effect of metal coatings of Ti, Ni, Ag, Au, Pt, and Pd. Figure 3.26 shows the time dependence of relative resistance change of either metal-coated or uncoated multiple ZnO nanorods as the gas ambient is switched from air to 500 ppm of H_2 in N_2. These were measured at a bias voltage of 0.5 V. There is a strong enhancement in response to Pd, and to a lesser extent Pt coatings, but the other metals produce little or no change. This is consistent with the known catalytic properties of these metals for hydrogen dissociation. Pd has a higher permeability than Pt, but the solubility of H_2 is larger in the former [109]. Moreover, studies of the bonding of H to Ni, Pd, and Pt surfaces have shown that the adsorption energy is lowest on Pt [110].

The power requirements for the sensors were very low, which is a key requirement for a competitive marketable sensor. Figure 3.27 shows the I–V characteristics measured at 25°C in both a pure

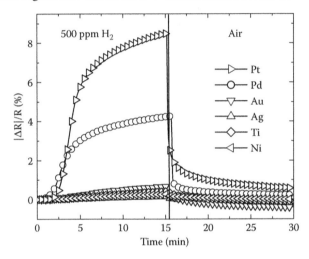

FIGURE 3.26 Time dependence of relative resistance response of metal-coated multiple ZnO nanorods as the gas ambient is switched from N_2 to 500 ppm of H_2 in air as time proceeds. There was no response to O_2.

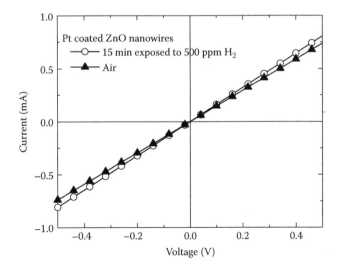

FIGURE 3.27 I–V characteristic of Pt-coated nanowires in air and after 15 min in 500 ppm H_2 in air.

N_2 ambient and after 15 min in a 500 ppm H_2 in N_2 ambient. Under these conditions, the resistance response is 8% and is achieved for a power requirement of only 0.4 mW. This compares well with competing technologies for hydrogen detection such as Pd-loaded carbon nanotubes [85,86].

3.4.2 GaN Nanowires

Figure 3.28 (top) shows scanning electron microscopy (SEM) micrographs of as-grown nanowires. A layer of 10 nm-thick Pd was deposited by sputtering onto the nanowires to verify the effect of catalyst on gas sensitivity. The bottom of Figure 3.28 shows the measured resistance at a bias of 0.5 V as a function of time from Pd-coated and uncoated multiple GaN nanowires exposed to a series of H_2 concentrations (200–1500 ppm) in N_2 for 10 min at room temperature. Pd-coating of the nanowires improved the sensitivity to ppm level H_2 by a factor of up to 11. The addition of Pd appears to be effective in catalytic dissociation of molecular hydrogen. Diffusion of atomic hydrogen to the metal–GaN interface alters the surface depletion of the wires and hence the resistance at fixed bias voltage [111]. The resistance change depended on the gas concentration, but the variations were small at H_2 concentration above 1000 ppm. The resistance after exposing the nanowires to air was restored to approximately 90% of initial level within 2 min [83,84].

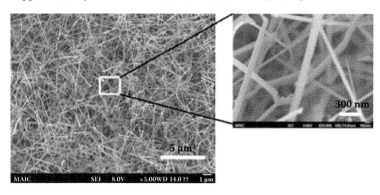

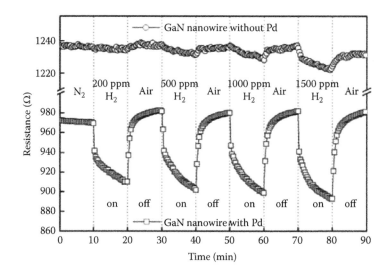

FIGURE 3.28   SEM images of as-grown GaN nanowires (top) and measured resistance at an applied bias of 0.5 V as a function of time from Pd-coated and uncoated multiple GaN nanowires exposed to a series of H_2 concentrations (200–1500 ppm) in N_2 for 10 min at room temperature (bottom).

3.4.3 InN Nanobelts

Similar results can be obtained with InN nanostructures. The hydrogen-sensing characteristics of multiple InN nanobelts grown by metalorganic chemical vapor deposition have been reported previously [112,113]. Pt-coated InN sensors could selectively detect hydrogen at the tens of ppm level at 25°C, while uncoated InN showed no detectable change in current when exposed to hydrogen under the same conditions. Upon exposure to various concentrations of hydrogen (20–300 ppm) in N_2 ambient, the relative resistance change increased from 1.2% at 20 ppm H_2 to 4% at 300 ppm H_2, as shown in Figure 3.29. Approximately 90% of the initial InN resistance was recovered within 2 min by exposing the nanobelts to air. Temperature-dependent measurements showed larger resistance change and faster response at high temperature compared to those at room temperature due to increase in catalytic dissociation rate of H_2 as well as diffusion rate of atomic hydrogen into the Pt–InN interface. The Pt-coated InN nanobelt sensors were operated at low power levels (~0.5 mW).

3.4.4 Single ZnO Nanowire

Figure 3.30 shows a schematic of single ZnO nanowire sensor and an SEM micrograph. The time dependence of resistance of Pt-coated ZnO nanowires as the gas ambient is switched from N_2 to

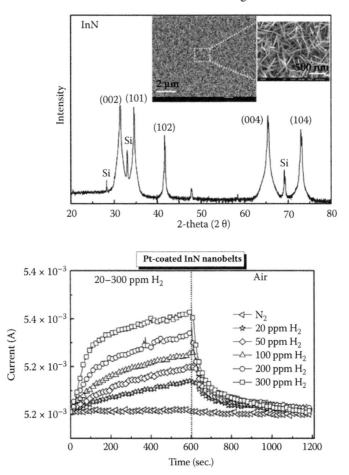

FIGURE 3.29 (Top) X-ray diffraction spectrum of MOCVD-grown InN nanobelts (the inset shows SEM images of the nanobelts) and change in current at fixed bias for switching from 20–300 ppm H_2 in air to pure air (bottom).

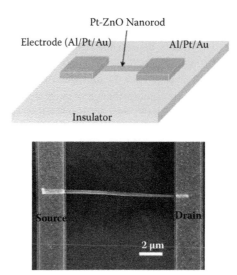

FIGURE 3.30   Schematic of ZnO nanowire sensor (top), SEM of completed device (bottom).

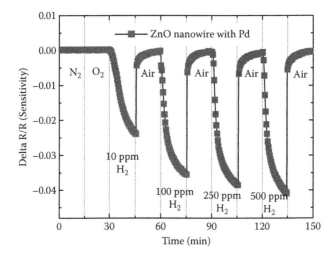

FIGURE 3.31   Change in resistance as a function of time when switching to H_2-containing ambient.

various concentrations of H_2 in N_2 (10–500 ppm) as time proceeds is illustrated in Figure 3.31. There are several aspects that are noteworthy. First, there is a strong increase (approximately a factor of 10) in the response of the Pt-coated nanowires to hydrogen relative to the uncoated devices. The addition of the Pt appears to be effective in catalytic dissociation of the H_2 to atomic hydrogen. Second, there was no response of either type of nanowires to the presence of O_2 in the ambient at room temperature. Third, the effective conductivity of the Pt-coated nanowires is higher due to the presence of the metal. Fourth, the recovery of the initial resistance is rapid (90%, <20 s) upon removal of the hydrogen from the ambient by either O_2 or air, while the nanowire resistance is still changing at least 15 min after the introduction of the hydrogen. The reversible chemisorption of reactive gases at the surface of metal oxides such as ZnO can produce a large and reversible variation in the conductance of the material. The gas-sensing mechanism suggested include the desorption of adsorbed surface hydrogen and grain boundaries in poly-ZnO, exchange of charges between adsorbed gas species and the ZnO surface leading to changes in depletion depth and changes in surface or grain boundary conduction by gas adsorption/desorption [113–117]. The detection mechanism is still not firmly established in these devices and needs further study. It should be remembered that hydrogen introduces a shallow donor state in ZnO, and this change in near-surface conductivity may also play a role.

3.5 SiC SCHOTTKY DIODE HYDROGEN SENSOR

SiC Schottky diodes with Pd or Pt contacts are also sensitive to the presence of hydrogen in the ambient, as shown in Figure 3.32. The advantage of the nitride system relative to SiC is the availability of a heterostructure and the strong piezoelectric and polarization fields present in the nitrides that enhance their capability for chemical sensing.

3.6 WIRELESS SENSOR NETWORK DEVELOPMENT

We have demonstrated a wireless hydrogen-sensing system using commercially available wireless components and GaN Schottky diodes as the sensing devices. The sensors used in the circuit have shown current stability for more than 1 year in an outdoor environment. The advantage of a wireless network sensing system is that it enables monitoring of independent sensor nodes and transmits wireless signals, to the sensor nodes can be placed anywhere within the range of a base station. This is especially useful in manufacturing plants and hydrogen-fueled automobile dealerships, where a number of sensors, possibly each detecting different chemicals, would be required. We have also developed an energy-efficient transmission protocol to reduce the power consumption of the remote

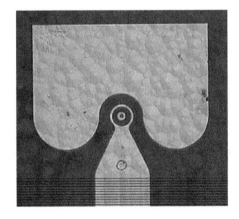

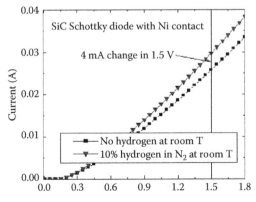

FIGURE 3.32 Pt/SiC Schottky diode (top) and change in current at fixed bias of 1.5 V when the ambient is switched from air to 10% H_2 in air (bottom).

sensor nodes. This enables very long lifetime operation using batteries, thus making the system truly wireless. Experimental results showed that a 150 m transmission distance can be achieved with 10 mW total power consumption. The entire sensor package can be built for less than $50, making it extremely competitive in today's market [62,63,118].

3.6.1 SENSOR MODULE

The sensor module was fully integrated on an FR4 PC board and packaged with a battery as shown in Figure 3.33 (left). The dimension of the sensor module package was $4.5 \times 2.9 \times 2$ in.^{3} The maximum line of sight range between the sensor module and the base station was 150 m. The base station of the wireless sensor network server was also integrated in a single module ($3.0 \times 2.7 \times 1.1$ in.^{3}) and ready to be connected to laptop by a USB cable, as shown in Figure 3.33 (center) and (right). The base station draws its power from the laptop's USB interface and does not require any battery or wall transformer [62,63,118].

3.6.1.1 Description of Wireless Transceiver

The sensor devices are based on the technology described in the previous sections and demonstrate comparable characteristics. An instrumentation amplifier is used for the detection circuit to sense the change of current in the device. The current variation, embodied as a change in the output voltage of the detection circuit, is fed into the microcontroller. The microcontroller calculated the corresponding current change and controlled the ZigBee transceiver to transmit the data to the wireless network server. The block diagram of the sensor module and the wireless network server are shown in Figure 3.34 [62,63,118].

FIGURE 3.33 Photo of sensor system (a) sensor with sensor device; (b) sensor and base station; (c) computer interface with base station.

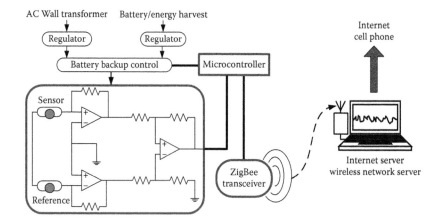

FIGURE 3.34 Block diagram of sensor module and wireless network server.

The ZigBee compliant wireless network supports the unique needs of low-cost, low-power sensor networks, and operates within the ISM 2.4 GHz frequency band. The transceiver module is completely turned down for most of the time, and is turned on to transmit data in extremely short intervals. The timing of the system is shown in Figure 3.35. When the sensor module is turned on, it is programmed to power up for the first 30 s. Following the initialization process, the detection circuit is periodically powered down for 5 s and powered up again for another 1 s, achieving a 16.67% duty cycle. The ZigBee transceiver is enabled for 5.5 ms to transmit the data only at the end of every cycle. This gives a RF duty cycle of only 0.09% [62,63,118].

3.6.1.2 Description of Web Server

A Web server was developed using MATLAB® to share the collected sensor data via the Internet. The interface of the server program, shown in Figure 3.28, illustrates three emulated sensors with different baseline currents. If any of the sensor's current increases to a level that indicates a potential hydrogen leakage, the alarm would be triggered. A client program was also developed to receive the sensor data remotely. As shown in Figure 3.36, the remote client was able to get a real-time log of the system for the past 10 min via the client program. In addition, a full data log obtained via accessing the server by an ftp client as the server program incorporates a full data logging functionality. When an alarm was triggered, the client was able to deactivate the alarm remotely by clicking a button on the interface. The server program for the wireless sensor network could also report a hydrogen leakage emergency through a phone line. When the current of any sensors exceeded a certain level, indicating a potential hydrogen leakage, the server would automatically call the phone-dial program, reporting the emergency to the responsible personnel [62,63,118].

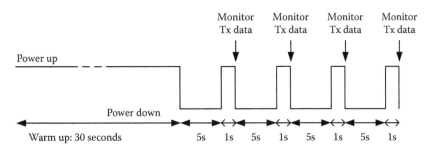

FIGURE 3.35 System timing of a wireless sensor node.

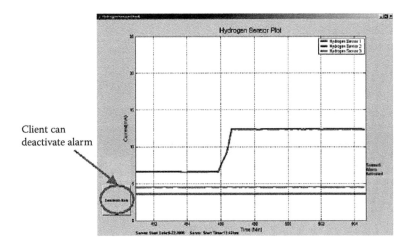

FIGURE 3.36 Interface of online hydrogen level monitoring.

3.6.2 FIELD TEST

Field tests have been conducted both at the University of Florida and at Greenway Ford in Orlando, FL. The setup at Greenway Ford was aimed at testing the stability of the sensor hardware and the server software under actual operational environment. Two sensor nodes were installed and the test was started on the 30th of August 2006. Six sensor modules and the server have been functioning to date [62,63,118].

3.6.2.1 Initial Field Test Results

The outdoor tests at University of Florida have been conducted several times to test the sensor's response to different concentrations of hydrogen at different distances. The tested hydrogen concentrations include 1%, 4%, and 100%, and the distance from the outlet of hydrogen to the sensor ranges from 1 ft to 6 ft. Hydrogen leakage was successfully detected for all of these cases, triggering the program to send an alarm to a cell phone. Figure 3.37 shows the test results with four running sensor modules and 4% hydrogen at 3 ft away from the hydrogen outlet. The sensors were tested sequentially, so the effects of each sensor can be isolated. The test results also displayed the reliability of the wireless network as it was able to collect the data from each individual sensor.

Initial results of field testing indicated that the reliability of the sensors could be a concern, as the single-diode sensors showed a periodic rise and fall in the current level, which can be attributed to a temperature effect. There was also a long-term current degradation, which was attributed to ohmic contact degradation within the device due to the continuous bias.

3.6.2.2 Improved Field Test Results

A couple of problems that arise when considering marketable applications are false alarms and stability. These can be caused by voltage swings in the device or simply by temperature changes altering the current level. A differential pair configuration of AlGaN/GaN HEMT diodes with a built-in control diode has been shown to reduce false alarms [72]. To avoid the thermal effects, the sensing

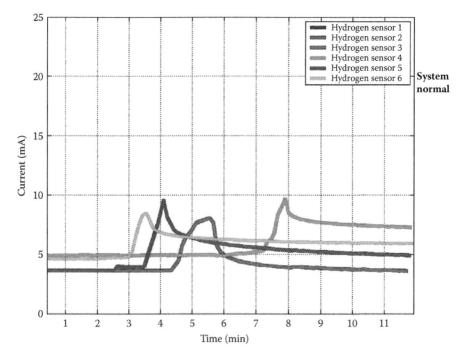

FIGURE 3.37 Field test results of four sensors.

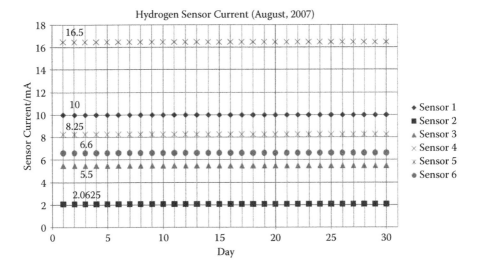

FIGURE 3.38 Field test results over a month period for differential diode sensors with boride-based ohmic contacts.

device was redesigned to employ a differential detection scheme described in previous sections. This involved a reference diode, which was encapsulated, and an active diode, which was open to the ambient. Detection is achieved by monitoring the difference in current between the two devices as opposed to measuring an absolute current level.

An additional key needed for long-term monitoring applications is the availability of stable ohmic contacts. It has been shown that boride-based ohmic contacts on HEMTs have lower contact resistance than Ti/Al/Pt/Au after extended aging at 350°C [64]. Also, TiB_2-based ohmic contacts show improved stability for long-term operation [65]. The combination of differential diode sensors and boride-based ohmic contacts has significantly improved stability and reliability. Recent field data is shown in Figure 3.38, and the data can be monitored in real time at the following Web site: http://ren.che.ufl.edu/app/realtimeSensing.htm [62,63,118].

3.7 SUMMARY

AlGaN/GaN HEMT, GaN diodes, and nanostructured wide-bandgap materials appear well suited for combustion gas sensing applications. The changes in forward current of Ga-faced AlGaN/GaN HEMTs are approximately double those of simple Ga-faced GaN Schottky diode gas sensors tested under similar conditions, and suggest that integrated chips involving gas sensors and HEMT-based circuitry for off-chip communication are feasible in the AlGaN/GaN system. A differential pair configuration of AlGaN/GaN HEMT diodes with a built-in control diode has been shown to reduce false alarms. This involved a reference diode, which was encapsulated, and an active diode, which was open to the ambient. An additional key needed for long-term monitoring applications was the availability of stable ohmic contacts. Lastly, the effect of humidity on the AlGaN/GaN HEMT diode was investigated.

New development in N-faced GaN Schottky diodes shows much higher sensitivity than Ga-faced diodes due to high polarity formed on surface termination. Due to the amplification of the dipole layer at the interface under reverse bias conditions instead of screening under forward bias conditions, the sensitivity of hydrogen detection is higher under the reverse bias conditions. By using reverse bias condition combined with the improved stability from boride contacts, the overall stability of the GaN system is very attractive for long-term applications requiring high sensitivity.

Nanostructured wide-bandgap materials functionalized with Pd or Pt have the potential for more sensitive detection of hydrogen than their thin-film counterparts due to the increase in surface-to-volume ratio. 1-D semiconductor nanomaterials, such as GaN nanowires, InN nanobelts, and ZnO nanowires have been demonstrated as good candidates to replace 2-D semiconductors.

A low-power wireless sensor network has been demonstrated. This type of network can accommodate different sensors at each node, and has been engineered to have a long battery life, making the system truly wireless. These sensors could conceivably be placed at strategic locations on a large scale and have been field-tested. In the anticipated mass deployment of hydrogen fuel cell vehicles, city-level monitoring and networking will be necessary for safety. This work shows that the technology exists for a low-cost large-scale sensor deployment.

ACKNOWLEDGMENTS

The hydrogen sensor work at UF is partially supported by AFOSR grant under grant number F49620-03-1-0370 (T. Steiner), NSF (CTS-0301178, monitored by Dr. M. Burka and Dr. D. Senich), by NASA Kennedy Space Center Grant NAG 3-2930 monitored by Daniel E. Fitch, ONR (N00014-98-1-02-04, H. B. Dietrich), and NSF DMR 0400416.

REFERENCES

1. Zhang, A.P., Rowland, L.B., E. Kaminsky, B., Tucker, J.B., Kretchmer, J.W., Allen, A.F., Cook, J., and Edward, B.J. 9.2 W/mm (13.8 W) AlGaN/GaN HEMTs at 10 GHz and 55 V drain bias. *Electron. Lett.* 2003, 39, 245–247.

2. Saito, W., Takada, Y., Kuraguchi, M., Tsuda, K., Omura, I., Ogura, T., and Ohashi, H. High breakdown voltage AlGaN-GaN Power-HEMT design and high current density switching behavior. *IEEE Trans. Electron. Dev.* 2003, 50, 2528–2531.

3. Zhang, A.P., Rowland, L.B., Kaminsky, E.B., Tilak, V., Grande, J.C., Teetsov, J., Vertiatchikh, A., and Eastman, L.F. Correlation of device performance and defects in AlGaN/GaN high-electron mobility transistors. *J. Electron. Mater.* 2003, 32, 388–394.

4. Lu, W., Kumar, V., Piner, E.L., and Adesida, I. DC, RF, and, microwave noise performance of AlGaN-GaN field effect transistors dependence of aluminum concentration. *IEEE Trans. Electron. Dev.* 2003, 50, 1069–1074.

5. Valizadeh, P. and Pavlidis, D. Investigation of the impact of Al mole-fraction on the consequences of RF stress on AlxGa1-xN/GaN MODFETs. *IEEE Trans. Electron. Dev.* 2005, 52, 1933–1939.

6. Pearton, S.J., Zolper, J.C., Shul, R.J., and Ren, F. GaN: Processing, defects, and devices. *J. Appl. Phys.* 1999, 86, 1–78 and references therein.

7. Hikita, M., Yanagihara, M., Nakazawa, K., Ueno, H., Hirose, Y., Ueda, T., Uemoto, Y., Tanaka, T., Ueda, D., and Egawa, T. AlGaN/GaN power HFET on silicon substrate with source-via grounding (SVG) structure. *IEEE Trans. Electron. Dev.* 2005, 52, 1963–1968.

8. Nakazawa, S., Ueda, T., Inoue, K., Tanaka, T., Ishikawa, H., and Egawa, T. Recessed-gate AlGaN/GaN HFETs with lattice-matched InAlGaN quaternary alloy capping layers. *IEEE Trans. Electron. Dev.* 2005, 52, 2124–2128.

9. Palacios, T., Rajan, S., Chakraborty, A., Heikman, S., Keller, S., DenBaars, S.P., and Mishra, U.K. Influence of the dynamic access resistance in the g(m) and f(T) linearity of AlGaN/GaN HEMTs. *IEEE Trans. Electron. Dev.* 2005, 52, 2117–2123.

10. Mishra, U.K., Parikh, P., and Wu, Y.F. AlGaN/GaN HEMTs—An overview of device operation and applications. *Proc. IEEE* 2002, 90, 1022–1031.

11. Eastman, L.F., Tilak, V., Smart, J., Green, B.M., Chumbes, E.M., Dimitrov, R., Kim, H., Ambacher, O.S., Weimann, N., Prunty, T., Murphy, M., Schaff, W.J., and Shealy, J.R. Undoped AlGaN/GaN HEMTs for microwave power amplification. *IEEE Trans. Electron Dev.* 2001, 48, 479–485.

12. Keller, S., Wu, Y.-F., Parish, G., Ziang, N., Xu, J.J., Keller, B.P., DenBaars, S.P., and Mishra, U.K. Gallium nitride based high power heterojunction field effect transistors: Process development and present status at UCSB. *IEEE Trans. Electron. Dev.* 2001, 48, 552–559.

13. Adivarahan, V., Gaevski, M., Sun, W.H., Fatima, H., Koudymov, A., Saygi, S., Simin, G., Yang, J., Khan, M.A., Tarakji, A., Shur, M.S., and Gaska, R. Submicron gate Si_3N_4/AlGaN/GaN-metal-insulator-semiconductor heterostructure field-effect transistors. *IEEE Electron. Device Lett.* 2003, 24, 541–543.

14. Tarakji, A., Fatima, H., Hu, X., Zhang, J.P., Simin, G., Khan, M.A., Shur, M.S., and Gaska, R. Large-signal linearity in III-N MOSDHFETs. *IEEE Electron Dev. Lett.* 2003, 24, 369–371.

15. Iwakami, S., Yanagihara, M., Machida, O., Chino, E., Kaneko, N., Goto, H., and Ohtsuka, K. AlGaN/GaN heterostructure field-effect transistors (HFETs) on Si substrates for large-current operation. *Jpn. J. Appl. Phys.* 2004, 43, L831–L833.

16. Mehandru, R., Kim, S., Kim, J., Ren, F., Lothian, J., Pearton, S.J., Park, S.S., and Park, Y.J. Thermal simulations of high power, bulk GaN rectifiers. *Solid-State Electron.* 2003, 47, 1037–1043.

17. Luther, B.P., Wolter, S.D., and Mohney, S.E. High temperature Pt Schottky diode gas sensors on n-type GaN. *Sens. Actuat. B* 1999, 56, 164–168.

18. Baranzahi, A., Spetz, A.L., and Lundström, I. Reversible hydrogen annealing of metal-oxide-silicon carbide devices at high-temperatures. *Appl. Phys. Lett.* 1995, 67, 3203–3205.

19. Schalwig, J., Muller, G., Ambacher, O., and Stutzmann, M. Group-III-nitride based gas sensing devices. *Phys. Status Solidi. A* 2001, 185, 39–45.

20. Schalwig, J., Muller, G., Eickhoff, M., Ambacher, O., and Stutzmann, M. Gas sensitive GaN/AlGaN-heterostructures. *Sens. Actuat. B* 2002, 87, 425–430.

21. Kim, J., Ren, F., Gila, B., Abernathy, C.R., and Pearton, S.J. Reversible barrier height changes in hydrogen-sensitive Pd/GaN and Pt/GaN diodes. *Appl. Phys. Lett.* 2003, 82, 739–741.

22. Wang, H.T., Kang, B.S., Ren, F., Fitch, R.C., Gillespie, J., Moser, N., Jessen, G., Dettmer, R., Gila, B.P., Abernathy, C.R., and Pearton, S.J. Comparison of gate and drain current detection of hydrogen at room temperature with AlGaN/GaN high electron mobility transistors. *Appl. Phys. Lett.* 2005, 87, 172105.

23. Kouche, A.E.L., Lin, J., Law, M.E., Kim, S., Kang, B.S., Ren, F., and Pearton, S.J. Remote sensing system for hydrogen using GaN Schottky diodes. *Sens. Actuat. B: Chem.* 2005, 105, 329–333.

24. Kang, B.S., Mehandru, R., Kim, S., Ren, F., Fitch, R., Gillespie, J., Moser, N., Jessen, G., Jenkins, T., Dettmer, R., Via, D., Crespo, A., Gila, B.P., Abernathy, C.R., and Pearton, S.J. Hydrogen-induced reversible changes in drain current in Sc$_2$O$_3$/AlGaN/GaN high electron mobility transistors. *Appl. Phys. Lett.* 2004, 84, 4635–4637.

25. Kang, B.S., Ren, F., Gila, B.P., Abernathy, C.R., and Pearton, S.J. AlGaN/GaN-based metal-oxide-semiconductor diode-based hydrogen gas sensor. *Appl. Phys. Lett.* 2004, 84, 1123–1125.

26. Kim, J., Gila, B., Chung, G.Y., Abernathy, C.R., Pearton, S.J., and Ren, F. AlGaN/GaN-based metal–oxide–semiconductor diode-based hydrogen gas sensor. *Solid-State Electron.* 2003, 47, 1069–1073.

27. Huang, J.R., Hsu, W.C., Chen, Y.J., Wang, T.-B., Lin, K.W., Chen, H.-I., and Liu, W.-C. Comparison of hydrogen sensing characteristics for Pd/GaN and Pd/Al0.3Ga0.7As Schottky diodes. *Sens. Actuat. B* 2006, 117, 151–158.

28. Kang, B.S., Kim, S., Ren, F., Gila, B.P., Abernathy, C.R., and Pearton, S.J. Comparison of MOS and Schottky W/Pt–GaN diodes for hydrogen detection. *Sens. Actuat. B* 2005, 104, 232–236.

29. Matsuo, K., Negoro, N., Kotani, J., Hashizume, T., and Hasegawa, H. Pt Schottky diode gas sensors formed on GaN and AlGaN/GaN heterostructure. *Appl. Surf. Sci.* 2005, 244, 273–276.

30. Song, J., Lu, W., Flynn, J.S., and Brandes, G.R. AlGaN/GaN Schottky diode hydrogen sensor performance at high temperatures with different catalytic metals. *Solid-State Electron.* 2005, 49, 1330–1334.

31. Weidemann, O., Hermann, M., Steinhoff, G., Wingbrant, H., Spetz, A.L., Stutzmann, M., and Eickhoff, M. Influence of surface oxides on hydrogen-sensitive Pd:GaN Schottky diodes. *Appl. Phys. Lett.* 2003, 83, 773–775.

32. Ali, M., Cimalla, V., Lebedev, V., Romanus, H., Tilak, V., Merfeld, D., Sandvik, P., and Ambacher, O. Pt/GaN Schottky diodes for hydrogen gas sensors. *Sens. Actuat. B* 2006, 113, 797–804.

33. Voss, L., Gila, B.P., Pearton, S.J., Wang, H.-T., and Ren, F. Characterization of bulk GaN rectifiers for hydrogen gas sensing. *J. Vac. Sci. Technol. B* 2005, 3, 2373–2377.

34. Song, J., Lu, W., Flynn, J.S., and Brandes, G.R. Pt-AlGaN/GaN Schottky diodes operated at 800°C for hydrogen sensing. *Appl. Phys. Lett.* 2005, 87, 133501.

35. Yun, F., Chevtchenko, S., Moon, Y.-T., Morkoç, H., Fawcett, T.J., and Wolan, J.T. GaN resistive hydrogen gas sensors. *Appl. Phys. Lett.* 2005, 87, 073507.

36. Schalwig, J., Muller, G., Eickhoff, M., Ambacher, O., and Stutzmann, M. Gas sensitive GaN/AlGaN-heterostructures. *Sens. Actuat. B* 2002, 81, 425–430.

37. Eickhoff, M., Schalwig, J., Steinhoff, G., Weidmann, O., Gorgens, L., Neuberger, R., Hermann, M., Baur, B., Muller, G., Ambacher, O., and Stutzmann, M. Electronics and sensors based on pyro-electric AlGaN/GaN heterostructures—Part B: Sensor applications *Phys. Stat. Solidi C* 2003, 6, 1908–1918.

38. Svenningstorp, H., Tobias, P., Lundström, I., Salomonsson, P., Mårtensson, P., Ekedahl, L.-G., and Spetz, A.L. Influence of catalytic reactivity on the response of metal-oxide-silicon carbide sensor to exhaust gases. 1999, *Sensors and Actators*, B57, 159–165.

39. Chen, L., Hunter, G.W., and Neudeck, P.G. X-ray photoelectron spectroscopy study of the heating effects on Pd/6H-SiC Schottky structure. *J. Vac. Sci. Technol A* 1997, 15, 1228–1234.

40. Baranzahi, A., Spetz, A.L., and Lundström, I. Reversible hydrogen annealing of metal-oxide-silicon carbide devices at high temperatures. *Appl. Phys. Lett.* 1995, 67, 3203–3205.

41. Kang, B.S., Mehandru, R., Kim, S., Ren, F., Fitch, R., Gillespie, J., Moser, N., Jessen, G., Jenkins, T., Dettmer, R., Via, D., Crespo, A., Gila, B.P., Abernathy, C.R., and Pearton, S.J. Hydrogen-induced reversible changes in drain current in Sc_2O_3/AlGaN/GaN high electron mobility transistors. *Appl. Phys. Lett.* 2004, 84, 4635–4637.

42. Kang, B.S., Heo, Y.W., Tien, L.C., Norton, D.P., Ren, F., Gila, B.P., and Pearton, S.J. Electrical transport properties of single ZnO nanorods. *Appl. Phys. A* 2005, 80, 1029–1032.

43. Kouche, A.El., Lin, J., Law, M.E., Kim, S., Kim, B.S., Ren, F., and Pearton, S.J. Remote sensing system for hydrogen using GaN Schottky diodes. *Sens. Actuat. B: Chem.* 2005, 105, 329–333.

44. Pearton, S.J., Kang, B.S., Kim, S., Ren, F., Gila, B.P., Abernathy, C.R., Lin, J., and Chu, S.N.G., GaN-based diodes and transistors for chemical, gas, biological and pressure sensing. *J. Phys: Condensed Matter* 2004, 16, R961–R994.

45. Kim, S., Kang, B., Ren, F., Ip, K., Heo, Y., Norton, D., and Pearton, S.J. Sensitivity of Pt/ZnO Schottky diode characteristics to hydrogen. *Appl. Phys. Lett.* 2004, 84, 1698–1700.

46. Kang, B.S., Wang, H.T., Ren, F., Gila, B.P., Abernathy, C.R., Pearton, S.J., Johnson, J.W., Rajagopal, P., Roberts, J.C., Piner, E.L., and Linthicum, K.J. pH sensor using AlGaN/GaN high electron mobility transistors with Sc_2O_3 in the gate region. *Appl. Phys. Lett.* 2007, 91, 012110.

47. Kang, B.S., Wang, H.T., Ren, F., Pearton, S.J., Morey, T.E., Dennis, D.M., Johnson, J.W., Rajagopal, P., Roberts, J.C., Piner, E.L., and Linthicum, K.J. Enzymatic glucose detection using ZnO nanorods on the gate region of AlGaN/GaN high electron mobility transistors. *Appl. Phys. Lett.* 2007, 91, 252103.

48. Kang, B.S., Wang, H.T., Lele, T.P., Tseng, Y., Ren, F., Pearton, S.J., Johnson, J.W., Rajagopal, P., Roberts, J.C., Piner, E.L., and Linthicum, K.J. Prostate specific antigen detection using AlGaN/GaN high electron mobility transistors. *Appl. Phys. Lett.* 2007, 91, 112106.

49. Wang, H.T., Kang, B.S., Ren, F., Pearton, S.J., Johnson, J.W., Rajagopal, P., Roberts, J.C., Piner, E.L., and Linthicum, K.J. Electrical detection of kidney injury molecule-1 with AlGaN/GaN high electron mobility transistors. *Appl. Phys. Lett.* 2007, 91, 222101.

50. Mitra, P., Chatterjee, A.P., and Maiti, H.S. ZnO thin film sensor. *Mater. Lett.* 1998, 35, 33–38.

51. Hartnagel, H.L., Dawar, A.L., Jain, A.K., and Jagadish, C. *Semiconducting Transparent Thin Films*. IOP Publishing: Bristol, England, U.K., 1995.

52. Chang, J.F., Kuo, H.H., Leu, I.C., and Hon, M.H. The effects of thickness and operation temperature on ZnO: Al thin film CO gas sensor. *Sens. Actuat. B* 1994, 84, 258–264.

53. See the databases at http://www.rebresearch.com/H2perm2.htm and http://www.rebresearch.com/H2sol2.htm.

54. Eberhardt, W., Greunter, F., and Plummer, E.W. Bonding of H to Ni, Pd, and Pt Surfaces. *Phys. Rev. Lett.* 1981, 46, 1085–1088.

55. Voss L., Gila, B.P., Pearton, S.J., Wang, H.-T., and Ren, F. Characterization of bulk GaN rectifiers for hydrogen gas sensing. *J. Vac. Sci. Technol. B* 2005, 23, 2373–2377.

56. Wang, H.-T., Kang, B.S., Ren, F., Fitch, R.C., Gillespie, J.K., Moser, N., Jessen, G., Jenkins, T., Dettmer, R., Via, D., Crespo, A., Gila, B.P., Abernathy, C.R., and Pearton, S.J. Comparison of gate and drain current detection of hydrogen at room temperature with AlGaN/GaN high electron mobility transistors. *Appl. Phys. Lett.* 2005, 87, 172105.

57. Anderson, T., Wang, H.T., Kang, B.S., Li, C., Low, Z.N., Lin, J., Pearton, S.J., and Ren, F. *NHA Conference Proceedings* 2007.

58. Anderson, T., Wang, H.T., Kang, B.S., Li, C., Low, Z.N., Lin, J., Pearton, S.J., Ren, F., Dabiran, A., Osinsky, A., and Painter, J. Hydrogen Safety Report, July 2007 <http://www.hydrogenandfuelcellsafety.info/2007/jul/sensors.asp>.

59. Anderson, T., Wang, H.T., Kang, B.S., Li, C., Low, Z.N., Lin, J., Pearton, S.J., Ren, F., Dabiran, A., and Osinsky, A. *J. Electrochem. Soc.* (submitted October 2007).

60. Hunter, G.W., Neudeck, P.G., Okojie, R.S., Beheim, G.M., Thomas, V., Chen, L., Lukco, D., Liu, C.C., Ward, B., and Makel, D. Development of. SiC gas sensor systems. *Proc. ECS* 2002, 03, 93–111.

61. Neuberger, R., Muller, G., Ambacher, O., and Stutzmann, M. High electron mobility AlGaN/GaN transistors for fluid monitoring applications. *Phys. Status Solidi. A* 2001, 185, 85–89.

62. Schalwig, J., Muller, G., Ambacher, O., and Stutzmann, M. Group-III-nitride based sensing devices. *Phys. Status Solidi. A* 2001, 185, 39–45.

63. Steinhoff, G., Hermann, M., Schaff, W.J., Eastmann, L.F., Stutzmann, M., and Eickhoff, M. pH response of GaN surfaces and its application for pH-sensitive field-effect transistors. *Appl. Phys. Lett.* 2003, 83, 177–179.

64. Khanna, R., Pearton, S.J., Ren, F., and Kravchenko, I.I. Comparison of electrical and reliability performances of TiB2-, CrB2-, and W2B5-based Ohmic contacts on n-GaN. *J. Vac. Sci. Technol.* 2006, B24, 744–748.

65. Wang, H.T., Anderson, T.J., Kang, B.S., Ren, F., Li, Changzhi, Low, Zhen-Ning, Lin, Jenshan, Gila, B.P., Pearton, S.J., Dabiran, A., and Osinsky, A. *Appl. Phys. Lett.* 2007, 90, 252109.

66. http://www.cityrating.com/relativehumidity.asp.

67. Conway, B.E., Barnett, B., Angerstein-Kozlowska, H., and Tilak, B.V. *J. Chem. Phys.* 1990, 93, 8361.

68. Harrington, D.A. *J. Electroanal. Chem.* 1997, 420, 101.

69. Marković, N.M. and Ross, P.N. *Surf. Sci. Rep.* 2002, 45, 121.

70. Paik, C.H., Jarvi, T.D., and O'Grady, W.E., *Electrochem. Solid-State Lett.* 2004, 7, A82.

71. Xu, H., Kunz, R., and Fenton, J.M., *Electrochem. Solid-State Lett.* 2007, 10, B1.

72. Wang, H.T., Anderson, T.J., Ren, F., Li, C., Now, Z.N., Lin, J., Gila, B.P., Pearton, S.J., Osinsky, A., and Dabiran, A. Robust detection of hydrogen using differential AlGaN/GaN high electron mobility transistor sensing diodes. *Appl. Phys. Lett.* 2006, 89, 242111.

73. Huang, F.C., Chen, Y.Y., and Wu, T.T. *Nanotechnology* 2009, 20, 065501.

74. Zhao, Z., Knight, M., Kumar, S., Eisenbraun, E.T., and Carpenter, M.A. *Sens. Actuat. B* 2008, 129, 726.

75. Starke, U., Sloboshanin, S., Tautz, F.S., Seubert, A., and Schaefer, J.A. Polarity, Morphology and reactivity of epitaxial GaN films on Al$_2$O$_3$ (0001), *Phys. Status Solidi A.* 2000, 177, 5–14.

76. Northrup, J.E., and Neugebauer, J., Strong affinity of hydrogen for the GaN(000-1) surface: Implications for molecular beam epitaxy and metalorganic chemical vapor deposition, *Appl. Phys. Lett.* 2004, 85, 3429–3431.

77. Mayumi, M., Satoh, F., Kumagai, Y., Takemoto, K., and Koukitu, A. Influence of polarity on surface reaction between GaN{0001} and hydrogen, *Phys. Status Solidi B.* 2001, 228, 537–541.

78. Ambacher, O., Smart, J., Shealy, J.R., Weimann, N.G., Chu, K., Murphy, M., Schaff, W.J., Eastman, L.F., Dimitrov, R., Wittmer, L., Stutzmann, M., Reiger, W., and Hilsenbeck, J. Two-dimensional electron gases induced by spontaneous and piezoelectric polarization charges in N- and Ga-face AlGaN/GaN heterostructures, *J. Appl. Phys.* 1999, 85, 3222–3233.

79. Wang, Yu-Lin, Ren, F., Zhang, Y., Sun, Q., Yerino, C.D., Ko, T.S., Cho, Y.S., Lee, I.H., Han, J., and Pearton, S.J. Improved hydrogen detection sensitivity in N-polar GaN Schottky diodes, *Appl. Phys. Lett.* 2009, 94, 212108-1–212108-3.

80. Wang, Yu-Lin, Chu, B.H., Chang, C.Y., Chen, K.H., Zhang, Y., Sun, Q., Lee, I.-H., Han, J., Pearton, S.J., and Ren, F. Hydrogen sensing of N-polar and Ga-polar GaN Schottky diodes. *Sens. Actuat. B: Chem.* 2009, 142, 175.

81. Sun, Q., Cho, Y.S., Kong, B.H., Cho, H.K., Ko, T.S., Yerino, C.D., Lee, I.-H., and Han, J. N-face GaN growth on c-plane sapphire by metalorganic chemical vapor deposition, *J. Cryst. Growth* 2009, 311, 2948–2952.

82. Sun, Q., Cho, Y.S., Lee, I.-H., Han, J., Kong, B.H., and Cho, H.K. Nitrogen-polar GaN growth evolution on c-plane sapphire, *Appl. Phys. Lett.* 2008, 93, 131912-1–131912-3.

83. Johnson, J.L., Choi, Y., Ural, A., Lim, W., Wright, J.S., Gila, B.P., Ren, F., and Pearton, S.J. *J. Electron. Mater.* 2009, 38, pp. 490–494.

84. Lim, W., Wright, J.S., Gila, B.P., Johnson, J.L., Ural, A., Anderson, T., Ren, F., and Pearton, S.J. *Appl. Phys. Lett.* 2008, 93, pp. 072110–072112.

85. Lu, Y., Li, J., Ng, H.T., Binder, C., Partridge, C., and Meyyapan, M. Room temperature methane detection using palladium loaded single-walled carbon nanotube sensors. *Chem. Phys. Lett.* 2004, 391, 344–348.

86. Sayago, I., Terrado, E., Lafuente, E., Horillo, M.C., Maser, W.K., Benito, A.M., Navarro, R., Urriolabeita, E.P., Martinez, M.T., and Gutierrez, J. Hydrogen sensors based on carbon nanotubes thin films. *Syn. Metals* 2005, 148, 15–19.

87. Kang, B.S., Ren, F., Heo, Y.W., Tien, L.C., Norton, D.P., and Pearton, S.J. pH measurements with single ZnO nanorods integrated with a microchannel. *Appl. Phys. Lett.* 2005, 86, 112105.

88. Wan, Q., Li, Q.H., Chen, Y.J., Wang, T.H., He, X.L., Li, J.P., and Lin, C.L. Fabrication and ethanol sensing characteristics of ZnO nanowire gas sensors. *Appl. Phys. Lett.* 2004, 84, 3654–3656.

89. Wan, Q., Li, Q.H., Chen, Y.J., Wang, T.H., He, X.L., Gao, X.G., Li, J.P. Positive temperature coefficient resistance and humidity sensing properties of Cd-doped ZnO nanowires. *Appl. Phys. Lett.* 2004, 84, 3085–3087.

90. Keem, K., Kim, H., Kim, G.T., Lee, J.S., Min, B., Cho, K., Sung, M.Y., and Kim, S. Photocurrent in ZnO nanowires grown from Au electrodes. *Appl. Phys. Lett.* 2004, 84, 4376–4378.

91. Huang, M.H., Mao, S., Feick, H., Yan, H., Wu, Y., Kind, H., Weber, E. Russo, R., and Yang, P. Room-temperature ultraviolet nanowire nanolasers. *Science* 2001, 292, 1897–1899.

92. Wang, Z.L. Nanostructures of zinc oxide. *Materials Today*, June 2004, pp. 26–33.

93. Heo, Y.W., Norton, D.P., Tien, L.C., Kwon, Y., Kang, B.S., Ren, F., Pearton, S.J., and LaRoche, J.R. ZnO nanowire growth and devices. *Mat. Sci. Eng. R* 2004, 47, 1–47.

94. Liu, C.H., Liu, W.C., Au, F.C.K., Ding, J.X., Lee, C.S., and Lee, S.T. Electrical properties of zinc oxide nanowires and intramolecular p–n junctions. *Appl. Phys. Lett.* 2003, 83, 3168–3170.

95. Park, W.I., Yi, G.C., Kim, J.W., and Park, S.M. Schottky nanocontacts on ZnO nanorod arrays. *Appl. Phys. Lett.* 2003, 82, 4358; *Phys. Lett.* 2003, 82, 4358–4360.

96. Ng, H.T., Li, J., Smith, M.K., Nguygen, P., Cassell, A., Han, J., and Meyyappan, M. Growth of epitaxial nanowires at the junctions of nanowalls. *Science* 2003, 300, 1249.

97. Park, W.I., Yi, G.C., Kim, M.Y., and Pennycook, S.J. Quantum confinement observed in ZnO/ZnMgO nanorod heterostructures. *Adv. Mater.* 2003, 15, 526–529.

98. Poole, P.J., Lefebvre, J., and Fraser, J. Spatially controlled, nanoparticle-free growth of InP nanowires. *Appl. Phys. Lett.* 2003, 83, 2055–2057.

99. He, M., Fahmi, M.M.E., Mohammad, S.N., Jacobs, R.N., Salamanca-Riba, L., Felt, F., Jah, M., Sharma, A., and Lakins, D. InAs nanowires and whiskers grown by reaction of indium with GaAs. *Appl. Phys. Lett.* 2003, 82, 3749–3751.

100. Wu, X.C., Song, W.H., Huang, W.D., Pu, M.H., Zhao, B., Sun, Y.P., and Du, J.J. Crystalline gallium oxide nanowires: Intensive blue light emitters. *Chem. Phys. Lett.* 2000, 328, 5–9.

101. Zheng, M.J., Zhang, L.D., Li, G.H., Zhang, X.Y., and Wang, X.F. Ordered indium-oxide nanowire arrays and their photoluminescence properties. *Appl. Phys. Lett.* 2001, 79, 839–841.

102. Zhang, B.P., Binh, N.T., Segawa, Y., Wakatsuki, K., and Usami, N. Optical properties of ZnO rods formed by metalorganic chemical vapor deposition. *Appl. Phys. Lett.* 2003, 83, 1635–1637.

103. Park, W.I., Jun, Y.H., Jung, S.W., and Yi, G. Excitonic emissions observed in ZnO single crystal nanorods. *Appl. Phys. Lett.* 2003, 82, 964–966.

104. Pan, Z.W., Dai, Z.R., and Wang, Z.L. Nanobelts of semiconducting oxides. *Science* 2001, 291, 1947–1949.

105. Lao, J.Y., Huang, J.Y., Wang, D.Z., and Ren, Z.F. ZnO nanobridges and nanonails. *Nano Lett.* 2003, 3, 235–238.

106. Kim, J., Ren, F., Gila, B., Abernathy, C.R., and Pearton, S.J. Reversible barrier height changes in hydrogen-sensitive Pd/GaN and Pt/GaN diodes. *Appl. Phys. Lett.* 2003, 82, 739–741.

107. Vasiliev, A., Moritz, W., Fillipov, V., Bartholomäus, L., Terentjev, A., and Gabusjan, T. High temperature semiconductor sensor for the detection of fluorine. *Sens. Actuat. B* 1998, 49, 133–138.

108. Kim, J., Gila, B., Abernathy, C.R., Chung, G.Y., Ren, F., and Pearton, S.J. Comparison of Pt/GaN and Pt/4H-SiC gas sensors. *Solid State Electron.* 2003, 47, 1487–1490.

109. Spetz, A.L., Tobias, P., Unéus, L., Svenningstorp, H., Ekedahl, L.-G., and Lundström, I. High temperature catalytic metal field effect transistors for industrial applications. *Sens. Actuat. B* 2000, 70, 67–76.

110. Connolly, E.J., O'Halloran, G.M., Pham, H.T.M., Sarro, P.M., and French, P.J. Comparison of porous silicon, porous polysilicon and porous silicon carbide as materials for humidity sensing applications. *Sens. Actuat. A* 2002, 99, 25–30.

111. L. Voss, B.P. Gila, S.J. Pearton, H. Wang, and F. Ren, *J. Vac. Sci. Technol.*, 2005, B23, pp. 6–10.

112. J.S. Wright, W. Lim, B.P. Gila, S.J. Pearton, F. Ren, W. Lai, L.C. Chen, M. Hu, and K.H. Chen, *J. Vac. Sci. Technol.* 2009, B 27, L8–L10.

113. W. Lim, J.S. Wright, B.P. Gila, S.J. Pearton, F. Ren, W. Lai, L.C. Chen, M. Hu, and K.H. Chen, *Appl. Phys. Lett.* 2008, 93, pp. 202109–202111.

114. K.D. Mitzner, J. Sternhagen, and D.W. Galipeau, *Sens. Actuat. B* 2003, 93, pp. 92.

115. P. Mitra, A.P. Chatterjee, and H.S. Maiti, *Mater. Lett.* 1998, 35, pp. 35.

116. Hartnagel, H.L., Dawar, A.L., Jain, A.K., and Jagadish, C. *Semiconducting Transparent Thin Films* (IOP Publishing, Bristol, 1995).

117. Chang, J.F., Kuo, H.H., Leu, I.C., and Hon, M.H. *Sens. Actuat. B* 1994, B 84, 258.

118. Eickhoff, M., Neuberger, R., Steinhoff, G., Ambacher, O., Muller, G., and Stutzmann, M. Wetting behaviour of GaN-surfaces with Ga- or N-face polarity. *Phys. Status Solidi B* 2001, 228, 519–522.

4 InN-Based Chemical Sensors

Yuh-Hwa Chang
Institute of Nanoengineering and Microsystems,
National Tsing Hua University, Hsinchu 30013, Taiwan

Yen-Sheng Lu
Institute of Electronics Engineering, National Tsing
Hua University, Hsinchu 30013, Taiwan

J. Andrew Yeh
Institute of Nanoengineering and Microsystems,
National Tsing Hua University, Hsinchu 30013, Taiwan

Yu-Liang Hong
Department of Physics, National Tsing Hua
University, Hsinchu 30013, Taiwan

Hong-Mao Lee
Department of Physics, National Tsing Hua
University, Hsinchu 30013, Taiwan

Shangjr Gwo
Department of Physics, National Tsing Hua
University, Hsinchu 30013, Taiwan

CONTENTS

4.1 INTRODUCTION

Open-gate Si-based ion-sensitive field-effect transistors (ISFETs) were proposed for use as pH sensors by Bergveld in 1970. These transistors are promising device structures for integrated chemical and biological sensors [1]. A gate voltage induced by ions adsorbed onto the open-gate region modulates the surface space-charge layer underneath and results in the change of the source-drain current, thus realizing the operation of ISFETs. To improve the relatively insensitive response and poor stability of the original ISFET structure, various sensing materials, such as Si_3N_4 [2], Al_2O_3 [2], or Ta_2O_5 [3], were investigated to substitute the role of silicon oxide. Moreover, materials with more robust surface properties are particularly required for chemical and biological sensors against possible damages and contaminations from harsh sensing environments.

III-nitrides, including gallium nitride (GaN), aluminum nitride (AlN), indium nitride (InN), and their alloys have recently emerged as promising materials for next-generation chemical and biological sensors because of their high sensitivity, biocompatibility, thermal stability, and robust surface properties [4–18]. The majority of GaN-based sensors utilize the configuration of high electron mobility transistor (HEMT) structure, featuring a polarization-induced two-dimensional electron gas (2DEG) with the sheet density of the order of 10^{13} cm^{-2} at the interface of heterostructure [4–12]. The sensing mechanism is similar to the case of Si-based ISFETs. The electron density in the 2DEG for a HEMT-based sensor is modulated by the induced gate voltage due to ion adsorption onto the open-gate region, giving rise to the change of the source-drain current. In recent years, InN becomes another appealing candidate for chemical and biological sensing applications mainly because of its unusually strong electron accumulation occurring in the near-surface region [19–24]. This chapter will discuss the InN surface properties briefly and demonstrate some advances of chemical sensors based on InN.

4.2 SURFACE PROPERTIES OF InN

4.2.1 ELECTRONIC PROPERTIES

Recently, due to the advances in epitaxial growth techniques of InN, high-quality single-crystalline InN films have been routinely obtained. Meanwhile, the study of their electronic properties has attracted much attention. It has been shown that the as-grown InN films exhibit a narrow bandgap (0.6−0.7 eV), excellent electron transport characteristics, and a background high electron density (typically in excess of 1×10^{18} cm^{-3}) for the nominally undoped InN films [25]. Moreover, the unusual phenomenon of a surface electron accumulation layer with a sheet density of the order of 10^{13} cm^{-2} at the InN surfaces has been found and confirmed by various experimental techniques, such as capacitance−voltage (C–V) profiling and high-resolution electron energy loss spectroscopy (HREELS) [19–24]. As characterized by C–V profiling, a gradient of carrier concentration ranging from 10^{20} to 10^{18} cm^{-3} within 6 nm in depth at the InN surface was observed, indicating a strong surface electron accumulation [19]. HREELS spectra on clean InN surfaces also revealed similar results. The broad peak of electron energy loss observed at ~250 meV was explained as the conduction band electron plasmon excitations. This peak shows a downward dispersion of ~40 meV as the probing electron energy increases from 15 to 60 eV, which results from a surface layer of higher plasma frequency than that of the bulk. This provides a clear evidence for the existence of an electron accumulation layer at the InN surface [22].

The phenomenon of the surface electron accumulation has been found to be universal for all as-grown InN surfaces except on the in situ cleaved nonpolar a-plane InN surface, on which the absence of surface electron accumulation has been demonstrated [26]. This unusual electronic property has been explained as a result of the extraordinarily low conduction band minimum (CBM) at the Γ-point (5.8 eV below the vacuum level), which is the largest value among known semiconductors. This ultralow CBM is located far below the universal Fermi-level stabilization energy (E_{FS},

4.9 eV below the vacuum level), which implies that the surface states tend to become donors, thus pinning the Fermi-level at 0.9 eV above the CBM at the InN surfaces [23]. As a result, the band is bended downward and a high density of electrons is accumulated near the surface. Recently, based on their first-principles calculations, Segev and Van de Walle suggested that the microscopic origin of donor-type surface states is mostly associated with In–In bonding states within the surface In adlayers [27,28].

In recent years, some studies for eliminating the surface electron accumulation of InN have been reported, but none of them succeeded and only the level of surface electron accumulation can be reduced [29–32]. For example, the surface sulfur treatment, which has been widely used in passivating the surface states of group IV, III–V, and II–VI semiconductors, could reduce the surface electron density by ~30%, corresponding to a decrease of surface band bending by ~0.3 eV [32]. Although the surface electron accumulation effect is detrimental to the applications of InN intended for electronic and optoelectronic devices, it is, on the contrary, attractive for chemical sensing applications. The charge accumulation layer forms a natural 2DEG at InN surfaces. Unlike the 2DEG occurring at the buried interface of the heterostructure for a HEMT-based sensor, the natural 2DEG of InN is located immediately under the surface, where the sensing functionality occurs. Adsorbates on the InN surface could effectively affect the surface band bending, thereby modulating the electron density in the surface electron accumulation layer. This can give rise to a significant surface conductance change.

4.2.2 CHEMICAL SENSING PROPERTIES

The electronic properties of InN have been well developed in the last decade, but the study of chemical sensing properties was rarely reported. The liquid-phase chemical response of bare InN surfaces was first demonstrated by Lu et al. [33]. It was found that the treatments using methanol, water, and isopropanol (IPA) at InN surfaces could increase both the sheet carrier density and Hall mobility. Despite the difference in thickness for the measured four samples, the similar magnitude of increase was observed with the same treatment. The treatments using water and methanol showed a higher and comparable response, while a relatively weaker response was observed for IPA. The mechanism of InN surface sensing effect to solvents might combine dipolar and unipolar van der Waals interactions, which determines the response of InN surface to different chemical exposures. This preliminary result reveals the potential of InN for chemical and biological sensing applications.

4.3 InN-BASED CHEMICAL SENSORS DEVELOPMENT

4.3.1 ION-SELECTIVE ELECTRODE (ISE)

Potentiometric ion-selective electrode (ISE) sensors have been developed for chemical, biomedical, physiological, and clinical applications for many decades. Ligands, functionalized groups, or ionophores absorbed on the ISE selectively responses to surrounding specific ions, thus inducing a surface potential change. To achieve high sensitivity, good repeatability, and fast response, the use of solid-state materials have been proposed [34–37]. For example, a Ga-face GaN-based ISE has been reported to show a good selectivity for anions, such as chlorine (Cl^-), sulfate (SO_4^{2-}), and hydroxide (OH^-) ions [36,37]. Here, an N-face InN-based ISE is demonstrated for selectively determining anion concentrations, such as chlorine (Cl^-) and hydroxide (OH^-) ions in aqueous solutions by using potentiometric measurements.

4.3.1.1 Device Fabrication

A wurtzite N-face InN($000\bar{1}$) film was used in the ISE fabrication. The heteroepitaxial growth of InN on the Si(111) substrate was conducted in a molecular-beam epitaxy (MBE) system (DCA-600) equipped with a radio-frequency nitrogen plasma source. The base pressure was maintained at ~6 × 10⁻¹¹ Torr during processing. Silicon was chosen as the substrate for the InN(000$\overline{1}$) epitaxial growth instead of the most widely used sapphire substrate due to the smaller lattice mismatch between InN(000$\overline{1}$) and Si(111) (~8%) than InN(000$\overline{1}$) and Al$_2$O$_3$(0001) (~25%). However, the reported film quality of InN grown on Si substrates using a single AlN buffer layer was poor. The unintentional nitridation of Si forms an amorphous SiN$_x$ layer during the buffer layer growth, leading to. To overcome this problem, a double-buffer layer (comprised of β-Si$_3$N$_4$ and AlN) technique was proposed [38,39]. First, a single-crystal β-Si$_3$N$_4$ ultrathin layer, which was 1:2 coincident lattice matched with the Si(111) substrate, was formed on the Si(111) substrate in N$_2$ plasma prior to the InN growth to prevent the formation of amorphous SiN$_x$ layer. Then, AlN (0001) was deposited on β-Si$_3$N$_4$ (0001) with 5:2 coincident lattices formed at the interface to facilitate the double-buffer layer for InN-on-Si heteroepitaxial growth. By using this technique, autodoping from Si outdiffusion could also be effectively reduced compared with the growth of InN with a single AlN buffer layer.

The InN film used here was 400 nm in thickness. The room temperature Hall measurement on the as-grown InN sample revealed an electron density of 4.1 × 10¹⁸ cm⁻³ with the mobility of 1012 cm²/V×s. The as-grown InN surfaces were first immersed in an HCl:H$_2$O (1:3) solution to remove excessive surface indium residues and surface oxide prior to the device fabrication. The sensing window of an InN-based ISE is 2 × 2 mm² in size. The ISE was bonded with a platinum wire via the gold contact pad on the InN surface, followed by the encapsulation using epoxy resin. The use of wire bonding and encapsulation can reduce the electronic noise and prevent some possible chemical erosion.

4.3.1.2 Testing Procedure and Results

First, the changes of Helmholtz voltage—that is the potential buildup between the InN surface and the reference electrode, with respect to the variation of the anion concentrations in analyte solutions—were measured. The open-circuit potentiometric signals of the InN-based ISE were recorded using a multimeter (Agilent 34401A). The multimeter with the input impedance of 1 MΩ ± 2% and capacitance of 100 pF in parallel can permit an open-circuit voltage for sensing measurements. The potentiometric measurements were performed by immersing the ISE into analyte solutions with an Hg/HgCl electrode (Hanna HI-5412), which was used for establishing a reference potential in solutions. The analyte solutions included KCl, NaCl, NH$_4$Cl, CaCl$_2$, BaCl$_2$, and NaOH in deionized water, respectively. Figure 4.1 shows the semi-log plot of potential buildup for the

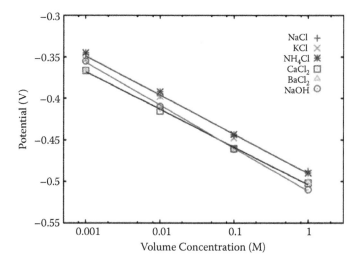

FIGURE 4.1 Potentiometric responses of an InN-based ISE to various volume concentrations of the IA/IIA chloride and hydroxyl solutions.

volume concentrations of the chloride and the hydroxyl solutions ranging from 10^{-3} M to 1 M. The fitting lines indicate the sensing responses to the IA, IIA chloride, and hydroxyl solutions, respectively. The measured potentials of the InN-based ISE are linearly proportional to the logarithm of volume concentration, satisfying the Nernst equation. The sensitivities indicated by the slopes of the fitted lines were determined to be −47 mV and −45 mV per decade at 25°C for the IA and IIA chlorides, respectively, and a similar sensitivity of −49 mV per decade was observed for hydroxyl. The repeatability measurement of the InN-based ISE yielded the potential variation of ±1 mV. The long-term stability was found to be 9 mV for 6 h, which is close to that of GaN-based ISE sensors [36]. The negative sensitivity implies that the potential shifted to more negative as the chlorine ion concentration increased. This result suggests that the InN-based ISE selectively respond to chlorine ions instead of cations, which might result from the chlorine ions attracted to the positively charged donor sites on the InN surface. Furthermore, note that the theoretical sensitivity of single-charged cations (e.g., K^+, Na^+, and NH_4^+) is twice as much as that of double-charged cations (e.g., Ca^{2+} and Ba^{2+}), according to the Nernst equation.

The dynamic response of the InN-based ISE for potassium chloride (KCl) of 0.01 M–0.05 M–0.1 M and 0.01 M–0.1 M is plotted in Figure 4.2. The response (fall) time and the recovery (rise) time indicate the times elapsed after the relative potential change from 10% (T_{10}) to 90% (T_{90}) and from 90% to 10%, respectively. The response and the recovery times of the one-decade volume concentration change (e.g., 0.01 M–0.1 M) are found to be 12 s and 78 s for the IA chloride solutions, and to be 18 s and 108 s for IIA chloride solutions. The response time is faster than the recovery time for the tested chloride solutions by roughly six times. Overall, both the short response times and the long-term potential stability suggest that the potential buildup mechanism results from purely a surface mechanism without bulk transition into InN crystals.

Charged donors from both extrinsic donor impurities and donor-type native surface/bulk defects at and near the surface region may provide charged sites that selectively adsorb anions onto the InN surface, inducing a Helmholtz voltage at the electrolyte–InN interface. The adsorption of anions to positive-charged donors expels the accumulated electrons in the near-surface region. Consequently, the N-face InN film grown on silicon substrate has been demonstrated to be an excellent sensing membrane that can selectively react to anions in aqueous solutions with fast-response, long-term stability, and high repeatability.

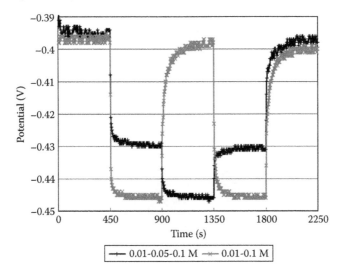

FIGURE 4.2 Dynamic response of an InN-based ISE for KCl concentration changes of 0.1 M→0.05 M→0.1 M and 0.01 M→0.1 M. The Cl⁻ concentration measurement using the InN surface potential shows a repeatability of ±1 mV.

4.3.2   Ion-Sensitive Field-Effect Transistor (ISFET)

An open-gate Si-based field-effect transistor (FET) with silicon oxide as a sensing material was first proposed as a pH sensor by Bergveld in 1970 [1]. The structure of an ISFET is similar to that of a conventional FET but without the gate metal. During the usage, the ISFET along with a reference electrode is immersed into the analyte solution. The analyte ions in the solution absorbed onto the open-gate region will induce a Helmholtz voltage, similar to that of an ISE, thus modulating the carrier distribution beneath and the conductivity between the source and the drain. Hence, the performance of an ISFET is primarily determined by the properties of the sensing material at the open gate region and the transconductance of the FET. In the practical use, the voltage bias between the reference electrode and the substrate (V_{GS}) can be varied for achieving the higher current variation with respect to the same ion concentration change. Later, we will introduce an InN-based ISFET and its applications in anion, pH, and polarity sensing.

4.3.2.1   Device Fabrication

The conventional ISFETs employ open-gate Si-based transistors, which require relatively complex process, including well formation, device isolation, source/drain formation, and contact kickups as well as a dielectric layer, such as metal oxides or III-nitrides, as a sensing membrane. An InN-based ISFET developed here is comprised of only an InN epilayer grown on an Si substrate and a pair of metal electrodes deposited on InN surface. The schematic cross section of an ultrathin (~10 nm) InN-based ISFET and its package photo are shown in Figure 4.3. The metal electrode was composed of a composite layer of 500 A Au/2000 A Al/400 A Ti, which was deposited with e-beam evaporation system and patterned with liftoff process. The channel length of the ISFETs was 100 μm wide. The entire device, except the open-gate region, was encapsulated with polyimide and polydimethylsiloxane (PDMS) for preventing the chemical erosion from sensing environments. During the operation, the InN epilayer acts as a conductive layer and the native indium oxide on the surface functions as a sensing membrane. The source-drain current (I_{DS}) flowing through the two electrodes can be modulated via the open gate by electrolyte-gate-biasing or chemical sensing.

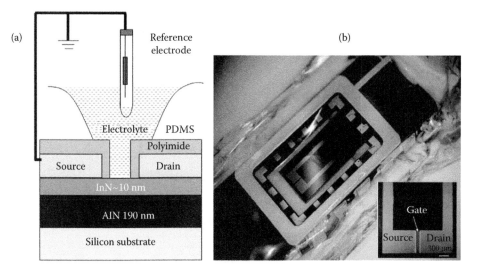

FIGURE 4.3 (See color insert).   (a) Schematic cross-sectional structure of an ultrathin (~10 nm) InN ISFET. The native oxide at InN surfaces functions as the sensing material, which is in direct contact with the sensing environment. The adsorption of analyte ions onto the InN surface builds up a Helmholtz potential, resulting in an output current variation. The source and drain are encapsulated by polyimide and PDMS to prevent the contact with electrolyte.

4.3.2.2 Testing Procedure and Results

The source-drain current (I_{DS}) of an InN-based ISFET was measured using a multimeter (Agilent 34401A) and a DC power supply (Agilent 6654A). The gate-source voltage (V_{GS}) was controlled through an Hg/HgCl reference electrode (Hanna HI-5412) immersed in the electrolyte, as shown in Figure 4.3a. First, to investigate the electrical performance of an InN-based ISFET, the I_{DS}–V_{GS} characteristics were measured through the electrolyte-gate biasing. Then, the responses of I_{DS} with respect to the various concentrations of analyte ions, including anion and hydrogen ion (pH), in aqueous solutions and to the polar liquids were studied.

4.3.2.2.1 Electrolyte-Gate Biasing Characterization

The I_{DS}–V_{GS} characteristics of InN-based ISFETs were investigated using the electrolyte-gate-biasing technique, that is, varying the applied voltage on the electrolyte gate through the reference electrode. The InN epilayers studied here include undoped N-polar–c-plane InN ($-c$-InN) of 1 μm and 10 nm in thickness, and 1.2-μm-thick Mg-doped a-plane InN (a-InN:Mg). Figure 4.4 shows three investigated ISFET structures where carriers flowing through the surface, interface, and bulk channels are schematically illustrated. The ISFETs fabricated with 10-nm-thick and 1-μm-thick undoped $-c$-InN epilayers are primarily employed to study the thickness effect on the current variation ratio. The ISFET using an 1.2-μm-thick a-InN:Mg epilayer is applied to demonstrate the Mg-doping effect on the current variation enhancement for an ISFET.

The I_{DS}–V_{GS} characteristics of three investigated InN ISFETs are shown in Figure 4.5. The drain-source currents (I_{DS}), normalized to the current at zero bias, are plotted as functions of gate bias (V_{GS}) with a fixed drain-source voltage (V_{DS}) of 0.25 V, while the gates are biased through a pH 7 buffer solution. The 1-μm-thick $-c$-InN ISFET shows an insensitive response (<0.1%) to the change of gate bias from −0.3 to 0.3 V. The small current modulation ability is attributed to the existence of the bulk channel. For the thick epilayer, the strong electron accumulation at the InN surface with a large electron density of the order of 10^{20} cm^{-3} reduces the field effects, which allows the field-effect affected region only within a few nanometers from the surface (small electrostatic screening length) [40]. Therefore, the carriers in the bulk channel and in the interface channel cannot be efficiently induced or suppressed (i.e., the channel current can only be slightly varied). On the contrary, the 10-nm-thick $-c$-InN ISFET exhibits a current variation of 18% (15%) under the gate bias of 0.3 V (−0.3 V) with respect to the unbiased one, demonstrating the current

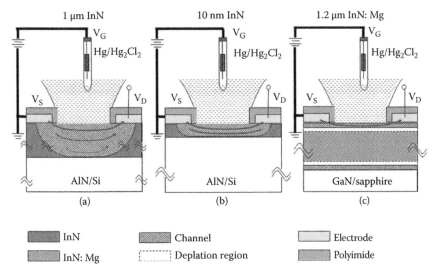

FIGURE 4.4 Schematic diagrams of three investigated ISFET structures, including 1-μm-thick $-c$-InN, 10-nm-thick $-c$-InN, and 1.2-μm-thick a-InN:Mg epilayers.

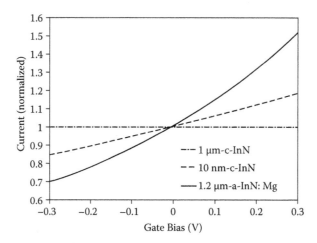

FIGURE 4.5 I_{DS}–V_{GS} characteristics of the open-gate InN-based ISFETs measured in a pH 7 buffer solution with a fixed $V_{DS} = 0.25$ V. The InN used in ISFET fabrication includes 1-μm-thick –c-InN, 10-nm-thick –c-InN, and 1.2-μm-thick a-InN:Mg epilayers. The I_{DS} is normalized with respect to the current at zero gate bias for each ISFET, respectively.

in the ultrathin channel is sensitive to the gate bias. In this case, the carriers are mainly from the surface and the interface channels, which have a much smaller value than that of the 1-μm-thick –c-InN ISFET. As a result, the I_{DS} of the 10-nm-thick –c-InN ISFET is smaller than that of the 1-μm-thick one (0.23 mA versus 2.52 mA at zero gate bias for $V_{DS} = 0.25$ V). When biased by a gate voltage, the 10-nm-thick –c-InN ISFET has a relatively large change of carrier number, and we can obtain a higher current variation ratio than that of the 1-μm-thick –c-InN ISFET. The 1.2-μm-thick a-InN:Mg ISFET shows the current response of 52% and 30% to gate bias of 0.3 and –0.3 V, respectively. Such a high current variation ratio may result from both the surface inversion layer and the depletion region in the near-surface region. The surface inversion carriers (electrons) isolated by the depletion region from the bulk channel are easily depleted upon sufficient negative gate biases. However, the current cannot be completely suppressed because the carriers at the InN–GaN heterointerface are still electrically connected to the surface possibly through the sidewalls or the dislocation threads in the bulk [41].

Compared with the ISFETs based on other III-nitrides, InN-based ISFETs exhibit the relatively high current variation ratio upon the same magnitude of gate-voltage change [5,42]. These results reinforce that the InN-based ISFET is very promising for the chemical sensing with high sensitivity. The chemical responses of the ISFETs fabricated with ultrathin (10 nm) –c-InN and a-InN:Mg to anions, pH, and polar liquids will be introduced in the following sections.

4.3.2.2.2   Anion Sensing

Ultrathin (10 nm) InN-based ISFETs were used to perform chlorine ion sensing in aqueous solutions. As shown in Figure 4.6, the I_{DS}–V_{DS} characteristics of an ultrathin InN-based ISFET reveal a difference between in deionized (DI) water and in 1 M potassium chloride (KCl) aqueous solution. The measured channel current increases with elevating the applied voltage up to 3 V, and it decreases upon exposure from DI water to the KCl aqueous solution at the same V_{DS}. The chlorine ions adsorbed onto the InN surface along with free electrons are balanced by the positively charged surface donors in the accumulation layer. To maintain charge neutrality in the layer, the free electrons within the channel redistribute throughout the InN layer. As a result, lower electron density in the channel decreases the current flowing through the channel. A maximum current difference of 510 μA is observed at 3 V. Besides, the measured current response in DI water shows no saturation phenomenon until breakdown at a current of 57 mA.

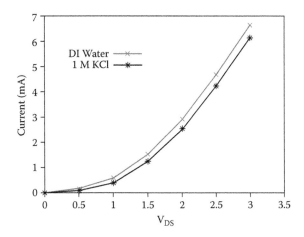

FIGURE 4.6 I_{DS}–V_{DS} characteristics of an ultrathin InN-based ISFET under exposure to DI water and to KCl solution, respectively. The output current (I_{DS}) of the ISFET shows a clear difference between in DI water and in KCl (1 M). A maximum difference of 510 μA is observed at V_{DS}= 3 V.

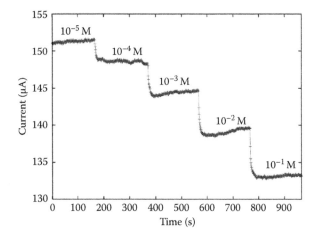

FIGURE 4.7 Dynamic response of an InN-based ISFET in a KCl solution with varying KCl concentration from 10^{-5} M to 10^{-1} M at a fixed V_{DS} = 0.2 V.

Figure 4.7 shows dynamic response of the InN-based ISFET for potassium chloride (KCl) with ion concentration adjusting from 10^{-5} to 10^{-1} M by titration without mechanical stirring. The chlorine ion concentration was calculated based on the molarity (M) assuming the nearly full dissociation of KCl (solubility of 34.0 g/100 cm³ at 20°C). In the measurement, V_{DS} remained unchanged at 0.2 V. The InN-based ISFET revealed an abrupt current change upon exposure to different ion concentrations. The current became smaller as the chlorine ion concentration was increased. The shift of current with respect to the per decade change of chlorine ion concentration was found to be 5 μA, indicating a sensitivity of 5 μA/pCl (pCl denotes –log[Cl⁻]). The response time defined here is the time elapsed after the ion concentrations change from 10% to 90% of interned increase. The response time of one-decade concentration change for the InN ISFET was estimated to be less than 10 s, which is limited by ion diffusion and equilibrium. The detected current in equilibrium fluctuated within 1.2 μA, implying that the tested sensor has a resolution limit of 0.25 pCl for the ISFET operation using the present detection circuitry.

Figure 4.8 shows the semi-log plot of I_{DS} for various volume concentrations of potassium chloride (KCl) solutions ranging from 10^{-5} M to 10^{-1} M at a fixed V_{DS} = 0.2 V. The measured current

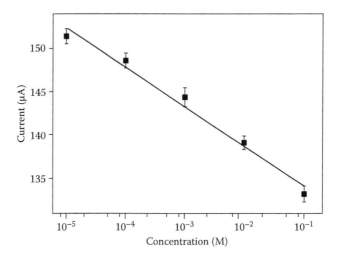

FIGURE 4.8 Current response of InN ISFET versus the logarithm of the concentration of KCl in aqueous solution with a fixed $V_{DS} = 0.2$ V. The tested device has a sensitivity of 5 μA/pCl.

of the InN-based ISFET decreases monotonically with the logarithm of the volume concentration of the chlorine ion, exhibiting a linear dependence on the logarithm of the volume concentration. The average response of the InN-based ISFET to KCl is 5 μA/pCl. At the concentration of 10^{-3} M, the maximum measurement error in our experiment corresponds to a standard deviation of 1.1 μA (i.e., ±0.7%). For the InN case, both charged donors from extrinsic donor impurities and donor-type native surface/bulk defects near the surface region may provide charged sites that selectively adsorb anions on the InN surface. The adsorption of anions to positive-charged donors is balanced by redistribution of free electrons in InN to maintain the charge neutrality condition in the accumulation layer of the ISFET channel. Also, it builds up the corresponding Helmholtz voltage between the InN gate region and the reference electrode. The change of the gate potential modulates the channel current.

In summary, we have shown that the ultrathin InN-based ISFET has a sensitivity of 5 μA per decade of chlorine ion concentration change and a response time of less than 10 s. The performance of the InN-based ISFET is primarily benefits from the unique surface electronic properties of the ultrathin InN film.

4.3.2.2.3 pH Sensing

ISFETs fabricated with ultrathin (10 nm) −c-InN and 1.2-μm-thick a-InN:Mg were used to investigate the chemical response of the bare InN and InN:Mg surfaces to the pH variation. The performance factors of static and dynamic chemical response of InN-based ISFETs such as gate sensitivity, current variation ratio, resolution, and response time are investigated here. The measurement using AC mode is further introduced to reduce the noise from the ambient environments, thus improving the sensing resolution.

The chemical response of InN-based ISFETs was performed by exposing to different pH buffers (Hanna) pH value was adjusted with hydrochloric acid (HCl) and sodium hydroxide (NaOH) in aqueous solution. Figure 4.9 shows the chemical response of InN-based ISFETs as a function of pH value of the aqueous solution with V_{DS} fixed at 0.2 V. The current monotonically decreased with the increase of pH value ranging from 2 to 10 in a step of 1. The average current variation of InN-based ISFETs is 17.1 μA/pH, corresponding to a current variation ratio of 4.12% with respect to the current at pH 7. The high current variation ratio ($\Delta I_{DS}/I_{DS}$) of ultrathin InN ISFETs to pH change suggests the high resolution of the sensor, which could be employed to detect slight pH change. The pH response of the ultrathin InN ISFET was also performed by titration using

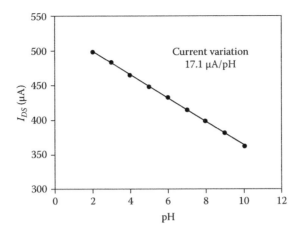

FIGURE 4.9 Current response of an ultrathin ISFET to various pH buffers ranging from pH 2 to pH 10 at a fixed $V_{DS} = 0.2$ V, showing an average current variation of 17.1 μA/pH (variation ratio of 4.12%).

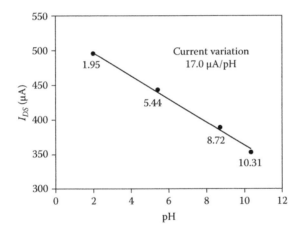

FIGURE 4.10 Current response of an ultrathin ISFET using titration at a fixed $V_{DS} = 0.2$ V, showing an average current variation of 17.0 μA/pH (variation ratio of 4.08%).

HCl and NaOH aqueous solutions, as shown in Figure 4.10. The ISFET was first immersed in the HCl aqueous solution, and then NaOH was added to change the pH value of the solution from acid to base. The pH value was in situ monitored by a pH meter (Hanna HI-111) for the comparison with the measured I_{DS}. The result is in good agreement with that using buffer solutions. The average current variation is 17.0 μA/pH, corresponding to a current variation ratio of 4.08%.

The gate sensitivity of the bare InN surface to pH changes in the aqueous solution was measured using the constant current method, that is, adjusting the gate bias V_{GS} via the reference electrode to compensate the influence of pH changes and return the I_{DS} to one constant value. The magnitude of adjustment on V_{GS} can be used to estimate the induced Helmholtz voltage at the liquid-InN interface in different pH buffer solutions. The constant current set here is the current at pH 7 and, consequently, the derived Helmholtz voltage in the solution of various pH is relative to the one of pH 7. Figure 4.11 shows the V_{GS} as a function of pH value ranging from 2 to 10 for an ultrathin InN-based ISFET. A linear relationship over the entire investigated range was observed, and a gate sensitivity of 58.3 mV/pH was derived. The gate sensitivity of the InN-based ISFET is close to the theoretical Nernst response of 59 mV/pH and similar to those of III-nitride-based pH sensors (57.3 mV/pH for a

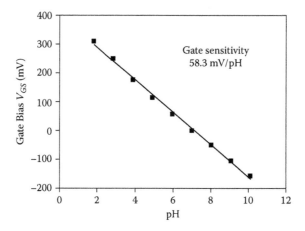

FIGURE 4.11   Gate bias V_{GS} as a function of pH value ranging from 2 to 10 for an ultrathin InN ISFET using the constant current measurement (constant current is set to the current at pH 7). The sensitivity is estimated to be 58.3 mV/pH.

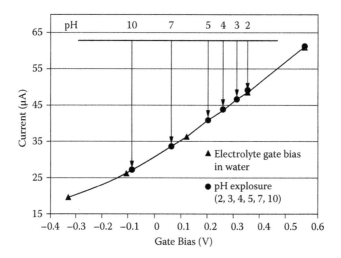

FIGURE 4.12 I_{DS}–V_{GS} characteristics of an a-InN:Mg ISFET in buffer solutions of various pH values ranging from 2 to 10 and in a pH 7 buffer solution with varying gate bias ($V_{DS} = 0.1$ V). The gate sensitivity is calculated to be 56.5 mV/pH using the experimental data.

GaN:Si/GaN:Mg ISFET [5], 56.0 mV/pH for a GaN-capped AlGaN/GaN HEMT [5], and 57.5 mV/pH for a bare AlGaN/GaN HEMT [7]).

The current response of an a-InN:Mg ISFET to different pH buffer solutions and, in comparison, to electrolyte gate bias in a pH 7 buffer solution are shown in Figure 4.12. During the measurement, V_{DS} was fixed at 0.1 V. Upon exposure to pH buffer solutions, a monotonic current increase in the pH range from 2 to 10 was observed. The average current variation with respect to the pH change per decade can be found to be 2.7 µA. With the relation of I_{DS} versus V_{GS} under electrolyte gate bias in DI water, a gate potential change (i.e., gate sensitivity) can be deduced to 56.5 mV/pH, which is similar to those of III-nitride-based pH sensors as mentioned earlier. The dynamic response of the a-InN:Mg ISFET in the pH range from 5 to 2 using titration with HCl solution is shown in Figure 4.13, which revealed abrupt current changes in pH transition. The response time (T_{90}–T_{10}) was estimated to be less than 10 s.

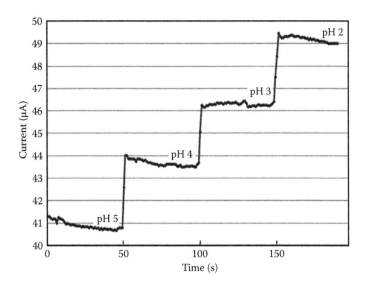

FIGURE 4.13 Dynamic response of an a-InN:Mg ISFET in the pH range from 5 to 2 using titration with HCl solution with a fixed $V_{DS} = 0.1$ V.

To reduce the signal fluctuation coming from the environmental noise and thus improve the resolution, the chemical response of an ultrathin InN-based ISFET was measured in AC mode using a lock-in amplifier (Stanford Research System, SR830). During the measurement, a standard resistor of ~50.0 Ω was connected in a series with an InN-based ISFET, and the voltage drop across the ISFET (V_{FET}) was measured. The frequency and the amplitude of the AC signal were set at 1 kHz and 250 mV_{rms}, respectively. The V_{FET} reveals a linear relationship to buffer solutions of various pH values ranging from 2 to 12, as shown in Figure 4.14. The current fluctuation in dynamic equilibrium is largely reduced, and the resolution of the ISFET can be estimated using the signal-to-noise ratio in this range, which was less than 0.05 pH. The predicted resolution was demonstrated in practice, as shown in Figure 4.15. The response of an ultrathin InN-based ISFET to pH variation in the range from 7.83 to 8.55 and in a step of ~0.05 was measured under the AC mode. The pH change was achieved by using titration with dilute NaOH solution. The resolution obtained by AC-mode measurement, in comparison with that by DC mode, was improved nearly by one order.

4.3.2.2.4 Polarity Sensing

The chemical response of the InN surface to polar liquids, including methanol, isopropyl alcohol (IPA), and acetone was investigated using the ultrathin InN-based ISFETs. The source-drain currents (I_{DS}) at a fixed V_{DS} of 0.25 V were measured for three samples upon exposure to polar liquid. As shown in Figure 4.16, the average I_{DS} of methanol, IPA, and acetone are 466, 462, and 405 μA, respectively, which shows a linear relationship with the value of p/ε of the polar liquid. The result can be explained using the simple Helmholtz model. From the model, the potential drop (ΔV) at the liquid–solid interface due to the molecular dipole of the polar liquid is given as [43]:

$$\Delta V = \frac{Ns\, p\, \cos\theta}{\varepsilon\varepsilon_0} \tag{3.1}$$

where Ns is the density of absorbed dipole on the solid surface per unit area, p is the dipole moment, θ is the angle between the dipole and the surface normal, ε is the dielectric constant of the liquid, and ε_0 is the permittivity of free space. The value of p, ε, and p/ε for the polar liquids investigated are shown in Table 4.1. Given the same absorbed dipole density on the InN surface (Ns) and dipole angle (θ) with

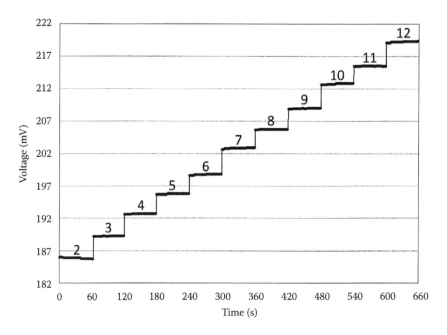

FIGURE 4.14 Response of an InN ISFET to buffer solutions with pH values ranging from 2 to 12 under AC-mode measurement conditions.

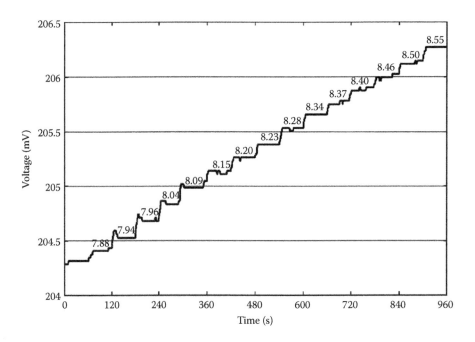

FIGURE 4.15 Transient behavior of the voltage drop across an ultrathin InN-based ISFET to pH variation ranging from 7.83 to 8.55 in a step of ~0.05 pH under AC-mode measurement conditions. The pH change was achieved by titration with a dilute NaOH solution, and which was in situ recorded using a commercial pH meter. The recorded pH value of each step is shown in the plot.

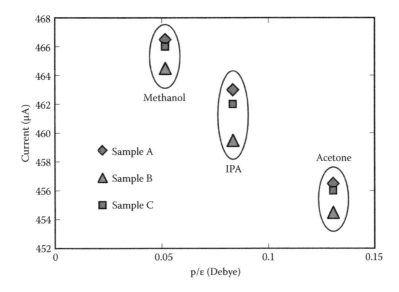

FIGURE 4.16 I_{DS} as a function of p/ε for three ultrathin InN-based ISFETs measured in polar liquids, including methanol, IPA, and acetone.

TABLE 4.1
The dipole moment (p), dielectric constant (ε), and p/ε for methanol, IPA, and acetone

	Methanol	IPA	Acetone
Dipole moment (p)	1.7 ± 0.02	1.58 ± 0.03	2.88 ± 0.03
Dielectric constant (ε)	33	20.18	21.01
p/ε	0.052	0.078	0.137

Source: After *CRC Handbook of Chemistry and Physics* **2008**, 88th ed., Taylor & Francis, London.

the surface normal, the potential drop at the liquid–InN interface should be proportional to p/ε of the polar liquid. Hence, due to the nearly linear relationship of I_{DS} versus V_{GS}, the I_{DS} of the ultrathin InN-based ISFET in the polar liquid should be also proportional to the value of p/ε. The result is consistent with those obtained using III-nitride-based or III-arsenide-based ISFETs [6,7,43].

4.4 SUMMARY

The unusual phenomenon of strong surface electron accumulation layer results in a natural 2DEG at InN surfaces, which makes InN attractive for chemical and biological sensing applications. The anion (Cl$^-$)-sensing capability for bare InN surfaces was demonstrated using an InN-based ISE, which could achieve a lower detection limit of 10^{-3} M. To improve the detection limit, sensitivity, and stability, ISFETs fabricated with the ultrathin (10 nm) InN and InN:Mg were demonstrated. Both types of InN-based ISFETs exhibit the relatively high current variation ratio upon the same magnitude of electrolyte-gate-bias change. Using the ultrathin InN-based ISFET, the lower detection limit of anion (Cl$^-$) sensing was improved to 10^{-5} M. In addition, pH sensing ranging from 2 to 10 was demonstrated, showing a linear relationship between the source-drain current and pH value. A gate sensitivity of 58.3 mV/pH is derived using the constant current measurement, which is close to the theoretical Nernstian response of 59 mV/pH while a gate sensitivity of 56.5 mV/pH is observed for the a-InN:MgISFET. The average current variation of InN-based ISFETs is 17.1 µA/pH, corresponding to a current variation ratio of 4.12%. The resolution of an ultrathin InN-based ISFET can be further improved to less than 0.05 pH using the AC-mode measurement at 1 kHz. For polarity sensing, the current variation of ISFETs shows a linear relationship to the value of p/ε of the polar liquid, which is attributed to the potential drop at the liquid–solid interface due to the molecular dipole of the polar liquid. In comparison with the conventional ISFET using an open-gate Si-based FET, the InN-based ISFET possesses the advantages of simple structure, easy fabrication, and the potential of integration with the read-out circuitry in a same chip for cost-effective sensing applications.

ACKNOWLEDGMENTS

The InN-based chemical sensor investigated at Tsing Hua University was partly supported by National Science Council in Taiwan through the projects NSC 98-3011-P-007-001 and NSC 97-3114-E-007-002.

REFERENCES

1. Bergveld, P. *IEEE Trans. Bio-Med. Eng.* **1970**, *BM17*, 70.
2. Matsuo, T., Esashi, M., and Abe, H. *IEEE Trans. Electron Devices.* **1979**, *26*, 1856–1857.
3. Gimmel, P., Schierbaum, K.D., Gopel, W., Van den Vlekkert, H.H., and De Rooy, N.F. *Sens. Actuators B* **1990**, *1*, 345–349.
4. Neuberger, R., Müller, G., Ambacher, O., and Stutzmann, M. *Phys. Status Solidi A* **2001**, *185*, 85–89.
5. Steinhoff, G., Hermann, M., Schaff, W.J., Eastman, L.F., Stutzmann, M., and Eickhoff, M. *Appl. Phys. Lett.* **2003**, *83*, 177–179.
6. Kokawa, T., Sato, T., Hasegawa, H., and Hashizume, T. *J. Vac. Sci. Technol. B*, **2006**, *24*, 1972–1976.
7. Mehandru, R., Luo, B., Kang, B., Kim, J., Ren, F., Pearton, S.J., Pan, C., Chen, G., and Chyi, J. *Solid State Electron.* **2004**, *48*, 351–353.
8. Kang, B.S., Ren, F., Kang, M.C., Lofton, C., Tan, W., Pearton, S.J., Dabiran, A., Osinsky, A., and Chow, P.P. *Appl. Phys. Lett.* **2005**, *86*, 173502.
9. Kang, B., Pearton, S.J., Chen, J., Ren, F., Johnson, J., Therrien, R., Rajagopal, P., Roberts, J., Piner, E., and Linthicum, K. *Appl. Phys. Lett.* **2006**, *89*, 122102.
10. Kang, B.S., Wang, H.T., Ren, F., and Pearton, S.J. *J. Appl. Phys.* **2008**, *104*, 031101.
11. Alifragis, Y., Georgakilas, A., Konstantinidis, G., Iliopoulos, E., Kostopoulos, A., and Chaniotakis, N. *Appl. Phys. Lett.* **2005**, *87*, 253507.
12. Chaniotakis, N. and Sofikiti, N. *Anal. Chim. Acta*, **2008,** *615*, 1–9.
13. Lu, H., Schaff, W., and Eastman, L. *J. Appl. Phys.* **2004**, *96*, 3577–3579.
14. Kryliouk, O., Park, H.J., Wang, H.T., Kang, B.S., Anderson, T.J., Ren, F., and Pearton, S.J. *J. Vac. Sci. Technol. B* **2005**, *23*, 1891–1894.
15. Chen, C.-F., Wu, C.-L., and Gwo, S. *Appl. Phys. Lett.* **2006**, *89*, 252109.
16. Lu, Y.-S., Huang, C.-C., Yeh, J.A., Chen, C.-F., and Gwo, S. *Appl. Phys. Lett.* **2007**, *91*, 202109.
17. Lu, Y.-S., Ho, C.-L., Yeh, J.A., Lin, H.-W., and Gwo, S. *Appl. Phys. Lett.* **2008**, *92*, 212102.
18. Lu, Y.-S., Chang, Y.-H., Hong, Y.-L., Lee, H.-M. , Gwo, S., and Yeh, J.A. *Appl. Phys. Lett.* **2009**, *95*, 102104.
19. Lu, H., Schaff, W.J., Eastman, L.F., and Stutz, C.E. *Appl. Phys. Lett.* **2003**, *82*, 1736–1738.
20. Rickert, K.A., Ellis, A.B., Himpsel, F.J., Lu, H., Schaff, W., Redwing, J.M., Dwikusuma, F., and Kuech, T.F. *Appl. Phys. Lett.* **2003**, *82*, 3254–3256.
21. Veal, T.D., Mahboob, I., Piper, L.F.J., McConville, C.F., Lu, H., and Schaff, W.J., *J. Vac. Sci. Technol. B* **2004**, *22(4)*, 2175–2178.
22. Mahboob, I., Veal, T.D., Piper, L.F.J., McConville, C.F., Lu, H., Schaff, W.J., Furthmüller, J., and Bechstedt, F. *Phys. Rev. B* **2004**, *69*, 201307.
23. Mahboob, I., Veal, T.D., McConville, C.F., Lu, H., and Schaff, W.J. *Phys. Rev. Lett.* **2004**, *92*, 036804.
24. Li, S.X., Yu, K.M., Wu, J., Jones, R.E., Walukiewicz, W., Ager III, J.W., Shan, W., Haller, E.E., Lu, H., and Schaff, W.J. *Phys. Rev. B.* **2005**, 71, 161201.
25. Bhuiyan, A.G., Hashimoto, A., and Yamamoto, A. *J. Appl. Phys.* **2003**, *94*, 2779–2808.
26. Wu, C.-L., Lee, H.-M., Kuo, C.-T., Chen, C.-H., and Gwo, S. *Phys. Rev. Lett.* **2008**, *101*, 106803.
27. Segev, D. and Van de Walle, C.G. *Europhysics Lett.* **2006**, *76(2)*, 305–311.
28. Van de Walle, C.G. and Segev, D. *J. Appl. Phys.* **2007**, *101*, 081704.

29. Cimalla, V., Lebedev, V., Wang, C.Y., Ali, M., Ecke, G., Polyakov, V.M., Schwierz, F., Ambacher, O., Lu, H., and Schaff, W.J. *Appl. Phys. Lett.* **2007**, 90, 152106.

30. Lebedev, V., Wang, C.Y., Cimalla, V., Hauguth, S., Ali, M., Himmerlich, M., Krischok, S., Schaefer, J.A., Ambacher, O., Polyakov, V.M., and Schwierz, F. *J. Appl. Phys.* **2007**, *101*, 123705.

31. Denisenko, A., Pietzka, C., Chuvilin, A., Kaiser, U., Lu, H., Schaff, W.J., and Kohn, E. *J. Appl. Phys.* **2009**, *105*, 033702.

32. Chang, Y.-H., Lu, Y.-S., Hong, Y.-L., Kuo, C.-T., Gwo, S., and Yeh, J.A. *J. Appl. Phys.* **2010**, *107*, 043710.

33. Lu, H., Schaff, W.J., and Eastman, L.F. *J. Appl. Phys.* **2004**, *96*, 3577.

34. Schalwig, J., Muller, G., Ambacher, O., and Stutzmann, M. *Phys. Status Solidi A* **2001**, *185*, 39–45.

35. Stutzmann, M., Steinhoff, G., Eickhoff, M., Ambacher, O., Nebel, C.E., Schalwig, J., Neuberger, R., and Muller, G. *Diamond Relat. Mater.* **2002**, *11*, 886–891.

36. Chaniotakis, N.A., Alifragis, Y., Konstantinidis, G., and Georgakilas, A. *Anal. Chem.* **2004**, *76*, 5552–5556.

37. Chaniotakis, N.A., Alifragis, Y., Georgakilas, A., and Konstantinidis, G. *Appl. Phys. Lett.* **2005**, *86*, 164103.

38. Wu, C.-L., Wang, J.-C., Chan, M.-H., Chen, T.T., and Gwo, S. *Appl. Phys. Lett.* **2003**, *83*, 4530–4532.

39. Gwo, S., Wu, C.-L., Shen C.-H., Chang, W.-H., Hsu, T.-M., Wang, J.-S., and Hsu, J.-T. *Appl. Phys. Lett.* **2004**, *84*, 3765–3767.

40. Ahn, C.H., Triscone, J.M., and Mannhart, J. *Nature* **2003**, *424 (6952)*, 1015–1018.

41. Brown, G.F., Ager III, J.W., Walukiewicz, W., Schaff, W.J., and Wu, J. *Appl. Phys. Lett.* **2008**, *93*, 262105.

42. Chiang, J.-L., Chen, Y.-C., Chou, J.-C., and Chen, C.-C. *Jpn. J. Appl. Phys.* **2002**, *41*, 541–545.

43. Bastide, S., Butruille, R., Cahen D., Dutta, A., Libman, J., Shanzer, A., Sun, L., and Vilan A. *J. Phys. Chem. B* **1997**, *101 (14)*, 2678–2684.

44. *CRC Handbook of Chemistry and Physics* **2008**, 88th ed., Taylor & Francis, London.

5 ZnO Thin-Film and Nanowire-Based Sensor Applications

Y.W. Heo
School of Materials Science and Engineering, Kyungpook
National University, Daegu, 702-701, Korea

S.J. Pearton
Department of Materials Science and Engineering,
University of Florida, Gainesville, FL 32611

D.P. Norton
Department of Materials Science and Engineering,
University of Florida, Gainesville, FL 32611

F. Ren
Department of Chemical Engineering, University
of Florida, Gainesville, FL 32611

CONTENTS

149

5.1   INTRODUCTION

ZnO has a long history as a material for gas sensing, varistors, piezoelectric transducers, as a light-transmitting electrode in optoelectronic devices, electro-optic modulators, and as a sunscreen.[1] In the last few years, major advances have been made in the areas of conductivity control and availability of high-quality bulk ZnO substrates. This work has refocused attention on ZnO for UV light emitters and transparent electronics. ZnO can be grown at relatively low temperatures on cheap substrates such as glass and has a larger exciton binding energy (~60 meV) than GaN (~25 meV).[2–22] In addition, the advances in growth of ZnO nanorods and nanowires have suggested that these may have applications to biodetection,[22] because of their large surface areas. Finally, incorporation of percent levels of transition metal impurities such as Mn leads to ferromagnetism in ZnO with practical Curie temperatures. This suggests that ZnO may be an ideal host as a spintronic material.

The purpose of this chapter is to provide a summary of recent advances in developing improved control of doping and unintentional impurities during growth of ZnO and also advances in device processing methods. In addition, we review the use of ZnO as a dilute magnetic semiconductor with an ordering temperature above 300 K. Finally, we focus on the development of ZnO nanowires for gas sensing, transparent transistors, low-power signal processing, and UV photodetection.

5.2   BASIC PROPERTIES OF ZnO

Another key advantage of the ZnO system is the availability of heterostructures, which are necessary to produce efficient light-emitting devices with low threshold voltages and electronic devices with enhanced mobility relative to bulk material. Cd substitution on the cation site leads to a reduction in the bandgap to ~3.0 eV.[23] By contrast, substituting Mg on the Zn site can increase the bandgap to approximately 4.0 eV while still maintaining the wurtzite structure. ZnO has a lower electron mobility than GaN, but a slightly higher saturation velocity.[23,24] Electron doping in nominally undoped ZnO has been attributed to Zn interstitials, oxygen vacancies, or hydrogen.[25–30] The intrinsic defect levels that lead to n-type doping lie approximately 0.01–0.05 eV below the conduction band. The optical properties of ZnO, studied using photoluminescence, photoconductivity, and absorption, reflect the intrinsic direct bandgap, a strongly bound exciton state, and gap states due to point defects.[31–34] A strong room temperature near band-edge UV photoluminescence peak at ~3.2 eV is attributed to an exciton state.[35] In addition, visible emission is also observed due to defect

TABLE 5.1
Properties of wurtzite ZnO

Property	Value
Lattice parameters at 300 K	
a_0	0.32495 nm
c_0	0.52069 nm
a_0/c_0	1.602 (ideal hexagonal structure shows 1.633)
u	0.345
Density	5.606 g/cm^3
Stable phase at 300 K	Wurtzite
Melting point	1975°C
Thermal conductivity	0.6, 1–1.2
Linear expansion coefficient(/C)	a_0: 6.5×10^{-6}
	c_0: 3.0×10^{-6}
Static dielectric constant	8.656
Refractive index	2.008, 2.029
Energy gap	3.4 eV, direct
Intrinsic carrier concentration	<10^6 cm^{-3} (max n-type doping >10^{20} cm^{-3} electrons; max p-type doping<10^{17} cm^{-3} holes
Exciton binding energy	60 meV
Electron effective mass	0.24
Electron Hall mobility at 300 K for low n-type conductivity	200 cm^2/V-s
Hole effective mass	0.59
Hole Hall mobility at 300 K for low p-type conductivity	5–50 cm^2/V-s

states. Other transitions, especially near 500 nm are thought to involve various types of intrinsic defects.[36–43]

Table 5.1 shows a compilation of basic physical parameters for ZnO.[22,44] There is still an uncertainty about some of these values. For example, there have been few reports of p-type ZnO, and therefore the hole mobility and effective mass are still being debated. The values for carrier mobility will undoubtedly increase as more control is gained over compensation and defects in the material. In addition, the maximum doping levels achievable will most likely increase for the same reason. These are all important parameters for their effect on device performance.

5.3 DOPANTS IN ZnO

Without question, the most significant problem to overcome in ZnO-related materials for electronic and photonic applications is the difficulty in achieving high p-type doping levels. For ZnO, n-type conductivity is relatively easy to realize via excess Zn or with Al, Ga, or In doping. However, ZnO displays significant resistance to the formation of shallow acceptor levels. Difficulty in achieving bipolar (n- and p-type) doping in a wide-bandgap material is not unusual. There have been several explanations put forward to explain doping difficulties in wide-bandgap semiconductors.[45–47] First, there can be compensation by native point defects or dopant atoms located on interstitial sites. The defect compensates for the substitutional impurity level through the formation of a deep-level trap. In some cases, strong lattice relaxations can drive the dopant energy level deeper within the gap. In other systems, one may simply have a low solubility for the chosen dopant, limiting the accessible extrinsic carrier density.[47] In ZnO, most candidate p-type dopants introduce deep acceptor levels. Copper doping introduces an acceptor level with an energy ~0.17 eV below the conduction band.[48] Silver behaves as an acceptor with a deep level ~0.23 eV below the conduction band.[49] Lithium introduces a deep acceptor, and induces ferroelectric behavior.[50–56]

It appears that the most promising dopants for p-type material are the group V elements, although theory suggests some difficulty in achieving shallow acceptor states.[57,58] For instance, while ab initio electronic band structure calculations for ZnO based on the local density approximation indicate that the Madelung energy decreases with group III cation substitution for n-type doping,[58] the Madelung energy increases with group V anion substitution, indicating significant localization of these acceptor states. However, there have been several reports suggesting acceptor doping levels with group V substitution. Photo-induced paramagnetic resonance studies of N-doped ZnO crystals indicate the presence of an acceptor state due to nitrogen substitution.[59] Minegishi et al. reported the growth of p-type ZnO by the simultaneous addition of NH_3 in hydrogen carrier gas with excess Zn.[60] The resistivity of these films was high, with $\rho \sim 100$ Ω-cm, suggesting that the acceptor level is relatively deep with a subsequent low mobile hole concentration. P-type ZnO has also been reported for films grown by PLD, in which a N_2O plasma is used for doping.[61] Rouleau et al.[62] have addressed nitrogen doping in epitaxial ZnO films in which an RF plasma source was used to crack N_2 in conjunction with Zn evaporation during pulsed-laser deposition from a ZnO target. Clearly, the films contain nitrogen. However, no definitive p-type behavior was observed as determined by Hall measurements, suggesting that compensating complexes are incorporated in the growing film. Similar results have been reported by others.[63] The lack of p-type behavior in nitrogen-doped ZnO has been addressed theoretically within the context of the N-N complex formation.[64] In particular, it appears that the formation of N-N related complexes introduces compensating centers. Isolated substitutional N is assumed to be required to realize the desired acceptor state. As such, the use of source species that contain only one nitrogen atom per entity (NO, N, NO_2) should be more amenable to acceptor state formation due to the large dissociation energy of N_2 (9.9 eV). Recently, promising results were reported on the synthesis of p-type ZnO using molecular beam epitaxy.[65] In this case, homoepitaxial nitrogen-doped ZnO was grown on semi-insulating Li-doped ZnO crystals from a high-purity Zn evaporation source, combined with atomic O and N flux created via an RF plasma. Hall measurements yielded p-type behavior with a hole mobility of 2 cm^2/V-s and a hole concentration of 9×10^{16}/cm^3. The acceptor level was estimated using low-temperature photoluminescence to be 170–200 meV. Clearly, achieving this result was dependent on minimizing the concentration of compensating donor levels from defects or complex formation.

A hole mobility of 0.5 to 3.5 cm^2/V-s and a carrier density of 10^{17} to 10^{19}/cm^3 upon annealing was reported in phosphorus-doped ZnO thin films deposited by sputter deposition.[66] Similarly, annealing of P-doped ZnO films yielded semi-insulating behavior, consistent with activation of a deep acceptor level.[67]

While most efforts on p-type doping of ZnO have focused on nitrogen doping, a few studies have considered other group V elements for substitutional doping on the O site. Given the mismatch in ionic radii seen in Table 5.2 for P (2.12 Å), As (2.22 Å), and Sb (2.45 Å) as compared to O (1.38 Å), solubility of these elements in ZnO should be limited. Nevertheless, p-n junction-like behavior has been reported between an n-type ZnO substrate and a surface layer that was heavily doped with phosphorus.[68] Activation of the P dopant was achieved via laser annealing of a zinc phosphide-coated ZnO single crystal. A related result was reported for epitaxial ZnO films on GaAs subjected to annealing.[69] In this case, a p-type layer was reportedly produced at the GaAs–ZnO interface. Both of these reports are promising, but several unresolved issues related to the solid solubility of the dopant and possible secondary phase formation in the doped region need to be overcome. It should be noted that p-type doping in other II–VI compound semiconductors using dopants with highly mismatched radii has been reported. In particular, a shallow acceptor level is realized in ZnSe doped with nitrogen, despite the large difference in radii between Se (1.98 Å) and N (1.46 Å).

Significant issues remain in p-type doping of ZnO and related alloys. First, reproducibility of reported p-type conduction in ZnO has yet to be demonstrated. Also, a definitive signature of p-type doping in the form of a light-emitting p-n junction has yet to be reported. The role of single N versus multi-N bearing dopants in the incorporation of activated nitrogen must be better understood.[70]

TABLE 5.2
Valence and ionic radii of candidate dopant atoms

Atom	Valence	Radius (Å)
Zn	+2	0.60
Li	+1	0.59
Ag	+1	1.00
Ga	+3	0.47
Al	+3	0.39
In	+3	0.62
O	−2	1.38
N	−3	1.46
P	−3	2.12
As	−3	2.22
F	−1	1.31

This entails a comparison of nitrogen source molecules that produce useful acceptor levels. Since Zn interstitials, O vacancies, and hydrogen complexes have been reported as contributors to compensating electrons,[71–74] this investigation must include understanding the role of oxidizing species in yielding low-native-defect thin-film material. Third, the dopant solubility (likely metastable) and acceptor state location in the gap should be determined for the various group V dopants under consideration. In addition, the effectiveness of codoping in lowering the energy level of the acceptor state has yet to be definitively demonstrated. The background impurity density during growth must be minimized so as to observe the presence of acceptors in transport measurements.

We have recently examined the behavior of phosphorus in ZnO epitaxial films, focusing on the effects of annealing on doping.[73] Pulsed laser deposition was used for film growth. Phosphorus-doped ZnO targets were fabricated using high-purity ZnO (99.9995%), with P_2O_5 (99.998%) serving as the doping agent. The targets were pressed and sintered at 1000°C for 12 h in air. Targets were fabricated with phosphorus doping levels of 0, 1, 2, and 5 at%. A KrF excimer laser was used as the ablation source. A laser repetition rate of 1 Hz was used, with a target-to-substrate distance of 4 cm and a laser pulse energy density of 1–3 J/cm^2. The ZnO growth chamber exhibits a base pressure of 10^{-6} Torr. Single-crystal (0001) Al_2O_3 (sapphire) was used as the substrate material in this study. The substrates were attached to the heater platen using Ag paint. Despite a relatively large lattice mismatch, sapphire is the most commonly used substrate for ZnO epitaxial growth. Film thickness ranged from 300 to 1000 nm. Film growth was performed over a temperature range of 300°C–500°C in an oxygen pressure of 20 mTorr. The transport properties of annealed P-doped ZnO epitaxial films were compared with undoped films grown and annealed under the same conditions. Annealing was carried out at temperatures ranging from 400°C to 700°C in a 100 Torr O_2 ambient for 60 min. Four-point van der Pauw Hall measurements were performed to determine transport properties.

With pulsed-laser deposition, stoichiometric transfer of target material to the substrate is generally accomplished, irrespective of the vapor pressure of individual constituent atoms in the material. Despite the anticipated stoichiometric transfer of doped material, measurements were performed to confirm the phosphorus content in the films. Both P and P_2O_5 sublime at relatively low temperatures (416°C and 360°C, respectively), making it possible that the films deposited and/or annealed at elevated temperatures would not retain the phosphorus content of the ablation targets. To examine this issue, the phosphorus content of the films was measured using energy dispersive spectrometry (EDS) and x-ray photoelectron spectroscopy. Despite the relatively high growth and annealing temperature, the phosphorus content in the ablation target was, in fact, replicated in the film within the experimental accuracy of the two techniques.

X-ray diffraction was used to determine the crystallinity of the phosphorus-doped ZnO films relative to undoped ZnO films. Based on the bulk phase diagram for ZnO-P_2O_5, the solid solubility of phosphorus in ZnO should be limited and metastable. Indications that the solid solubility has been exceeded in the films would be the appearance of impurity phases or a saturation of dopant-relevant material properties with increasing doping. The diffraction data shows only ZnO (000ℓ) and substrate peaks. The films are oriented in a plane as determined by a four-circle x-ray diffraction. Note that some degradation in x-ray diffraction intensity is observed with increasing phosphorus content. No discernible shift in x-ray diffraction peak location was observed upon annealing.

Hall measurements were performed on both the as-deposited and annealed films to delineate the doping behavior of phosphorus in ZnO. For the as-deposited films, the inclusion of phosphorus yields a significant increase in electron density, resulting in ZnO that is highly conductive and an n-type.[73] The shallow donor behavior in the as-deposited films is inconsistent with P substitution on the O site, and presumably originates from either substitution on the Zn site or the formation of a phosphorus-bearing complex. Previous work has shown that the defect-related carrier density in nominally undoped ZnO can be reduced via high-temperature annealing in oxygen or air. In the case of undoped material, the reduction in donor density is presumed due to either a reduction in oxygen vacancies, Zn interstitials, or perhaps out-diffusion of hydrogen that is incorporated in the ZnO lattice during synthesis. In an effort to reduce electron density and elucidate the electronic nature of phosphorus doping in ZnO, annealing studies were performed. Anneals were performed on films with a nominal phosphorus content of 0, 1, 2, and 5 at%. The resistivity of the as-deposited phosphorus doped films was significantly lower than that for the nominally undoped film. For as-deposited films, a shallow donor state dominated transport. As the films were annealed at increasing temperatures, the resistivity of the phosphorus-doped films increased rapidly. This was particularly evident for films subjected to annealing temperatures of 600°C or higher. The increase in resistivity with annealing was more rapid in the films with higher phosphorus content. When annealed at 700°C, the phosphorus-doped ZnO films became semi-insulating with a resistivity approaching 10^4 Ω-cm. The conversion of transport behavior from highly conducting to semi-insulating with annealing should be attributed to at least two factors. First, the defect associated with the shallow donor state in as-deposited films appears to be relatively unstable. This would explain the increase in resistivity, but would alone predict a saturation of resistivity at the value given by the undoped material. This does not explain why the phosphorus-doped films become significantly more resistive than undoped films subjected to the same annealing treatment. The dependence of postannealed resistivity on phosphorus content suggests that a deep level associated with phosphorus dopant is present. This is, in fact, consistent with the expected results that P substitution on the oxygen site yields a deep acceptor.

In addition to resistivity, the Hall coefficients were measured at room temperature for both the unannealed and annealed films. A van der Pauw geometry was used for extracting Hall mobility and carrier concentration. For the experimental Hall measurement system used, the lower limit sensitivity for Hall mobility magnitude was on the order of 1 cm^2/V-s. For all of the samples, the Hall sign, if measurable, was negative, indicating n-type material. For phosphorus-doped films with resistivity greater than approximately 50 Ω-cm, an unambiguous Hall voltage could not be measured. From the measurements yielding an unambiguous Hall voltage, both the carrier density and mobility in phosphorus-doped films are observed to decrease with annealing. This is consistent with a reduction in the shallow donor state density and activation of a deep (acceptor) level in the gap. Note that in all cases where an unambiguous Hall voltage was determined, the sign was negative, indicating an n-type conduction.

5.4 ION IMPLANTATION

Ion implantation is a precise method of introducing dopants into a semiconductor and can also be used to create high-resistance regions for interdevice isolation. Implant doping is in its infancy in

ZnO, and there has been no clear demonstration of activation of an implanted donor or acceptor. The residual implant damage remaining after annealing appears to have a donor-like character. To minimize this damage, it may be necessary to adopt techniques used for other compound semiconductors, such as elevated temperatures during the implant step to take advantage of so-called dynamic annealing, in which vacancies and interstitials created by the nuclear stopping process are annihilated before they can form stable complexes. It may also be necessary to co-implant O in order to maintain the local stoichiometry of the implanted region since the O recoil is much more significant than the Zn recoil.

Kucheyev et al.[74,75] have recently reported a systematic study of implant isolation in n-type ZnO epi layers grown on sapphire. At 350°C, the maximum sheet resistance only reached values of around 10^5 ohms per square, while at 25°C the maximum sheet resistance was as high as 10^{11} ohms per square for samples also implanted at 25°C. This type of defect isolation is stable only to ~300°C–400°C, and the sheet resistance is returned to its preimplantation value by ~600°C. There was no significant chemical effect noted for Cr, Fe, or Ni implantation relative to O, indicating that these elements do not introduce a large concentration of deep acceptors into ZnO. Arrhenius plots of sheet resistance of the O-implanted material yielded activation energies of 15–47 meV. The thermal stability of the defect isolation is much lower than is the case in GaN, where the high resistance persists to anneal temperatures >900°C.

5.5 ETCHING OF ZnO

ZnO is readily etched in many acid solutions, including HNO_3/HCl and HF.[76–79] In most cases, the etching is reaction-limited, with activation energies of >6 kCal.mol.$^{-1}$ Initial results have also appeared on plasma etching of sputter-deposited thin films,[79–82] while plasma-induced damage from high-ion-density Ar or H_2 discharges was found to increase the conductivity of the near surface of similar samples and lead to improved n-type ohmic contact resistivities.[83]

Figure 5.1 shows the ZnO etch rate at 150°C in inductively coupled plasmas of Cl2/Ar as a function of the substrate bias, V_b. The x-axis is plotted as the square root of the average ion energy, which is the plasma potential of ~25 V minus the dc self-bias.[84] A commonly accepted model for an etching process occurring by ion-enhanced sputtering in a collision-cascade process predicts that the etch rate will be proportional to $E^{0.5}-E_{TH}^{0.5}$, where E is the ion energy and E_{TH} is the threshold energy.[85] Therefore, a plot of etch rate versus $E^{0.5}$ should be a straight line with an x-intercept equal to E_{TH}.[86–90] For Cl_2/Ar, this threshold is ~170 eV. In the case of $CH_4/H_2/Ar$, the value of E_{TH} is ~96 eV. The fact that both Cl_2/Ar and $CH_4/H_2/Ar$ exhibit an ion-assisted etch mechanism is consistent with

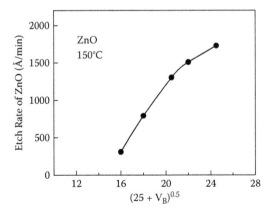

FIGURE 5.1 Etch rate of ZnO in Cl_2/Ar at 150°C as a function of the average ion kinetic energy (plasma potential of +25 V minus the measured dc bias voltage).

the moderate vapor pressure for the expected group II etch product, namely, $(CH_3)_2Zn$, with a vapor pressure of 301 mTorr at 20°C.[83-90] and the high bond strength of ZnO. To form the etch product, the Zn-O bonds must first be broken by ion bombardment. The $ZnCl_2$ etch product has a lower vapor pressure (1 mTorr at 428°C),[82] consistent with the slower etch rates for this chemistry. Given the higher etch rates for $CH_4/H_2/Ar$, we then focused on results with that plasma chemistry.

Figure 5.2 shows 300 K PL spectra from the samples etched in $CH_4/H_2/Ar$ at different rf chuck powers. The overall PL intensity is decreased from both the band edge (3.2 eV) and the deep-level emission bands (2.3–2.6 eV). The magnitude of this decrease scales with ion energy up to approximately −250 eV, and saturates for larger energies. This may be due to more efficient dynamic annealing of ion-induced point defects at higher defect production rates, producing essentially a saturation damage level as is observed in III–V compound semiconductor etching.[90] We did not observe the increase in band-edge intensity and suppression of the deep-level emission reported for H_2 plasma exposure of ZnO,[82,83] suggesting that the Ar ion bombardment component dominates during $CH_4/H_2/Ar$ etching at room temperature.

The near-surface stoichiometry of the ZnO was unaffected by $CH_4/H_2/Ar$ etching, as measured by AES. This indicates that the $CH_4/H_2/Ar$ plasma chemistry is capable of equirate removal of the Zn and O etch products during ICP etching.

Given this result and the fact that the etching occurs through an ion-assisted mechanism, we would expect to observe smooth, anisotropic pattern transfer. Figure 5.3 shows an SEM micrograph

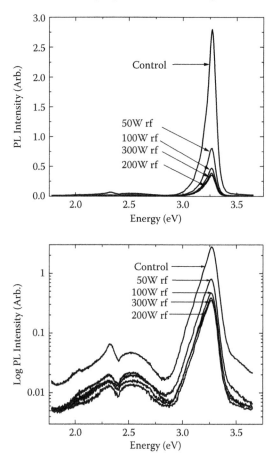

FIGURE 5.2   PL spectra at 300 K from ZnO before and after $CH_4/H_2/Ar$ etching at different rf chuck powers, shown on both log (top) and linear (bottom) scales.

FIGURE 5.3 SEM micrograph of features etched into ZnO using a $BCl_3/Cl_2/Ar$ plasma. The Cr/Ni mask is still in place.

of features etched in ZnO using a Ni/Cr mask and $BCl_3/Cl_2/Ar$ discharge. The vertical sidewalls are an indication of the fact that the etch products are volatile only with additional ion assistance. In addition, the etched field shows only a slight degree of roughening, consistent with the fact that the surface retains its stoichiometry.

The etch mechanism for ZnO in plasma chemistries of $CH_4/H_2/Ar$ and Cl_2/Ar is ion-assisted.[90–92] For both chemistries, the etch rate increases with ion energy as predicted from an ion-assisted chemical sputtering process. The near-surface stoichiometry is unaffected by $CH_4/H_2/Ar$ etching, but the PL intensity decreases, indicating the creation of deep-level recombination centers.

5.6 OHMIC CONTACTS

Low-specific-contact-resistance ohmic metallization is necessary for obtaining good electronic device performance.[93] In ZnO, the usual approaches involve surface cleaning to reduce barrier height or increase of the effective carrier concentration of the surface through preferential loss of oxygen.[93,94] Specific contact resistances of $\sim 3 \times 10^{-4}$ Ω-cm^{-3} were reported for Pt-Ga contacts on n-ZnO epitaxial layers,[93,95] 2×10^{-4} Ω-cm^{-3} for Ti/Au on Al-doped epitaxial layers,[96,97] 0.7 Ω-cm^{-3} for nonalloyed In on laser-processed n-ZnO substrates,[98] 2.5×10^{-5} Ω-cm^{-3} for nonalloyed Al on epitaxial n-type ZnO,[7] 7.3×10^{-3} Ω-cm^{-3} to 4.3×10^{-5} Ω-cm^{-3} for Ti/Au on plasma-exposed, Al-doped n-type epitaxial ZnO[99] and 9×10^{-7} Ω-cm^{-3} for Ti/Al on n$^+$-epitaxial ZnO.[100] Several points are clear from the past works, namely, that the minimum contact resistance generally occurs for postdeposition annealing temperatures of 200°C to 300°C on doped samples, which must be treated so as to further increase the near-surface carrier concentration.

We have found that Ti/Al/Pt/Au contacts on n-type ZnO layers deposited by pulsed laser deposition produce excellent ohmic behavior. The phosphorus-doped ZnO epitaxial films in this study were grown by pulsed-laser deposition (PLD) on single-crystal (0001) Al_2O_3 substrate, using a ZnO:$P_{0.02}$ target and a KrF excimer laser ablation source. The laser repetition rate and laser pulse energy density were 1 Hz and 3 J-cm^{-2}, respectively. The films were grown at 400°C under an oxygen pressure of 20 mTorr. The samples were annealed in the PLD chamber at temperatures ranging from 425°C to 600°C in O_2 ambient (100 mTorr) for 60 min. The resulting film thickness ranged from 350 to 500 nm. Four-point van der Pauw Hall measurements were performed to obtain the carrier concentration and mobility in the films. The carrier concentrations ranged from 7.5×10^{15} to 1.5×10^{20} cm^{-3}, with corresponding mobilities in the range 16–6 cm^2/V-s.[38]

Transmission line method (TLM) patterns, consisting of 100 μm^2 contact pads and gap spacings varying from 5 to 80 μm, were created by dry etching of mesa and lift-off of e-beam evaporated metals. The samples were then deposited with Ti (200 Å)/Al (800 Å)/Pt (400 Å)/Au (800 Å) by e-beam evaporation. After metal deposition, the samples were annealed in a Heatpulse 610 T system at 200°C for 1 min in N$_2$ ambient.

Figure 5.4 shows the dependence of specific contact resistance for as-deposited Ti/Al/Pt/Au and also carrier concentration in the ZnO films on the postgrowth annealing temperature. In general, the contact resistance is lower for higher carrier concentrations, with a minimum value for 8×10^{-7} Ω-cm^2 at a carrier concentration of 1.5×10^{20} cm^{-3}. At this high carrier density, tunneling was found to be the dominant transport mechanism, with the specific contact resistance

$$\rho_C \sim \exp\left[\frac{2\sqrt{\varepsilon_S m^*}}{\hbar}\left(\frac{\phi_{Bn}}{\sqrt{N_D}}\right)\right],$$

where q is the electronic charge, ϕ_{Bn} the barrier height, ε_S the ZnO permittivity, m* the effective mass, $\hbar$ Planck's constant, and N_D the donor density. The strong influence on doping is attributable to the $N_D^{-1/2}$ term. At lower carrier concentrations, temperature-dependent measurements over a fairly limited range (30°C–200°C) showed that thermionic emission was the dominant transport mechanism, with the contact resistance being

$$\rho_C = \frac{k}{qA^*T}\exp\left(\frac{q\phi_{Bn}}{kT}\right),$$

where k is Boltzmann's constant, A* the Richardson constant, and T the measurement temperature. Figure 5.5 shows a summary of the contact resistance data as a function of carrier concentration, for different anneal conditions and measurement temperatures. The lowest contact resistances obtained in the annealed samples, minimum specific contact resistances of 3.9×10^{-7} Ω-cm^2 and 2.2×10^{-8} Ω-cm^2 were obtained in samples with carrier concentrations of 6.0×10^{19} cm^{-3} measured at 30°C and 2.4×10^{18} cm^{-3} measured at 200°C, respectively.

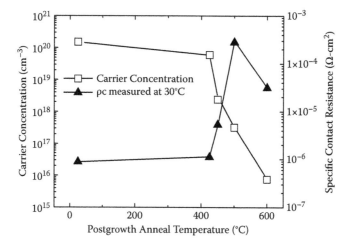

FIGURE 5.4   Carrier concentration of epi-ZnO and specific contact resistance of as-deposited Ti/Al/Pt/Au ohmic contacts measured at 30°C versus postgrowth anneal temperature.

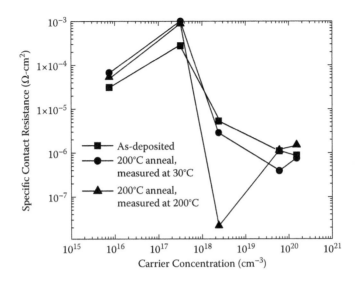

FIGURE 5.5   Specific contact resistance versus carrier concentration measured at 30°C at 200°C.

AES depth profiles of the Ti/Al/Pt/Au contact after annealing at 200°C showed that the initially sharp interfaces between the different metals are degraded by reactions, especially between the Ti and the ZnO to form Ti-O phases[29,30] and between the Pt and Al. We found that anneals at 600°C almost completely intermixed the contact metallurgy. Low thermal stability of both ohmic and Schottky contacts on ZnO appears to be a significant problem in this materials system, and clearly there is a need to investigate refractory metals with better thermal properties if applications such as high-temperature electronics or lasers operating at high current densities are to be realized.

In summary, specific contact resistances in the range 10^{-7}–10^{-8} Ω-cm^2 were obtained for Ti/Al/Pt/Au contacts on heavily n-type ZnO thin films, even in the as-deposited state. However, the contacts show significant changes in morphology even for low-temperature (200°C) anneals, and this suggests that more thermally stable contacts schemes should be investigated.

We have also done some initial work on ohmic contacts to p-type ZnMgO. The phosphorus-doped Zn$_{0.9}$Mg$_{0.1}$O oriented polycrystalline films were grown by PLD on top of a glass substrate, using a ZnO:P$_{0.02}$ target and a KrF excimer laser ablation source. The 600-nm-thick films were grown at 400°C in an oxygen pressure of 20 mTorr. The samples were annealed in the PLD chamber in O$_2$ ambient (100 mTorr) for 60 min to reduce the background n-type conductivity. Type conversion to p-type conductivity was achieved for anneals at 600°C, with hole concentrations in these p-type films on the order of 10^{16} cm^{-3}. Ohmic contact was made by lift-off of e-beam evaporated Ti (200 Å)/Au (800 Å) or Pt Au (800 Å). Circular top Schottky contacts with diameter 100 μm of either 200 Å thick Ti or Ni, topped by 800-Å-thick Au to reduce sheet resistance were patterned by lift-off.

The lowest specific contact resistivity of 3×10^{-3} Ω-cm^2 was obtained for Ti/Au annealed at 600°C for 30 s. Ni/Au was less thermally stable and showed severe degradation of contact morphology at this annealing temperature. Both Pt and Ti with of Au overlayers showed rectifying characteristics on p-ZnMgO, with barrier heights of ~0.55–0.56 eV and ideality factors of ~1.9. Comparison of these results with the same metals on n-type ZnO indicates that high surface state densities play a significant role in determining the effective barrier height.

5.7   SCHOTTKY CONTACT

High-quality Schottky diodes are necessary for some applications such as field-effect transistors or metal-semiconductor-metal photodetectors.[101] It has been found that low-reactive metals such as Au, Ag, and Pd form relatively high Schottky barriers of 0.6–0.8 eV with n-ZnO,[102–108] but the barrier heights do not seem to follow the difference in the work function values, indicating the nonnegligible impact of the interface defects states. The thermal stability of the Schottky diodes on n-ZnO has not been extensively studied, but many authors indicate that for Au, serious problems arise for temperatures higher than 330 K.[104,105,107] At least one study has shown the thermal stability of the Ag/n-ZnO Schottky diodes to be higher than that of Au Schottky diodes.[104] The effect of various surface treatments on the values of reverse currents and of the ideality factor of the forward current characteristics has also not been systematically studied. Neville and Mead[103] reported ideality factors very close to unity ($n = 1.05$) after depositing Au and Pd on the n-ZnO surface etched for 15 min in concentrated phosphoric acid, followed by 5 min etch in concentrated HCl and rinsed in organic solvents. However, similar surface treatments used in other reports led to unacceptably high leakage for Au diodes. Low ideality factors of 1.19 and excellent reverse currents of 1 nA at −1 V were obtained for the Au/n-ZnO Schottky diodes prepared on the O surface of undoped bulk crystals boiled in organic solvents and given a short etch in concentrated HCl.[105] Some authors have reported good electrical performance of the Schottky diodes after etching in concentrated nitric acid.[104] However, it should be noted that in the majority of papers, the ideality factors of the ZnO Schottky diodes are considerably higher than unity, which has been explained by the prevalence of tunneling, the impact of interface states, or the influence of deep recombination centers.

Figure 5.6 compares the 293 K current–voltage (I–V) characteristics of the Au Schottky diodes deposited on our n-ZnO crystals with the surface subjected to three different treatments. Curve 1 is for the surface given a standard organic solvents cleaning by boiling for 3 min in acetone, in trichloroethane TCE, and methanol with deionized DI water rinse and blowing dry with nitrogen (treatment I). Curve 2 was obtained for the Schottky diodes additionally etched in concentrated HCl for 3 min and given a rinse in DI water (treatment II). Curve 3 corresponds to the organic cleaning of the surface and additional etching for 3 min in concentrated HNO$_3$ (treatment III). capacitance–voltage (C–V) measurements yielded cutoff voltages of 0.65 V in each case, in good agreement with the reported 0.65 eV values of the Schottky barrier heights of Au on the (0001) ZnO surface.[102–108] The I–V characteristics of all studied samples showed the ideality factor n to be close to 2 (see Table 5.3) and almost the same in all cases. This is commonly observed for the n-ZnO Schottky diodes and is often ascribed to the prevalence of tunneling. The temperature dependence of the saturation current showed activation energies much lower than those expected from the barrier height obtained from the C–V plots. Thus, tunneling seems to be an important factor in determining the

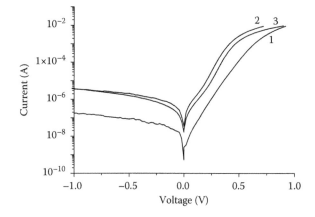

FIGURE 5.6 Room temperature I–V characteristics for Au Schottky diodes prepared on n-ZnO with only organic solvents cleaning (curve 1), with organic solvents cleaning and etching in HCl for 3 min (curve 2), and with organic solvents cleaning and etching in HNO$_3$ for 3 min (curve 3).

TABLE 5.3

Ideality factor n, reverse current at -0.5 V I_r, saturation current density J_s, the activation energy for the saturation current E_a, and the cutoff voltage V_c in C–V for various Au/ZnO and Ag/ZnO Schottky diodes measured at room temperature

Metal	Surface treatment	I_r (A)	n	J_s (A/cm²)	E_a (eV)	V_c (V)
Au	Organic solvents	8.6×10^{-8}	1.8	8.4×10^{-7}	0.35	0.65
Au	HCl	2.1×10^{-6}	1.6	8×10^{-6}	0.4	0.64
Au	HNO₃	1.6×10^{-6}	1.8	4.8×10^{-6}	0.29	0.65
Ag	HCl	2.5×10^{-6}	1.6	1×10^{-5}	0.3	0.69
Ag	HNO₃	2.0×10^{-6}	1.8	6×10^{-6}	0.35	0.68

current flow mechanism in our diodes. DLTS spectra measured on all these diodes showed similar spectra, with the presence of two electron traps with activation energies close to 0.2 and 0.3 eV and a low concentration of about 10^{14} cm^{-3} often observed in undoped n-ZnO.[104,105,107] Thus, the diodes prepared on epiready (0001)Zn surface of undoped bulk n-ZnO crystals without etching show a considerably lower reverse current while having virtually the same ideality factor, C–V characteristics, and DLTS spectra as the diodes with etched surfaces. Anneal temperatures near 370 K were sufficient to produce surface reactions involving Au, leading to the formation of the near-surface layer with reduced electron concentration and a high density of deep-level defects.

Ag Schottky diodes prepared with similar surface treatments showed no meaningful differences with the Au contacts in terms of I–V characteristics, C–V characteristics, DLTS spectra, or thermal stability. Once again, solvent cleaning produced the lowest reverse currents.

To examine the effect of polish damage, we used a diamond slurry to remove approximately 10 μm of material. The resulting surface was smooth and shiny, showing no scratches. This surface was first cleaned in organic solvents and etched for 3 min in concentrated HCl. The electron concentration deduced from C–V measurements gave the value of 6×10^{16} cm^{-3}, close to that in the control material. DLTS spectra showed that, in addition to the 0.2 and 0.3 eV electron traps, polishing introduces a high density of 0.55 and 0.65 eV electron traps whose signatures are similar to the defects created in n-ZnO by proton implantation.[107] Additional HNO₃ etching for 6 min resulted in an I–V characteristic close to that observed on the diode prepared on epiready surface with HNO₃ etching. The electron concentration measured on this diode was close to 10^{17} cm^{-3}. The C–f characteristic also almost returned to its "epiready" position, and the concentration of the 0.55 and 0.65 eV traps in DLTS was decreased by about an order of magnitude. Thus, we conclude that at least 9 min etch in HNO₃ is needed to remove the surface damage introduced by mechanical polishing.

Similar results were obtained for Pt contacts on n-type ZnO. Figure 5.7 shows the temperature dependence of barrier height for both as-deposited and 300°C annealed contacts. The barrier heights range from 0.46 to 0.62 eV for the as-deposited contacts over the temperature range 290–480 K.

The characteristics of device structures that employ phosphorus-doped (Zn,Mg)O have been examined in a effort to delineate the carrier type behavior in this material. The C–V properties of metal/insulator/P-doped (Zn,Mg)O diode structures were measured and found to exhibit a polarity consistent with the P-doped (Zn,Mg)O layer being p-type. In addition, thin-film junctions comprising n-type ZnO and P-doped (Zn,Mg)O display asymmetric I–V characteristics that are consistent with the formation of a p-n junction at the interface. Although Hall measurements of the P-doped (Zn,Mg)O thin films yielded an indeterminate Hall sign due to a small carrier mobility, these results are consistent with previous reports that phosphorus can yield an acceptor state and p-type behavior in ZnO materials. Figure 5.8 (top) shows the C–V characteristics of a structure using undoped n-type ZnO as the semiconductor. In this case, a heavily doped n-type indium-tim-oxide layer served as the bottom electrode. The polarity of the C–V characteristic for the device employing nominally

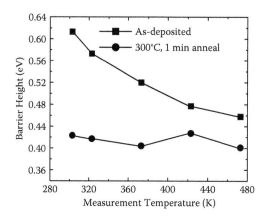

FIGURE 5.7 Pt barrier height on n-type ZnO as a function of measurement temperature for as-deposited and 300°C contacts.

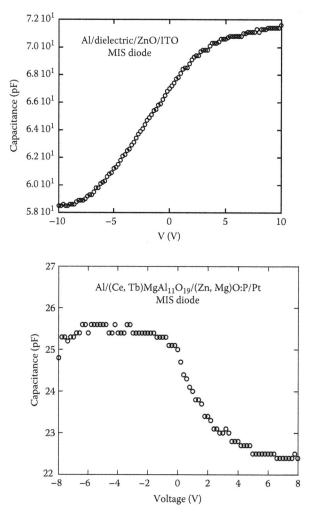

FIGURE 5.8 Capacitance–voltage characteristics of n-type ZnO MIS diode (top) and of P-doped (Zn,Mg) O MIS diode.

undoped ZnO is clearly n-type, with capacitance decreasing with an applied negative voltage. For an MIS diode, the net ionized dopant density, N_A-N_D, can be estimated from the C–V behavior, where N_A is the ionized acceptor density and N_D is the donor density. In particular, for a uniformly doped semiconductor, the high-frequency capacitance/area in the depletion region is given by

$$N_A - N_D = \frac{2}{q\varepsilon_s}\left[\frac{d}{dV}\left(\frac{1}{C^2}\right)\right]^{-1}$$

where q is the electron charge and ε_s is the permittivity of the semiconductor. A plot of $1/C^2$ as a function of V for the diode employing n-type ZnO yielded an ionized donor density is estimated to be $1.8 \times 10^{19}/\text{cm}^3$, which is in agreement with Hall measurements for similar polycrystalline ZnO films.

Similar device structures were then fabricated that employed phosphorus-doped (Zn,Mg)O as the semiconductor material. Figure 5.8 (bottom) shows the C–V behavior of these devices. The symmetry of the C–V curve indicates that the (Zn,Mg)O:P film is p-type. The net acceptor concentration is calculated to be on the order of $2 \times 10^{18}/\text{cm}^3$. If one assumes that all of the phosphorus dopant atoms are substitutional on an oxygen site, one can estimate the activation energy based on a simple hydrogenic model. With this, the activation energy is estimated to be 250–300 meV. In addition, given that the resistivity of similar thin-film material is on the order of 100–1000 Ω-cm, one can estimate the magnitude of the hole mobility to be on the order of 0.01 to 0.001 cm²/V-s. This should not be taken as an intrinsic value of hole mobility for P-doped (Zn,Mg)O, given that the films are polycrystalline with little optimization with respect to transport properties. Nevertheless, this does explain why Hall measurements of the carrier type is difficult in many samples.

The ability to grow ZnO thin films on c-plane Al₂O₃ via molecular beam epitaxy using ozone as an oxygen source with different carrier concentrations is also an advantage in obtaining low contact resistances. Under these conditions, epitaxial growth required high Zn and O₃/O₂ flux rates, with a limited temperature range for film growth. The ZnO films grown using ozone (O₃/O₂) were highly conductive and n-type (Figure 5.9).

The low thermal stability of Au and Ag Schottky contacts on ZnO limits their use in device applications. There is clearly much scope for work on improved rectifying contacts to ZnO, such as W and WSi$_X$.

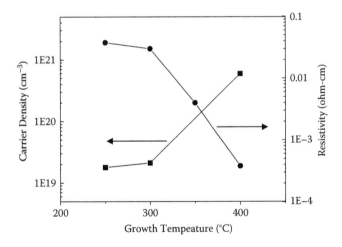

FIGURE 5.9 Carrier concentration and resistivity of ZnO grown on sapphire as a function of growth temperature with a Zn pressure of 2×10^{-6} mbar and O₃/O₂ pressure of 5×10^{-4} mbar.

5.8 PROPERTIES OF HYDROGEN IN ZnO

There is particular interest in the properties of hydrogen in ZnO, because of the predictions from density functional theory and total energy calculations that it should be a shallow donor.[28,30,109,110] The generally observed n-type conductivity, therefore, may at least in theory be explained by the presence of residual hydrogen from the growth ambient, rather than to native defects such as Zn interstitials or O vacancies. There is some experimental support for the predicted observations of its muonium counterpart[29,111] and from electron paramagnetic resonance of single-crystal samples.[112] There have been many other studies on the effects of hydrogen on the electrical and optical properties of ZnO.[70,113–128]

5.8.1 PROTON IMPLANTATION

An investigation into the retention of implanted hydrogen in single-crystal, bulk ZnO as a function of annealing temperature and the effects of the implantation on both the crystal quality and optical properties of the material showed that the temperatures at which implanted hydrogen is evolved from the ZnO are considerably lower then for GaN, which is a more developed materials system for visible and UV light-emitters.

Figure 5.10 shows the SIMS profiles of implanted ^{2}H in bulk ZnO as a function of subsequent annealing temperature. Note that the effects of the annealing are an evolution of ^{2}H out of the ZnO crystal, with the remaining deuterium atoms of each temperature decorating the residual implant damage. The peak in the as-implanted profile occurred at 0.96 μm, which is in good agreement with the projected range from Transfer-of-Ion-in-Matter (TRIM) simulations. The thermal stability of the implanted ^{2}H is considerably lower in ZnO than in GaN,[120,121] where temperatures of ~900°C are needed to remove deuterium to below the detection limit (~3 × 10^{15} cm^{-3}) of SIMS, and this suggests that slow-diffusing H$_2$ molecules or larger clusters do not form during the anneal. Since we did not observe conventional out-diffusion profiles, we were unable to estimate a diffusion coefficient for

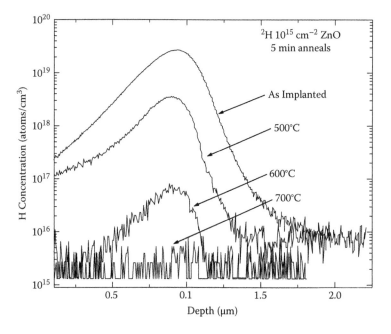

FIGURE 5.10 SIMS profiles of ^{2}H implanted into ZnO (100 keV, 10^{15} cm^{-2}) before and after annealing at different temperatures (5 min anneals).

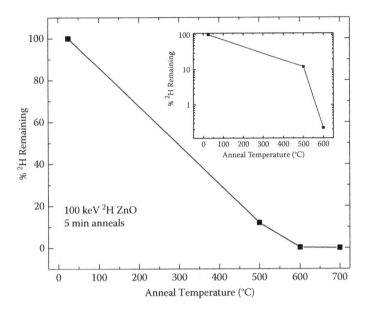

FIGURE 5.11 Percentage of retained ^{2}H implanted into ZnO (100 keV, 10^{15} cm^{-2}) as a function of annealing temperature (5 min anneals). The inset shows the data on a log scale.

the ^{2}H in ZnO. Our results are consistent with an implant-damaged trap-controlled release of ^{2}H from the ZnO lattice for temperatures >500°C.

Figure 5.11 shows the percentage of ^{2}H remaining in the ZnO as a function of anneal temperature. The ^{2}H concentrations were obtained by integrating the area under the curves in the SIMS data. It is evident that the thermal stability of implanted deuterium is not that high, with 12% of the initial dose retained after 500°C anneals and ~0.2% after 600°C anneals.

RBS/C showed that implantation of ^{1}H, even to much higher doses (10^{16} cm^{-2}), did not affect the backscattering yield near the ZnO surface. However, there was a small (but detectable) increase in scattering peak deeper in the sample, in the region where the nuclear energy loss profile of 100 keV H$^+$ is a maximum. The RBS/C yield at this depth was ~6.5% of the random level before H$^+$ implantation and ~7.8% after implantation to the dose of 10^{16} cm^{-2}.

While the structural properties of the ZnO were minimally affected by the hydrogen (or deuterium) implantation, the optical properties were severely degraded. Room-temperature cathodoluminescence showed that even for a dose of 10^{15} cm^{-2} ^{1}H$^+$ ions, the intensity of near-gap emission was reduced by more than 3 orders of magnitude as compared to control values. This is due to the formation of effective nonradiative recombination centers associated with ion-beam-produced defects. Similar results were obtained from PL measurements. The band-edge luminescence was still severely degraded even after 700°C anneals, where the ^{2}H has been completely evolved from the crystal. This indicates that point defect recombination centers are still controlling the optical quality under these conditions. Kucheyev et al.[74,75] have found that the resistance of ZnO can be increased by about 7 orders of magnitude as a result of trap introduction by ion irradiation.

The results of deep-level transient spectroscopy (DLTS) and capacitance–voltage (C–V) profiling studies of undoped n-ZnO[123,124] showed that the carrier removal rate in this material is considerably lower than in other wide-bandgap semiconductors such as GaN. The dominant electron traps introduced by such irradiation were the $E_c - 0.55$ eV and the $E_c - 0.78$ eV centers. Implant isolation studies have shown low thermal stability (350°C) for the electrical effects of implantation.[74,75]

Proton implantation was performed with an energy of 50 keV with doses of 5×10^{13}–5×10^{15} cm^{-2}. Figure 5.12 presents the room-temperature carrier concentration profile obtained from C–V measurements at 1 MHz on the virgin sample. The electron concentration is about 9×10^{16} cm^{-3}, which is in

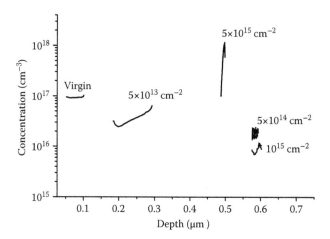

FIGURE 5.12   Room temperature electron concentration profiles deduced from 1 MHz C–V measurements on Au Schottky diodes prepared on the virgin sample, on the sample implanted with 5×10^{13} cm^{-2} 50 keV protons, on the samples implanted with 5×10^{14} cm^{-2} and 10^{15} cm^{-2} 100 keV protons (through the Au metal already in place), and on the sample implanted with the dose of 5×10^{15} cm^{-2} 50 keV protons.

reasonable agreement with the van der Pauw data. Implantation with 5×10^{13} cm^{-2} 50 keV protons led to a quite measurable decrease of the carrier concentration. For the dose of 5×10^{15} cm^{-2}, the capacitance dependence on voltage was very slight, and the apparent concentration deduced from the slope of the $1/C^2$ versus voltage curve was quite high, on the order of 10^{18} cm^{-3}. The depth of this region of high electron concentration was close to the estimated range of 50 keV protons in ZnO. The apparent room-temperature electron concentrations are close to 2×10^{16} cm^{-3} for the dose of 5×10^{14} cm^{-2} and close to 8×10^{15} cm^{-3} for the dose of 10^{15} cm^{-2}. At lower measurement temperatures (85 K), the profiles looked similar, with slightly lower absolute concentrations due to the freeze-out of some of the deeper shallow donors (admittance spectroscopy data suggests that these frozen-out donors have an activation energy of about 70 meV). The 5×10^{15} cm^{-2} dose sample still showed a high electron concentration wall near the end of the range of implanted protons, exceeding the initial shallow donors' concentration. This suggests a strong freeze-out of the major centers in the implanted region occurred at low temperature, leaving the sample depleted of carriers down to the high-electron concentration region near the end of the protons' range. Illumination of the proton-implanted samples with UV light generated by a deuterium lamp led to the persistent increase of capacitance of these samples and to persistent changes in the measured profiles as though some donors were activated within the part of the space charge region (SCR) that had been depleted in the dark (compare the dark and persistent photocapacitance [marked as PPC] profiles in Figure 5.13). The changes in capacitances persisted for many hours at 85 K and could still be observed even at room temperature. Two electron traps with apparent activation energy of 0.2 eV and 0.3 eV and very low concentration of 7×10^{13} cm^{-3} and 1.8×10^{14} cm^{-3}, respectively, were observed. Implantation of the sample with 50 keV protons to the dose of 5×10^{13} cm^{-2} had little effect on the density of these two traps but introduced two additional traps with activation energy 0.55 eV and 0.75 eV with concentrations of 3.7×10^{14} cm^{-3} and 3.75×10^{14} cm^{-3}. At higher temperatures, three traps were observed, namely, the 0.55 eV traps detected in the 5×10^{13} cm^{-2} dose implanted sample (however, the signal from this trap was very weak, most likely because the freeze-out of carriers was still a factor at the temperatures of the peak), the 0.75 eV traps with the peak near 280 K, and 0.9 eV traps with the peak near 370 K. The carrier removal rate deduced from room-temperature C–V profiling was significantly higher than the removal rate reported for 1.8 MeV protons[123,124] The range of the 50 keV protons is about 50 times shorter than the range of the 1.8 MeV protons and, consequently, the actual density of radiation defects causing the carriers to be removed should be much higher in the former case.

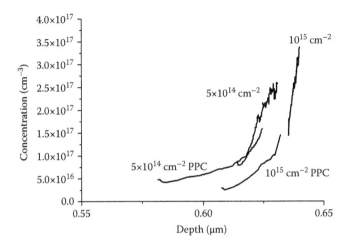

FIGURE 5.13 Electron concentration profiles measured at 85 K on two samples implanted with 100 keV protons through the Au Schottky diode metal with doses of 5×10^{14} cm^{-2} and 10^{15} cm^{-2}; the profiles were measured in the dark (the curves labeled with the dose values) and after exposure to the deuterium UV lamp light and a wait of 15 min (the curves marked by the dose value and PPC label).

5.8.2 HYDROGEN PLASMA EXPOSURE

Bulk ZnO samples were exposed to ^{2}H plasmas at temperatures of 100°C–300°C at 900 mTorr with 50 W of 13.56 MHz power. Some of these samples were subsequently annealed at temperatures up to 600°C under flowing N$_2$ ambients for 5 min. Figure 5.14 shows SIMS profiles of ^{2}H in plasma-exposed ZnO, for different sample temperatures during the plasma treatment. The profiles follow those expected for diffusion from a constant or semi-infinite source, that is,

$$C(x,t) = C_O \, erfc\left(\frac{X}{\sqrt{4Dt}}\right)$$

where $C(x,t)$ is the concentration at a distance x for diffusion time t, C_o is the solid solubility, and D is the diffusivity of ^{2}H in ZnO.[129] The incorporation depths of ^{2}H are very large compared to those in GaN or GaAs under similar conditions, where depths of 1–2 μm are observed.[121,122] It is clear that hydrogen must diffuse as an interstitial, with little trapping by the lattice elements or by defects or impurities. The position of H in the lattice after immobilization has not yet been determined experimentally, but from theory the lowest energy states for H$^+$ are at a bond-centered position forming an O-H bond, while for H$_2$ the antibonding Zn site is most stable.[30]

Using a simple estimate of the diffusivity D, from $D = X^2/4t$, and where X is taken to be the distance at which ^{2}H concentration has fallen to 5×10^{15} cm^{-3} in Figure 5.14, we can estimate the activation energy for diffusion. The extracted activation energy, E_a, is 0.17 ± 12 eV for ^{2}H in ZnO. Note that the absolute diffusivities of ^{1}H would be ~40% larger because of the relationship for diffusivities of isotopes, that is,[129]

$$\frac{D_{1_H}}{D_{2_H}} = \left(\frac{M_{2_H}}{M_{1_H}}\right)^{1/2}$$

The small activation energy is consistent with the notion that the atomic hydrogen diffuses in interstitial form.

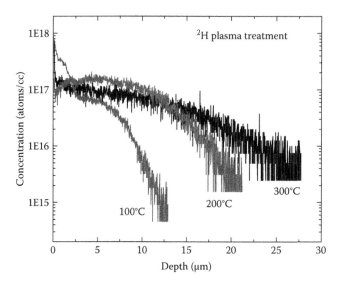

FIGURE 5.14   SIMS profiles of ^{2}H in ZnO exposed to deuterium plasmas for 0.5 h at different temperatures.

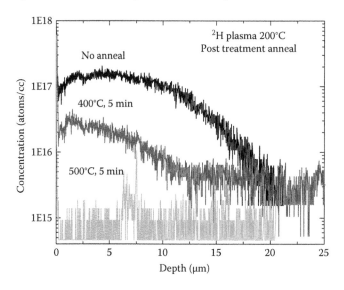

FIGURE 5.15   SIMS profiles of ^{2}H in ZnO exposed to deuterium plasma for 0.5 h at 200°C and then annealed at 400°C or 500°C for 5 min.

Figure 5.15 shows SIMS profiles of a ZnO sample exposed to a ^{2}H plasma of 0.5 h at 200°C, then annealed for 5 min under N$_2$ at different temperatures. There was significant loss of ^{2}H even after a short anneal at 400°C, with virtually all of it having evolved out of the crystal by 500°C. This is in sharp contrast to ^{2}H in GaN, where much higher temperatures ($\geq$800°C) are needed to evolve the deuterium out of the sample.[121,122]

Lavrov et al.[130] have identified two hydrogen-related defects in ZnO, by using local vibrational mode spectroscopy. The H-I center consists of a hydrogen atom at the bond-centered site, while the H-II center contains two inequivalent hydrogen atoms bound primarily to two oxygen atoms.[130]

The ^{2}H plasma treatment causes an increase in donor concentration, consistent with past reports.[29,111] In that case, the effect was attributed to hydrogen passivation of compensating acceptor impurities present in the as-grown ZnO epitaxial layers.[1] An alternative explanation is that the hydrogen induces a donor state and thereby increases the free electron concentration.[2] Subsequent annealing reduces the carrier density to slightly below the initial value in the as-received ZnO, which may indicate that it contained hydrogen as a result of the growth process. We emphasize that the n-type conductivity probably arises from multiple impurity sources,[131,132] and we cannot unambiguously assign all of the changes to hydrogen.

Hydrogen is found to exhibit a very rapid diffusion in ZnO when incorporated by plasma exposure, with a D of 8.7×10^{-10} cm²/V-s at 300°C. The activation energy for diffusion is indicative of interstitial motion. All of the plasma-incorporated hydrogen is removed from the ZnO by annealing at ≥500°C. When the hydrogen is incorporated by direct implantation, the thermal stability is somewhat higher, due to trapping at residual damage. The free electron concentration increases after plasma hydrogenation, consistent with the small ionization energy predicted for H in ZnO[1] and the experimentally measured energy of 60 ± 10 meV for muonium in ZnO.[30] The electrical activity and rapid diffusivity of H or ZnO must be taken into account when designing device fabrication processes such as deposition of dielectrics using SiH_4 as a precursor or dry etching involving the use of $CH_4/H_2/Ar$ plasmas since these could lead to significant changes in near-surface conductivity.

For many other semiconductor materials, it has been shown that the hydrogen introduced, for example, from a plasma at relatively low temperatures, readily forms complexes with various donor and acceptor species and often renders them electrically neutral (the phenomenon widely known as *hydrogen passivation*). There seems to be a general agreement that the hydrogen can be relatively easily introduced from a hydrogen plasma at moderate temperatures of 200°C–300°C. At least two papers report considerable increase in electron concentration in the near-surface regions of the hydrogen-plasma-treated n-ZnO bulk crystals and thin films.[70,125] In Reference 70, the authors present arguments in favor of this phenomenon being due to passivation of compensating acceptors rather than to the hydrogen-producing shallow donors. The authors of Reference 125 argue that there can be several types of shallow donors in n-ZnO, and hydrogen introduced from plasma can be one such shallow donor.

Figure 5.16 shows the room-temperature electron concentration profiles obtained from C–V measurements at 1 MHz on the virgin n-ZnO sample and on the sample after hydrogen plasma exposure. The profile for the control sample is flat and shows the concentration of uncompensated shallow donors as 9×10^{16} cm⁻³, which is in good agreement with van der Pauw measurements made on this sample (room-temperature electron concentration of 9×10^{16} cm⁻³ with a mobility of 190 cm²/V-s).

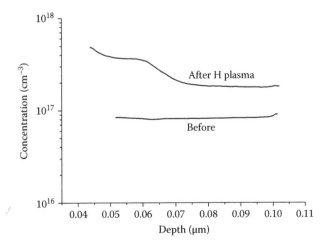

FIGURE 5.16 Room temperature electron concentration profiles measured on the n-ZnO sample by C–V profiling at 1 MHz before and after the hydrogen plasma treatment.

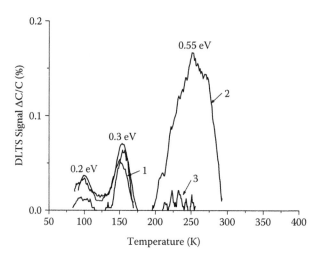

FIGURE 5.17   DLTS spectra measured on the n-ZnO sample before (curve 1) and after (curves 2 and 3) exposure to hydrogen plasma.

After the plasma exposure, the concentration of donors became quite high, close to 4×10^{17} cm^{-3}, near the surface and then showed a plateau at the level of 1.8×10^{17} cm^{-3} down to a depth of about 0.15 μm available to C–V measurements without causing an electrical breakdown. One can observe a striking similarity in the SIMS profile of hydrogen and the profile of the uncompensated shallow donors measured by C–V.

The DLTS spectrum (Figure 5.17) was dominated by two peaks corresponding to the electron traps with apparent activation energies of 0.2 and 0.3 eV. The concentrations of these traps were 8×10^{13} cm^{-3} and 1.4×10^{14} cm^{-3}. The second trap at $E_c^{-0.3}$ eV seems to be the same trap as the E2 center and is often dominant in DLTS spectra measured by several groups.[123–124]

One feature of the hydrogen plasma treatment results for our ZnO samples is the lack of the hydrogen passivation effect for deep electron and hole traps, which is the hallmark of hydrogen passivation phenomena in many other semiconductors. We have shown that hydrogen plasma treatment of good-quality n-ZnO crystals leads to about a factor of 2 increase of the electron concentration in the surface region of the samples, which is almost in one-to-one correspondence with the density of introduced hydrogen atoms according to SIMS measurements. This increase correlates with the similar increase of the intensity of the bound exciton 3.35 eV line, suggesting that the type of shallow donors introduced by hydrogen plasma treatment is the same as in the virgin sample. From admittance spectroscopy measurements and from comparison with the results of shallow donors studies in undoped ZnO, we associate these donors with the 37 meV donors and attribute them to either the hydrogen donors per se or to hydrogen donor complexes with native defects.

5.9   FERROMAGNETISM IN ZnO

The manipulation of spin in semiconductors (so-called spintronics) presents a new paradigm for functionality in electronic materials. Ideas currently under development offer intriguing opportunities to pursue novel device concepts based on the discrimination and manipulation of spin distributions. Of the semiconducting materials, ZnO offers significant potential in providing charge-, photonic-, and spin-based functionality. ZnO is a direct, wide-bandgap semiconductor with potential utility in UV photonics and transparent electronics. For spintronics, theoretical predictions suggest that room-temperature carrier-mediated ferromagnetism should be possible in ZnO, albeit for p-type material. Unfortunately, the realization of p-type ZnO has proved difficult. Ab initio calculations do predict ferromagnetism in n-type ZnO doped with most transition metal ions, including

Co and Cr, but predict no ferromagnetism for Mn-doped ZnO. This is consistent with experimental results, where ferromagnetism is not observed in Mn-doped ZnO that is n-type due to group III impurities. However, we recently observed ferromagnetism in n-type Mn-implanted, Sn-doped ZnO crystals, where Sn (a group IV element) serves as a doubly ionized donor impurity. The Curie temperature is quite high, approaching 250 K. It is unclear what special role the Sn dopant plays, as compared to group III elements, in enabling ferromagnetism in the Mn-doped ZnO system. The Sn may simply provide carriers, albeit electrons, that effectively mediate the spin interactions. The Sn dopant might alternatively form complexes with Mn, resulting in both Mn^{+2} and Mn^{+3} sites that could yield a ferrimagnetic ordering. Nevertheless, ferromagnetism is observed in Mn-doped ZnO (co-doped with Sn).

In recent years, the injection and manipulation of spin-polarized electrons in semiconductor materials has become the focus of intense research activity.[133–135] A functional spin offers an opportunity to develop new device concepts based on the discrimination and manipulation of spin distributions. Fundamental challenges related to the lifetime, manipulation, and detection of spin-polarized electrons in a semiconductor host must be addressed in order to realize spintronic technology based on semiconductors. Of particular importance is the realization of spin-polarized currents in a semiconductor matrix. Efforts directed at investigating the injection of electrons through a ferromagnetic metal–semiconductor junction have shown this structure to be ineffective. However, recent experiments indicate that injection of spin-polarized electrons from a ferromagnetic semiconductor into a nonferromagnetic semiconductor is possible without detrimental interface scattering. Unfortunately, ferromagnetism in semiconductors is rare and poorly understood, with ferromagnetic transition temperatures well below room temperature for most known materials. Clearly, the discovery of ferromagnetism above 300 K in a semiconductor would prove enabling in realizing practical spin-based electronic devices. More generally, the study of any semiconducting material that supports spin-polarized electron distributions would be useful in understanding and developing spintronic concepts.

5.9.1 Ferromagnetism in Semiconductors

Magnetism in semiconductor materials has been studied for many years, and includes spin glass and antiferromagnetic behavior in Mn-doped II–VI compounds, as well as ferromagnetism in europium chalcogenides and Cr-based spinels.[135–137] In recent years, ferromagnetism in semiconductors has received renewed attention, partly due to interest in spintronic device concepts.[133] Due to transition metal solubility and technological interest, contemporary research has primarily focused on II–VI and III–V materials. Strong ferromagnetic interaction between localized spins has been observed in Mn-doped II–VI compounds with high carrier densities. In $Pb_{1-x-y}Sn_yMn_xTe$ (y > 0.6), which possesses a hole concentration on the order of 10^{20}–10^{21} cm^{-3}, ferromagnetism has been achieved.[138] In addition, it has been shown that free holes in low-dimensional structures of II–VI dilute magnetic semiconductors and can induce ferromagnetic order.[139] For many semiconductor materials, the bulk solid solubility for magnetic and/or electronic dopants is not conducive to the coexistence of carriers and spins in high densities. However, the low solubility for transition metals in semiconductors can often be overcome via low-temperature epitaxial growth. This approach has been used with Mn-doped GaAs[140] in achieving ferromagnetism with a transition temperature of 110 K, which is remarkably high compare to traditional dilute magnetic semiconductor material. Behavior indicative of ferromagnetism at temperatures above 300 K has recently been reported for GaN and chalcopyrite semiconductors doped with transition metals, illustrating the potential of achieving room temperature spintronics technologies.[140,141]

Despite recent experimental success, a fundamental description of ferromagnetism in semiconductors remains incomplete. Recent theoretical treatments have yielded useful insight into fundamental mechanisms.[143] Dietl et al.[143,144] have applied Zener's model for ferromagnetism, driven by exchange interaction between carriers and localized spins, to explain the ferromagnetic transition temperature in III–V and II–VI compound semiconductors. The theory assumes that ferromagnetic correlations mediated by holes from shallow acceptors in a matrix of localized spins in a magnetically doped semiconductor. Specifically, Mn ions substituted on the group II or III site provide the local spin. In the case of III–V semiconductors, Mn also provides the acceptor dopant. High concentrations of holes are believed to mediate the ferromagnetic interactions among Mn ions. Direct exchange among Mn is antiferromagnetic, as observed in fully compensated (Ga,Mn)As that is donor-doped. In the case of electron-doped or heavily Mn-doped materials, no ferromagnetism is detected. Theoretical results suggest that carrier-mediated ferromagnetism in n-type material is relegated to low temperatures, if it occurs at all, while it is predicted at higher temperatures for p-type materials.[144] In p-type GaAs grown by low-temperature MBE, Mn doping in the concentration range $0.04 \leq x \leq 0.06$ results in ferromagnetism in GaAs. The model described has been reasonably successful in explaining the relatively high transition temperature observed for (Ga,Mn)As.

Carrier-mediated ferromagnetism in semiconductors is dependent on the magnetic dopant concentration as well as on the carrier type and carrier density. As these systems can be envisioned as approaching a metal–insulator transition when carrier density is increased and ferromagnetism is observed, it is useful to consider the effect of localization on the onset of ferromagnetism. As carrier density is increased, the progression from localized states to itinerant electrons is gradual. On the metallic side of the transition, some electrons populate extended states, while others reside at singly occupied impurity states. On crossing the metal–insulator boundary, the extended states become localized, although the localization radius gradually decreases from infinity. For interactions on a length scale smaller than the localization length, the electron wavefunction remains extended. In theory, holes in extended or weakly localized states could mediate the long-range interactions between localized spins. This suggests that for materials that are marginally semiconducting, such as in heavily doped semiconducting oxides, carrier-mediated ferromagnetic interactions may be possible.

This theoretical treatment presents several interesting trends and predictions. For the materials considered in detail (semiconductors with zinc-blende structure), magnetic interactions are favored in hole-doped materials due to the interaction of Mn^{+2} ions with the valence band. This is consistent with previous calculations for the exchange interaction between Mn^{+2} ions in II–VI compounds[135,145] showing that the dominant contribution is from two-hole processes. This superexchange mechanism can be viewed as an indirect exchange interaction mediated by the anions, thus involving the valence band.[146,147] Note that valence band properties are primarily determined by anions in II–VI compounds. The model by Dietl et al. predicts that the transition temperature will scale with a reduction in the atomic mass of the constituent elements due to an increase in p-d hybridization and a reduction in spin-orbit coupling. Most importantly, the theory predicts a T_c greater than 300 K for p-type GaN and ZnO, with T_c dependent on the concentration of magnetic ions and holes. Recent experimental evidence for ferromagnetism in GaN appears to substantiate the theoretical arguments.[142,146,147]

5.9.2 Spin Polarization in ZnO

As pointed out earlier, Dietl's theory predicts room temperature ferromagnetism for Mn-doped p-type ZnO. In addition to Dietl's prediction, ferromagnetism in magnetically doped ZnO has been theoretically investigated by ab initio calculations based on local density approximation.[148] Again, the results suggest that ferromagnetic ordering of Mn is favored when mediated by hole doping. However, for V, Cr, Fe, Co, and Ni dopants, ferromagnetic ordering in ZnO is predicted to occur without the need for additional charge carriers. Several groups have investigated the magnetic properties of TM-doped ZnO. In all of these studies, the ZnO material was n-type. The magnetic properties of Ni-doped ZnO thin films were reported.[149] For films doped with 3 to 25 at% Ni, ferromagnetism was observed at 2 K. Above 30 K, superparamagnetic behavior was observed. Fukumura et al. have shown that epitaxial thin films of Mn-doped ZnO can be obtained by pulsed-laser deposition, with Mn substitution as high as 35% while maintaining the wurtzite structure.[150] This is well above the equilibrium solubility limit of ~13%, and illustrates the utility of low-temperature epitaxial growth in achieving metastable solubility in thin films. Co-doping with Al resulted in n-type material with carrier concentration in excess of 10^{19} cm^{-3}. Large magnetoresistance was observed in the films, but no evidence for ferromagnetism was reported. However, Jung et al. recently reported ferromagnetism in Mn-doped ZnO epitaxial films, with a Curie temperature of 45 K.[151] The discrepancy appears to lie in differing film-growth conditions.

Recently, researchers have utilized ion implantation to survey the magnetic properties of a number of transition metal dopants in various semiconducting oxide materials. Among the system investigated, the researchers have observed high-temperature ferromagnetism in ZnO crystals that are implanted with transition metal dopants, including Co and Mn, co-doped with Sn. In the case of Mn co-doped with Sn, the Sn ions are provided as doubly ionized donors. This result differs from that reported for Mn-doped ZnO doped n-type with Al or Ga, and suggests that group IV dopants may behave differently from shallow group III donors in terms of interaction with magnetic dopants. If carrier-mediated mechanisms are responsible, one must explain why the behavior depends on the specific cation dopant species chosen (Sn versus Al,Ga). Insight into this issue may reside in the fact that doping via a multi-ionized impurity likely introduces relatively deep donor levels in the energy gap. Conduction from deep donors is often due to impurity band and/or hopping conduction, as opposed to conventional free electrons excited to the conduction band. Any carrier-mediated processes would be dependent on the relevant conduction mechanisms. This result appears to contradict the expectation that, in the absence of a shallow acceptor level, the dominant exchange mechanism is short-range superexchange which, for Mn^{+2} in ZnO, should favor antiferromagnetic ordering. The results will be discussed in detail in a later section. Despite the uncertainty in the mechanism, these results (high-temperature ferromagnetism in Co- and Mn, Sn-doped ZnO) indicate a pathway for exploring spintronics in ZnO materials.

Table 5.4 shows the valence and ionic radii for a number of dopant candidates. The ionic radius of Mn^{+2} (0.66 Å) is relative close to that of Zn (0.60 Å), suggesting moderate solid solubility without phase segregation. As such, the primary transition metal dopant of interest will be Mn. Chromium and cobalt present the possibility of achieving ferromagnetism in ZnO via doping with magnetic ions for which the net superexchange coupling is ferromagnetic. Low-temperature ferromagnetic behavior has been observed in Cr-based spinel semiconductors. Theoretical results by Blinowski, Kacman, and Majewski predict that Cr doping in II-VI semiconductors should result in ferromagnetism.[152] Ab initio calculations based on the local density of states approximation specifically predict ferromagnetism in Co- and Cr-doped ZnO without the need for additional doping.[148]

Figure 5.18 shows the magnetization versus field behavior at 10 K for Sn-doped ZnO samples implanted with 3 and 5 at% Mn. Hysteretic behavior is clearly observed, consistent with

TABLE 5.4
Valence and Ionic Radii for Candidate Dopant Atoms

Atom	Valence	Ionic radius (Å)
Zn	+2	0.60
Sn	+4	0.55
Li	+1	0.59
Ag	+1	1.00
Mn	+2	0.66
Cr	+3	0.62
Fe	+2	0.63
Co	+2	0.38
V	+3	0.64
Ni	+2	0.55
Mg	+2	0.57

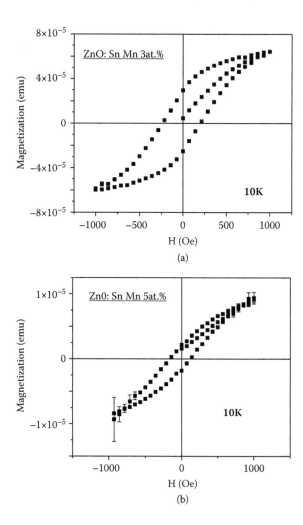

FIGURE 5.18 M versus H curve for Mn-implanted ZnO:Sn single crystals showing ferromagnetism in (a) 3 at% Mn and (b) 5 at% Mn implantation doses.

ferromagnetism. At 10 K, the coercive field in the 3 at% Mn-doped sample is 250 Oe. It must be noted that other possible explanations for hysteretic M versus H behavior that are remotely possible include superparamagnetism and spin-glass effects.[153–156] Magnetization measurements were also performed on Sn:ZnO crystals that were not subjected to the Mn implant. This was done to eliminate the possibility that spurious transition metal impurities might be responsible for the magnetic response. The Sn-doped ZnO crystals exhibit no magnetic hysteresis, showing that the Mn doping is responsible for the behavior. To track the hysteretic behavior in the implanted samples as a function of temperature, both field-cooled and zero field-cooled magnetization measurements were performed from 4.2 to 300 K. By taking the difference between these two quantities, the para- and diamagnetic contributions to the magnetization can be subtracted, leaving only a measure of the hysteretic ferromagnetic regime.

When assigning the origin of ferromagnetism in transition-metal-doped semiconductors, one must carefully consider the possibility that secondary phase formation is responsible. High-resolution transmission electron microscopy is the most direct means of exploring this issue. These activities are currently being pursued for the Mn-implanted ZnO samples. Nevertheless, one can also consider what known ferromagnetic impurity phases are possible.[153–156] First, metallic Mn is antiferromagnetic, with a Néel temperature of 100 K. In addition, nearly all of the possible Mn-based binary and ternary oxide candidates are antiferromagnetic. The exception to this is Mn_3O_4, which is ferromagnetic with a Curie temperature of 46 K in thin films.[157] X-ray diffraction measurements on the implanted samples show no evidence for Mn-O phases, although it is recognized that diffraction is limited in detecting secondary phases that may represent a fraction of a percent of the total volume in the implanted region. However, even if this phase were present, it could not account for the remarkably high ferromagnetic transition temperature of ~250 K observed for the Mn-implanted ZnO:Sn crystals. Note also that the Mn concentrations used in this study were well below the solid solubility limit of Mn in ZnO. It should also be noted that increasing the Mn content from 3 to 5 at% resulted in a significant decrease in the relative magnetization response, as is seen in Figure 5.18. This provides strong evidence, albeit indirect, that the magnetization is not due to any precipitating secondary phase. If the formation of a secondary Mn-related phase was responsible for the ferromagnetic behavior, an increase in Mn concentration would presumably increase the secondary phase volume fraction and related magnetization signature. Instead, the opposite behavior is observed.

One can postulate as to the mechanism by which Sn doping yields ferromagnetic interactions among the Mn ions. First, the Sn ions may simply provide carriers or bound donor states with extended wave functions that mediate interactions among the Mn ions. The Sn dopants may alternatively form complexes with the Mn ions, yielding a distribution of Mn^{+3} sites. In this case, the mixture of Mn^{+2}/Mn^{+3} sites could yield ferrimagnetic ordering through superexchange interaction. With this, the magnetic moment/Mn ion should increase with Sn doping even if the measured conductivity changes very little. It should be noted that very little is known about the behavior of Sn in ZnO. Identifying the location of the energy state associated with Sn in ZnO would be most useful in the interpretation of magnetic properties when co-doped with Mn. One could also look for shifts in the Sn or Mn dopant state location as the other dopant is added.

An additional means of elucidating the mechanism for ferromagnetism in ZnO co-doped with a deep donor is to consider an alternative transition metal. For this, Cr is particularly attractive. First, Cr provides an opportunity to realize ferromagnetism in ZnO via ions for which the net superexchange coupling is ferromagnetic. As with Mn, Cr itself is antiferromagnetic, thus eliminating any role of Cr precipitates in yielding spurious ferromagnetism. Third, theory predicts that Cr-doped n-type ZnO should be ferromagnetic. As such, experiments similar to those described earlier for ZnO co-doped with Mn and Sn should be performed with Cr replacing Mn. Ferromagnetism in Cr-doped semiconducting oxides has been reported for several materials, including ZnO and TiO_2.

The observation that a doubly ionized donor (Sn) leads to ferromagnetic interactions among Mn atoms in ZnO raises the question as to whether deep states are, in general, effective in mediating ferromagnetism. Of particular interest are the theoretical predictions that ferromagnetism above room temperature should be possible in Mn-doped ZnO that is a p-type. Given the difficulty in realizing a shallow acceptor in ZnO, an obvious experiment, given the results for the Mn co-doped with a deep donor (Sn), is to co-dope Mn with a deep acceptor. Attractive deep acceptors for addressing this question are Cu and As. Copper doping introduces an acceptor level with an energy ~0.17 eV below the conduction band. There is a large ionic radii mismatch for As (2.22 Å) on the O (1.38 Å) site, suggesting limited solid solubility for these anions.[158] Nevertheless, p-n junction-like behavior has been reported between an n-type ZnO films on GaAs subjected to annealing.[159] In this case, a p-type layer was reportedly produced at the GaAs–ZnO interface. Further understanding of group V substitution in ZnO requires the study of doped materials that are free from complications of reactive substrates or interfacial layers. The primary interest is to investigate whether the deep acceptor states from Cu or As mediate ferromagnetic interactions in TM-doped ZnO.

For practical applications in spintronic devices, the Curie temperature should be well above room temperature.[160,161] As discussed earlier, theory suggests that the Curie temperature will tend to increase with decreasing cation mass. In addition, there is phenomenological evidence that T_c increases with increasing gap. The observed trend is for T_c to increase with increasing bandgap. While the equilibrium solid solubility of Mg in wurtzite ZnO is only ~2%, epitaxial $Zn_{1-x}Mg_xO$ thin films have been realized with x as large as 0.35.[162]

A key requirement in understanding ferromagnetism in transition-metal-doped semiconductors, including ZnO, is to delineate whether the magnetism originates from substitutional dopants on cation sites, or from the formation of a secondary phase that is ferromagnetic. The importance of this issue cannot be understated. The concept of spintronics based on ferromagnetic semiconductors assumes that the spin polarization exists in the distribution of semiconductor carriers. Localized magnetic precipitates might be of interest in nanomagnetics, but is of little utility for semiconductor-based spintronics. The question of precipitates versus carrier-mediated ferromagnetism is complex, and is a central topic of discussion for other semiconducting oxides that exhibit ferromagnetism, in particular, the Co-doped TiO_2 system.[163,164] Several issues must be addressed in order to gain insight into the possible role of secondary phase precipitates in the magnetic properties of transition-metal-doped semiconductors, specifically for ZnO films. First, one should identify all candidate magnetic phases possible from the assemblage of elements. The coincidence of T_c with a known candidate secondary ferromagnetic phase indicates a likely source of at least part of the magnetic signature. The UF group has been investigating a number of TM-doped semiconducting oxide materials in order to understand ferromagnetism in semiconductors, and to identify promising candidates for spintronics. As an example, we recently investigated the magnetic properties of Mn-doped Cu_2O. Cu_2O is a p-type semiconductor with a bandgap of 2.0 eV and a hole mobility of 100 cm^2 $V^{-1}s^{-1}$. The growth of Mn-doped Cu_2O films was achieved via pulsed laser deposition from an Mn-doped Cu-O target. These epitaxial Cu_2O films doped with Mn are clearly ferromagnetic with a T_c of ~48 K. However, some question as to the origin of the ferromagnetism arises when it is recognized that the measured Curie temperature is close to that of Mn_3O_4, which has a T_c of ~46 K. Obviously, the simplest explanation for ferromagnetic behavior in this material is Mn_3O_4 precipitates. However, there is no evidence for this phase in X-ray diffraction. Additional work is needed in order to delineate the possible formation of Mn_3O_4 in the Cu_2O films. Nevertheless, knowledge of the candidate ferromagnetic secondary phases is invaluable in sorting out the ferromagnetic response. Fortunately, for Mn-doped ZnO:Sn, the only ferromagnetic secondary phase candidate is the Mn_3O_4 spinel. None of the other possible secondary phases involving combinations of Zn, Mn, O, and Sn yield a known ferromagnetic material. The high-temperature ferromagnetism in Mn-implanted ZnO:Sn crystals cannot be attributed to Mn_3O_4 as T_c is much higher in the Mn-doped ZnO than in the Mn_3O_4 phase.

In addition to TEM, one can also use x-ray diffraction to search for secondary phases within the films. Despite the obvious sensitivity limitations involved in detecting impurity phases that represent only 1%–5% vol% of a thin-film sample, we have been able to detect nanoscale precipitates in transition-metal-implanted samples. This is demonstrated specifically for another transition-metal-doped semiconducting oxide that we have investigated, namely, co-implanted ZnO. It has been reported that epitaxial $(Zn_{1-x}Co_x)O$ ($x = 0.05–0.25$) exhibits high-temperature ferromagnetism with a T_c greater than 300 K.[165] The ferromagnetic behavior is assigned to substitutional Co on the Zn site. Monte Carlo simulations on the indirect exchange interaction of Co-doped ZnO also predict ferromagnetism in these materials.[166] SQUID magnetometry measurements of Co-doped ZnO clearly indicate ferromagnetism. However, a careful examination of the x-ray diffraction θ-2θ scan along the surface normal clearly indicates the presence of Co precipitates. A Co(110) peak is clearly evident. From the width of the peak, the size of the cobalt precipitates can be estimated to be approximately 3.6 nm. Clearly, the possibility of precipitate formation must be carefully considered for each dopant investigated.

It should be noted that the presence of magnetic precipitates may mask an underlying carrier-mediated ferromagnetism due to substitutional doping. In this case, a more direct means of measuring the spin properties of the semiconductor carrier population is needed. Much of the research activity being presently pursued in spintronics is directed at demonstrating device structures (spin-LED, spin-FET) that differentiates spin polarization distribution among the electron/hole population in the semiconductor. Future work would employ these materials in such device structures. Some techniques, such as magnetic circular dichroism[167] and the magneto-optical Kerr effect,[168] also yield information regarding global spin distributions.

The successful realization of most spintronics applications depends critically on the ability to create spin-polarized charge carriers in a conventional semiconductor in a device structure. This can be accomplished under ambient conditions via optical pumping with an appropriately polarized laser light.[135] However, ultimate device integration will require electrical spin injection. Electrical spin injection can be accomplished either by injecting from a spin-polarized source or by spin-filtering unpolarized carriers at the interface. Despite persistent efforts by many groups, spin injection from a conventional ferromagnetic metal into a semiconductor has proved highly inefficient.[136,137] In contrast, efficient spin injection has recently been successfully demonstrated in all-semiconductor tunnel diode structures either by using a spin-polarized dilute magnetic semiconductor (DMS) as the injector or by using a paramagnetic semiconductor under a high magnetic field as a spin filter. These experiments, coupled with the earlier discoveries of ferromagnetic ordering in (Ga,Mn)As at an unprecedented temperature of 120 K and the long spin coherence length and time in a variety of semiconductors, offer the possibility of a breakthrough in spintronics. In order to implement these applications, however, DMS with an ordering temperature over 300 K must be synthesized.

5.9.3 NANORODS

Recently, many groups have demonstrated site-specific growth of ZnO nanorods using catalyst-driven molecular beam epitaxy (MBE) or vapor transport.[169,170] The large surface area of the nanorods makes them attractive for gas and chemical sensing, and the ability to control their nucleation sites makes them candidates for microlasers or memory arrays.

An SEM of a single rod is shown in Figure 5.19 (left). The growth time was ~2 h at 400°C. The typical length of the resultant nanorods was ~2 μm, with typical diameters in the range of 15–30 nm. It is also possible to grow cored nanorods with ZnMgO composition varying across the rod diameter (Figure 5.19, right). The samples were subsequently implanted with Mn^+ and Co^+ ions at a fixed energy of 250 keV and doses of $(1–5) \times 10^{16}\,cm^{-2}$, while the samples were held at ~350°C to avoid amorphization. The projected range for both types of ion is ~1500 Å, with peak transition metal concentrations corresponding to roughly 1–5 at%. The samples were then annealed at 700°C for 5 min under flowing N_2 to promote migration of the transition metal ions into substitutional positions. Even at this highest dose condition, the nanorods are stable to the implant/anneal cycle. A comparison of micrographs before and after this cycle did not show any observable change in the nanorods.

Figure 5.20 shows the magnetization versus field behavior at 300 K for an Mn-implanted $(5 \times 10^{16}\,cm^{-2}$ dose) nanorod sample after the 700°C, 5 min anneal. Hysteretic behavior is clearly present, with a coercive field at 100 K of ≤ 100 Oe. The possible explanations for this data include

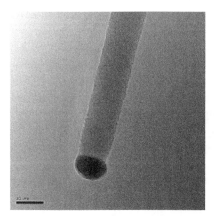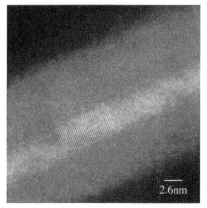

FIGURE 5.19 TEM image of single-crystal ZnO nanorod (left) and TEM micrographs showing cored $(Zn_{1-x}Mg_x)O$ nanorods having Zn-rich phase surrounded by another $(Zn_{1-x}Mg_x)O$ phase (right).

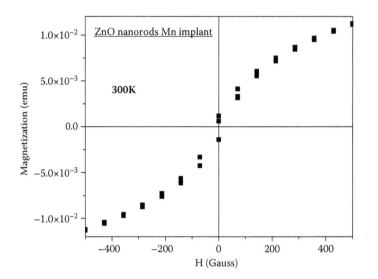

FIGURE 5.20   Magnetization versus field at 300 K for Mn-implanted ZnO nanorod.

ferromagnetism, superparamagnetism, or spin-glass effects.[56] The magnetization of unimplanted, annealed nanorods was as much as 3 orders of magnitude lower than in the implanted, annealed samples, demonstrating that the transition metals are responsible for the observed magnetic properties. The calculated moment from the saturation magnetization was ~2.2 Bohr magnetons per Mn ion, showing that a significant fraction of the implanted Mn is contributing to the magnetization.

None of the potential second phases can account for the observed magnetic behavior. Thus, for Mn-implanted nanorods, it does not appear that secondary phases play a significant role in the magnetic properties. Indeed, the magnetization results for the Mn-implanted nanorods are very similar to those reported previously for Mn-implanted, bulk n-type ZnO single crystals, in which high-resolution x-ray diffraction did not detect any secondary phases. We note also that solubility limits for Mn in ZnO are between 13% for equilibrium conditions to 35% for incorporation by pulsed laser deposition. These values are well beyond the concentrations employed here.

In sharp contrast, in the case of Co-implanted ZnO, macroscopic Co precipitates are ferromagnetic with Curie temperatures up to 1382 K when large enough to exhibit bulk properties. Bulk, single crystals of ZnO implanted with Co under the same conditions as employed here showed the presence of (110)-oriented hexagonal Co nanocrystals with an average diameter of ~35 Å, well below the superparamagnetic critical diameter of Co near room temperature of ~50 Å. This is consistent with the magnetization data for the Co-implanted nanorods, which do not show ferromagnetism at room temperature. Thus, in the case of the nanorods examined here, it appears the magnetic properties most likely result from the presence of Co nanocrystals. The results are similar to those obtained by implantation of these same species into bulk, single-crystal ZnO. One can envision implanting Mn into p-n junction nanorods to investigate the possibilities of spin-polarized UV light emission or using Mn-doped nanorods as magnetic storage elements.

The most effective measurement of the quality of the oxide-based ferromagnetic materials will be in the operation of device structures, such as spin-FETs or photoinduced ferromagnets. While promising advances in spin-injection efficiency from metal contacts into semiconductors have been made recently (using injection from ohmic contacts or ballistic point contacts, tunnel injection, or hot electron injection), in general, there are difficulties in reproducibility. The use of ferromagnetic semiconductors as the injection source in device structures should allow a more direct measurement of the efficiency and length scale of spin transport. There are two test structures based on ZnO that will demonstrate whether novel aspects of ferromagnetism can be exploited in devices.

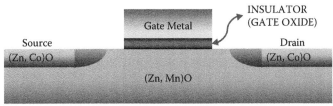

FIGURE 5.21   Schematic of ZnO-based spin-fet.

The first structure is shown in Figure 5.21 and was suggested by Sato and Katayama-Yoshida (*Mat. Res. Soc. Symp. Proc.* Vol. 666, F4.6.1, 2001) as an analogue of the spin-FET. The structure takes advantage of the fact that (Zn,Mn)O can be grown as an antiferromagnetic spin-glass insulator, while hole- and Mn-co-doped ZnO can be a half-metallic ferromagnet. In the structure of Figure 5.21, the application of negative gate bias brings holes into the (Zn,Mn)O and converts it to the half-metallic ferromagnetic state. Using ferromagnetic (Zn,Co)O as the source and drain contact material, it should be possible to have a 100% spin-polarized electron flow in the (Zn,Mn)O channel. The device can be fabricated by growing the source/drain materials on top of the (Zn,Mn) O and then etching away in the gate region for selective growth of the gate oxide and deposition of the gate metal.

The proposed photomagnet concept is based on the fact that ZnO:MnCr becomes a half-metallic ferromagnet upon hole doping while ZnO:FeMn is a half-metallic ferromagnet upon electron doping. For photons of appropriate energy, electrons and holes created in the GaAs substrate near the interface with the ZnO:MnCr or ZnO:FeMn can be drawn into those materials by biasing, causing them to become half-metallic ferromagnets. The presence of these ordered states can be detected using the magneto-optical effect from another probe beam of photons with energy lower than the ZnO bandgap. The device is readily grown on conducting GaAs to provide good ohmic contact to the substrate.

5.9.4 PROPERTIES OF MN- AND CO-IMPLANTED BULK ZNO

The properties of transition metal (TM) impurities, together with studies of the nature of shallow donors and shallow acceptors, are of particular interest due to the predicted room temperature ferromagnetism of heavily TM-doped ZnO. TM dopants can be incorporated into ZnO both during growth (from the vapor phase, by solid-state recrystallization, by the hydrothermal method, or, for example, from PbF_2 solutions [see, for example, a book[171] for early references and, for example, references 172 and 173 as a recent example]) and by ion implantation. The work done in the early 1980s was mostly driven by the general interest in the behavior of TMs in various semiconductors stimulated by the success of semi-insulating GaAs and InP. At that period, no group succeeded in observation of room-temperature magnetism. The results of electron spin resonance (ESR) and photoluminescence spectra studies of Fe, Ni, Mn, and Co-doped ZnO were rather confusing. Most thoroughly studied were the properties of Fe-doped samples. ESR of these samples showed clear indications of the presence of the Fe^{3+}-ionized donors in semi-insulating material with the Fermi level deep in the bandgap. The signal was photosensitive and could be quenched by light with photon energies peaking near 2.75 eV, which was explained by the optically stimulated transition from Fe^{3+} to Fe^{2+} state by capturing an electron from the valence band. Detailed studies placed the optical ionization transition threshold from the Fe^{3+} state to the conduction band at about 1.4 eV, suggesting a strong lattice coupling of the deep donor state of Fe in ZnO.[171]

A somewhat similar photosensitive ESR signal was also observed for Ni-doped ZnO.[169] At the same time, ESR signals in the Mn-doped and Co-doped ZnO were found not to be photosensitive. However, for the Co-doped material a strong optical band at 0.55–0.66 μm was detected in absorption and photocurrent spectra and attributed to an optical transition from the ground state to the excited states of the Co ion with subsequent thermal excitation of the electrons into the conduction band. Photoluminescence (PL) spectra studies in Fe-, Ni-, and Co-doped ZnO crystals revealed the presence of red luminescence bands at, respectively, 0.63 μm, 0.68 μm, and 0.6850 μm that were attributed to transitions involving excited states of the TM ions.[171]

In more recent papers, the electrical properties of Mn and Co were studied by means of I–V, C–V, and isothermal capacitance transients (ICTS) in ZnO crystals solution grown from PbF_2.[172] No indication of the presence of deep states that could be related to Mn and Co were observed in ICTS spectra, which could point to respective centers being deeper than the Schottky barrier height (about 0.7 eV). Some indirect manifestations of the influence of deep states due to Mn and Co could be observed in C–V and especially I–V measurements.[172] ZnFeO crystals co-doped with Cu (but not simple ZnFeO crystals) grown by solid-state recrystallization showed a Curie temperature of 500 K, and the authors suspected that some role in that was played by Cu acceptors.[173]

For Co- and Mn-implanted and 700°C annealed samples, the room-temperature van der Pauw characteristics were virtually the same as for the control sample. The temperature dependence of the conductivity showed the same activation energy of 13 meV. The optical transmission spectrum of the virgin ZnO crystal showed no strong absorption bands anywhere but at the band-edge region. The first band, observed with the threshold near 0.75 eV, and the second, with the threshold near 1.4 eV, are very similar for all samples. In addition to those, the Mn- and Co-implanted ZnO crystals show strong absorption bands with an optical threshold near 2 eV, and all samples show defect absorption bands with the threshold near 2.3–2.5 eV. The TM-related bands near 2 eV show the apparent threshold energies of about 2 eV for Mn and of about 1.9 eV for Co. The energy of the yellow-green defect absorption band is close to 2.5 eV for both the Mn- and the Co-implanted ZnO crystals and is somewhat lower, close to 2.35 eV for the heavily proton-implanted sample.

It is clear that no p-type layer is formed at the surface of Mn-, Co-implanted, and annealed samples and thus the near-room-temperature ferromagnetism observed in such implanted layers is not promoted by free holes as assumed in the model proposed by Dietl.[143,144] This correlates very well with the results obtained on high-Curie-temperature bulk ZnOFe(Cu) crystals that were shown to be an n-type.[173] Therefore, the mechanism of room-temperature ferromagnetism in ZnO doped with TM is unconventional, as for example, for GaN doped with TM.

The next question is whether or not any features in the absorption or MCL spectra can be associated with Mn and Co. The most likely candidates in that sense are the absorption bands at 1.9 and 2 eV and the red MCL bands at 1.84 and 1.89 eV. The absorption and the MCL bands seem to be closely related in the sense that their energies are reasonably close and the relative shift of the bands' energies when changing Mn to Co is similar though not quite equal in both bands. The positions of these MCL bands for different TM ions are very close to each other, which brings to mind the possibility that the bands are due to intracenter transitions between the states of TM ions. However, we do not see any fine structure confirming such an interpretation either in absorption or in MCL, and this question requires further study. For example, as mentioned earlier, a similar band near 2 eV observed in photocurrent and absorption spectra of Co-doped samples was attributed to intracenter transitions from the ground to excited states with subsequent thermal excitation of electrons into the conduction band.[171] The intensity of absorption in the red band is considerably higher, and the MCL peak is considerably broader for the Mn-implanted sample compared to the Co-implanted sample. Since the concentrations of implanted Mn and Co are virtually the same, this is most likely related to the higher density of substitutional Mn achieved after implantation and annealing compared to Co due to higher solubility of Mn at 700°C.

In summary, we have shown that the Mn- and Co-implanted and 700°C annealed regions of ZnO crystals remain an n-type and even show a higher electron concentration than the virgin samples. Therefore, the high Curie temperature of about 250 K measured on similarly implanted and annealed crystals is not due to hole-mediated spin alignment but rather due to some other mechanism. The implanted crystals show optical absorption and MCL bands near 2 eV that can be attributed to TM ions on the grounds that such bands are not observed in the virgin or proton-implanted samples. The position of the bands slightly shifts for various TM ions (e.g., corresponding MCL bands are at 1.84 eV for Co, 1.89 eV for Mn; the respective absorption bands are at 1.9 and 2 eV). The origin of the bands is not quite clear at the moment. We somewhat tentatively attribute them to internal transitions between the crystal-field-split states of substitutional TM ions. If so, one has to conclude that the concentration of such ions is considerably higher after implantation and annealing for the Mn-implanted samples compared to the Co-implanted samples. This could be due to the higher solubility of the Mn in ZnO at 700°C. This higher concentration of deep donor defects in the Mn-implanted samples stimulates formation of acceptor-type Zn vacancies, giving rise to the violet MCL band near 3 eV not observed in the Co-implanted or proton-implanted samples. In addition to these directly or indirectly TM-related bands, we also observed in Mn- and Co-implanted samples' absorption bands near 0.75, 1.4, and 2.5 eV that are due to radiation damage defects as follows from comparison with the results obtained on heavily proton-implanted samples.

5.10 GAS SENSING USING ZnO THIN FILM

5.10.1 ETHYLENE SENSING

ZnO has a long history of use as a gas-sensing material.[1,174] There is a strong interest in the development of wide-bandgap semiconductor gas sensors for applications, including detection of combustion gases for fuel leak detection in spacecraft, automobiles and aircraft, fire detectors, exhaust diagnosis, and emissions from industrial processes. Of particular interest are methods for detecting ethylene (C_2H_4), which offers problems because of its strong double bonds and hence the difficulty of dissociating it at modest temperatures. Wide-bandgap semiconductors such as ZnO are capable of operating at much higher temperatures than more conventional semiconductors such as Si. Diode or field-effect transistor structures are sensitive to gases such as hydrogen and hydrocarbons. Ideal sensors have the ability to discriminate between different gases and arrays that contain different metal oxides (e.g., SnO_2, ZnO, CuO, and WO_3) on the same chip can be used to obtain this result.[175] The gas-sensing mechanism suggested include the desorption of adsorbed surface oxygen and grain boundaries in poly-ZnO,[176] and exchange of charges between adsorbed gas species and the ZnO surface leading to changes in depletion depth[177] and changes in surface or grain boundary conduction by gas adsorption/desorption.[176]

Pt/ZnO Schottky diodes show changes in forward current of 0.3 mA at a forward bias of 0.5 V or, alternatively, a change of 50 mV bias at a fixed forward current of 8 mA when 5 ppm of H_2 is introduced into a N_2 ambient at 25°C, as shown in Figure 5.22. The rectifying I–V characteristic

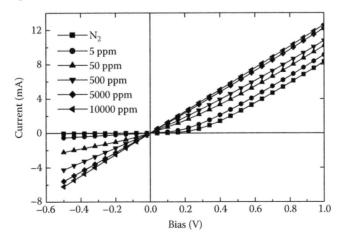

FIGURE 5.22 I–V characteristics at 25°C of Pt/ZnO diode measured in N_2- or H_2-containing ambients.

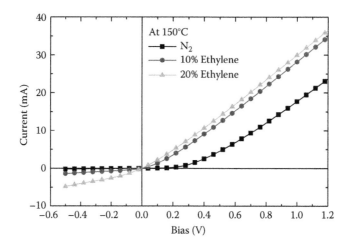

FIGURE 5.23 I–V characteristics at 150°C of Pt/ZnO diodes measured in different gas ambients.

shows a nonreversible collapse to ohmic behavior when as little as 50 ppm of H_2 is present in the N_2 ambient. At higher temperatures, the recovery is thermally activated with an activation energy of ~0.25 eV. This suggests that introduction of hydrogen shallow donors into the ZnO is a contributor to the change in the current of the diodes.

Figure 5.23 shows the I–V characteristics at 50°C and 150°C of the Pt/ZnO diode both in pure N_2 and in ambients containing various concentrations of C_2H_4. At a given forward or reverse bias, the current increases upon introduction of the C_2H_4, through a lowering of the effective barrier height. One of the main mechanisms is again the catalytic decomposition of the C_2H_4 on the Pt metallization, followed by diffusion to the underlying interface with the ZnO. In conventional semiconductor gas sensors, the hydrogen forms an interfacial dipole layer that can collapse the Schottky barrier and produce more ohmic-like behavior for the Pt contact. The recovery of the rectifying nature of the Pt contact was many orders of magnitude longer than for Pt/GaN or Pt/SiC diodes measured under the same conditions in the same chamber. For measurements over the temperature range 50°C–150°C, the activation energy for recovery of the rectification of the contact was estimated from the change in forward current at a fixed bias of 1.5 V. This was thermally activated through a relation of the type $I_F = I_0 \exp(-E_a/kT)$ with a value for E_a of ~0.22 eV, comparable for the value of 0.17 eV obtained for the diffusivity of atomic deuterium in plasma-exposed bulk ZnO. This suggests that at least some part of the change in the current upon hydrogen gas exposure is due to in-diffusion of hydrogen shallow donors that increase the effective doping density in the near-surface region and reduce the effective barrier height.

5.10.2 CO Sensing

Pt/ZnO bulk Schottky diodes are capable of detection of low concentrations (1%) of carbon monoxide at temperatures >100°C. Figure 5.24 (top) shows a schematic of the completed Pt/ZnO bulk Schottky diode, while Figure 5.24 (bottom) shows a packaged sensor. The bulk ZnO crystals from Cermet Inc. showed an electron concentration of 9×10^{16} cm^{-3} and an electron mobility of 200 cm^2/V-s at room temperature from van der Pauw measurements. The back (O-face) of the substrate was deposited with full-area Ti (200 Å)/Al (800 Å)/Pt (400 Å)/Au (800 Å) by e-beam evaporation. After metal deposition, the samples were annealed in a Heatpulse 610 T system at 200°C for 1 min in N_2 ambient. The front face was deposited with plasma-enhanced chemical-vapor-deposited SiN$_x$ at 100°C, and windows opened by wet etching so that a thin (20 nm) layer of Pt could be deposited by e-beam evaporation. After the final metal of e-beam-deposited Ti/Au(300 Å/1200 Å) interconnection contacts was deposited, the devices were bonded to electrical feedthroughs and exposed to different gas ambients in an environmental chamber, while the diode

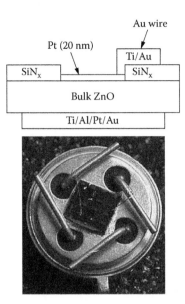

FIGURE 5.24 (See color insert).   Schematic of ZnO Schottky rectifier (top) and photograph of packaged device (bottom).

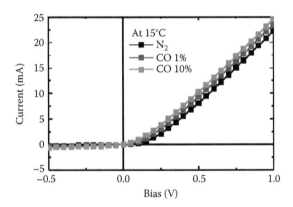

FIGURE 5.25   I–V characteristics of ZnO rectifier at 150°C in different ambients.

I–V characteristics were monitored, generally 8 min after introducing the CO and C_2H_4 partial pressure into the test chamber.

The I–V characteristics at 150°C of the ZnO rectifiers are shown in Figure 5.25, for N_2 and 1% and 10% CO in N_2 ambients. The forward and reverse currents increase as the concentration of CO increases. In SnO_2 conductance devices,[178–181] the detection of CO at 200°C–500°C was analyzed in terms of a surface reaction between absorbed CO and chemisorbed oxygen to produce CO_2.

This leads to a charge transfer to SnO_2 and a change in the conductance. In the case of ZnO, the mechanism may be the same, but must occur by diffusion of oxygen through the Pt contact since the rest of the device is covered with SiN_x, which is impermeable to oxygen. The net effect is a reduction in effective barrier height, Φ_B, of ~20 meV at 150°C for 10% CO in the N_2 ambient.

The changes in forward bias at fixed current (20 mA) and forward current at fixed bias (1V) from the rectifiers are shown in Figure 5.26 for different CO concentrations in the ambient, as a function of temperature.

The response is not detectable below 90°C, but increases significantly with temperature. In the model of Reference 181 for CO detection by SnO_2, if the temperature is too low, the CO product

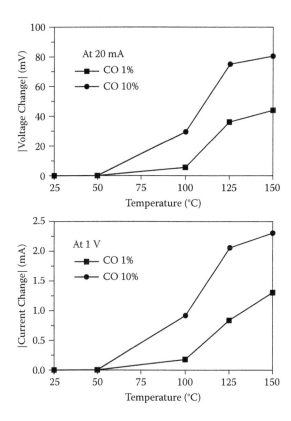

FIGURE 5.26   Change in voltage at fixed forward current (top) or change in current at fixed forward bias (bottom) for two different CO concentrations in N_2, as a function of measurement temperature.

species will not be desorbed and therefore poisons the site for future adsorption of oxygen. The specific on-state resistance of the rectifier RON is given by

$$R_{ON} = (4V^2_B/\varepsilon\mu E^3_M) + \rho_s W_s + R_C,$$

where ε is the ZnO permittivity, μ the carrier mobility, ρ_s and W_s the substrate resistivity and thickness, and RC the contact resistance. Over a relatively limited range of CO partial pressures in the N_2 ambient, the on-state resistance decreased according to $R_{ON} = (R_O + A(P_{CO})^{0.5})^{-1}$, where A is a constant and R_O is the resistance in pure N_2 ambient. This is a similar dependence as in the case of CO detection by SnO_2 conduction sensors, in which the effective conductance increased as the square root of CO partial pressure. The gas sensitivity can be calculated from the difference in resistances in CO-containing ambients, divided by the resistance in pure N_2, that is, $(R_{N2} - R_{CO})/R_{N2}$. At 150°C, the gas sensitivity was 4% for 1% CO in N_2 and 8% for 10% CO in N_2. These values are comparable to those reported for CO detection by SnO_2 conductance sensors.[178–181]

Figure 5.27 shows the time dependence of the changes in the forward current at fixed bias (0.4V) for different temperatures. The increase in the current shows a square-root dependence on time, suggesting that a diffusion process is occurring. The activation energy of this process was determined from the Arrhenius plot at the bottom of Figure 5.27, namely, 40.7 kJ/mol. This may represent the effective activation energy needed to catalytically dissociate the CO on the Pt metal contact surface and then have diffusion of the atomic oxygen to the interface with the ZnO. It is much lower than the bond strength of the CO molecule, which is 1076.5 kJ/mol.[175] The detection mechanism is clearly still not firmly established in these devices and needs further study.

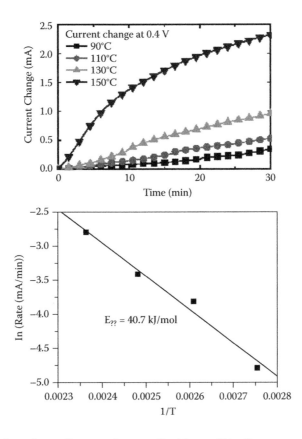

FIGURE 5.27 Time dependence of current change at fixed forward bias for measurement in 10% CO in N_2, for different measurement temperatures (top), and Arrhenius plot of time rate of change of current (bottom).

5.11 TRANSPORT IN ZnO NANOWIRES

ZnO is attractive for forming various types of nanorods, nanowires, and nanotubes for observation of quantum effects.[182–203] The large surface area of the nanorods makes them attractive for gas and chemical sensing, and the ability to control their nucleation sites makes them candidates for microlasers or memory arrays. To date, most of the work on ZnO nanostructures has focused on the synthesis methods, and there have been only a few reports of the electrical characteristics. The initial reports show a pronounced sensitivity of the nanowire conductivity to ultraviolet illumination and the presence of oxygen in the measurement ambient. The photoresponse of the nanowires shows that the potential barrier between the ZnO and the contact is lowered by illumination with above-bandgap light.

The preparation of the nanorods proceeded as follows: discontinuous Ag droplets were used as the catalyst for ZnO nanorod growth and formed by annealing e-beam-evaporated Ag thin films (~100 Å) on p-Si (100) wafers at 700°C. ZnO nanorods were deposited by MBE with a base pressure of 5×10^{-8} mbar using a high purity (99.9999%) Zn metal and an O_3/O_2 plasma discharge as the source chemicals. The Zn pressure was varied between 4×10^{-6} and 2×10^{-7} mbar, while the beam pressure of the O_3–O_2 mixture was varied between 5×10^{-6} and 5×10^{-4} mbar. The growth time was ~2 h at a temperature of 400°C ~ 600°C. The typical length of the resultant nanorods was 2 ~ 10 μm, with diameters in the range of 30–150 nm. Selected area diffraction patterns showed the nanorods to be single-crystal. They were released from the substrate by sonication in ethanol and then transferred to SiO_2-coated Si substrates. E-beam lithography was used to pattern sputtered Al/Pt/Au electrodes contacting both ends of a single nanorod. The separation of the electrodes was

~3.7 μm. A scanning electron micrograph of the completed device is shown in Figure 5.28. Au wires were bonded to the contact pad for I–V measurements performed over the range 25°C–150°C in a range of different ambients (C_2H_4, N_2O, O_2 or 10% H_2 in N_2).

The as-grown nanorods were resistive and passed very small currents (<10^{-10} A). To make them more conducting, we annealed them in hydrogen gas at 400°C. Figure 5.29 shows the I–V characteristics as a function of measurement temperature for these conducting nanorods. It is expected that the hydrogen increases the n-type doping in the ZnO. The hydrogen-annealed nanorods are indeed more conducting, with currents of ~3×10^{-8} A at 0.5 V bias voltage of either polarity. The current is thermally activated of the form $I = I_o \exp(-E_a/kT)$, where E_a is the activation energy, k is Boltzmann's constant, and T is the absolute measurement temperature. The nanorods showed a strong, reversible response to above-bandgap (366 nm) ultraviolet light, with the UV-induced current being approximately a factor of 7 larger than the dark current at a given voltage. The intensity of the photocurrent was independent of the illumination time for a fixed modulation frequency. Previous reports have suggested that photocurrents excited in ZnO nanowires by above-bandgap light are surface-related and do not represent true bulk conduction.

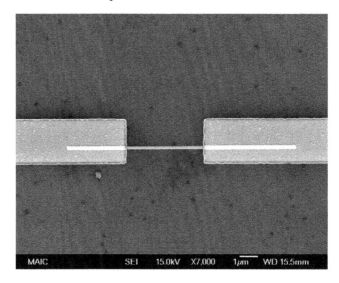

FIGURE 5.28   SEM micrograph of single ZnO nanorod bridging two Al/Pt/Au ohmic contact pads.

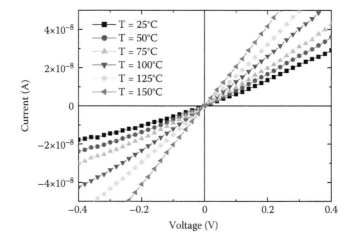

FIGURE 5.29   I–V characteristics of single nanorod measured at different substrate temperatures.

Figure 5.30 shows the resistivity the nanorods obtained from the I–V characteristics, in Arrhenius form. The apparent activation energy derived from this plot is 0.089 ± 0.02 eV. This does not correspond to any of the known donor dopant or native defect ionization energies in ZnO, and it is possible that the conduction is again surface-related, as in the UV photoresponse measurements. In the hydrogen-annealed nanorods, the currents were insensitive to the measurement ambient, as shown in Figure 5.31. One might expect the current to be dependent on this ambient if the conduction were truly dominated by the surface. Various gas-sensing mechanisms in ZnO-based gas sensors have been suggested, including the desorption of adsorbed surface oxygen and grain boundaries in poly-ZnO, exchange of charges between adsorbed gas species and the ZnO surface leading to changes in depletion depth, and changes in surface or grain boundary conduction by gas adsorption/desorption. In all of these cases, the mechanism is surface-related. In sharp contrast to the data for the hydrogen-annealed nanorods, the unannealed samples showed a strong sensitivity to the measurement ambient, with increased currents in the presence of hydrogen. Clearly, the control of both bulk and surface quality is a key area for practical applications of the nanorods in sensing of gases, chemicals, or UV light.

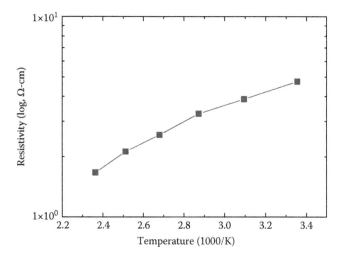

FIGURE 5.30 Arrhenius plot of single nanorod resistivity obtained from I–V measurements in air.

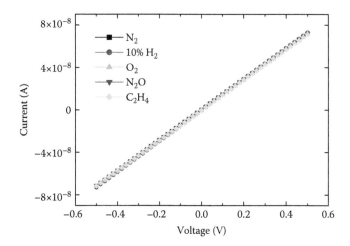

FIGURE 5.31 I–V characteristics of nanorod at 25°C as a function of the measurement ambient.

5.12 ZnO NANOWIRE SCHOTTKY DIODES

The typical length of the nanowires used was ~5 μm, with typical diameters in the range of 30–100 nm. Selected area diffraction patterns showed the nanowires to be single-crystal. They were released from the substrate by sonication in ethanol and then transferred to SiO_2-coated Si substrates. E-beam lithography was used to pattern sputtered Al/Pt/Au electrodes contacting both ends of a single nanowire. The separation of the electrodes was ~3 μm. E-beam-evaporated Pt/Au was used as the gate metallization by patterning a 1 μm wide strip orthogonal to the nanowire. A scanning electron micrograph of the completed device is shown in Figure 5.32. Au wires were bonded to the contact pads for I–V measurements at 25°C. In some cases, the diodes were illuminated with above-band-edge ultraviolet (UV) light during the measurement.

Figure 5.33 shows I–V characteristics from the Pt/ZnO nanowire diode, measured both in the dark and with UV illumination. While the measurement in the dark shows rectifying behavior, the

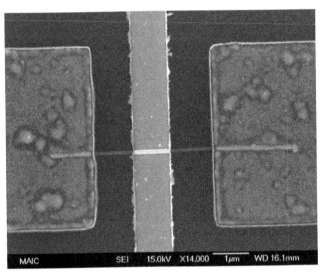

FIGURE 5.32   SEM micrograph of ZnO nanowire Schottky diode.

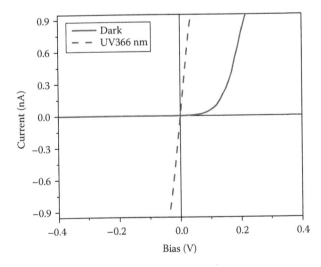

FIGURE 5.33   I–V characteristics from the ZnO Schottky nanowire diodes both in the dark and with UV illumination.

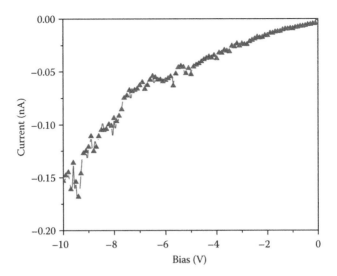

FIGURE 5.34 Reverse current of ZnO nanowire diode measured in the dark.

UV illumination produces an ohmic characteristic. This is similar to the result reported by Keem et al., who showed that the above-bandgap light lowered the potential barrier between the Schottky contact and the ZnO nanowire. In their case, however, the photoresponse recovery times were very long (>10⁴ s), which was ascribed to the dominance of surface states, whereas our recovery times were limited by turn-off of the Hg lamp and suggest the conduction in our nanowires is dominated by bulk transport.

Figure 5.34 shows the reverse current characteristic in more detail. The reverse current was still only 1.5×10^{-10} A at −10 V bias, corresponding to a current density of 2.35 A.cm⁻². If one defines the reverse breakdown voltage as the bias at which the reverse current density is 1 mA.cm⁻², this value is ~2 V for the nanowire diodes.

The forward I–V characteristics in the dark (shown in Figure 5.35) were fit to the relation for thermionic emission over a barrier

$$J_F = A^* \cdot T^2 \exp\left(-\frac{e\phi_b}{kT}\right)\exp\left(\frac{eV}{nkT}\right)$$

where J is the current density, A^* is Richardson's constant for n-GaN, T the absolute temperature, e is the electronic charge, Φ_b is the barrier height, k is Boltzmann's constant, n is the ideality factor, and V is the applied voltage.

We could not drive our nanowires to the saturation current condition because of fears the high current density would create self-heating problems and therefore were unable to extract the barrier height. However, previous measurements of the barrier height of Pt on ZnO have yielded values of 0.75 eV reported by Mead using internal photoemission yield spectroscopy to obtain the barrier height of the laterally homogeneous portions of the contacts and, more recently, 0.61 eV on bulk ZnO. From the measured data, we derived an ideality factor of $n = 1.1$. This shows that there is little recombination in our nanowires. The on-state resistance, R_{ON}, was 1.65 Ω·cm² (the resistivity of the nanowire was 22.6 Ω·cm) for an individual nanowire device, with an on/off current ratio of ~6 at 0.15/−5 V bias. These values can be further improved by optimizing the growth process of the nanowires and the subsequent contact metallization processes.

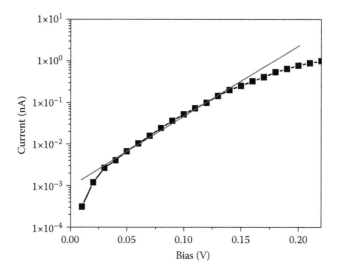

FIGURE 5.35   Forward current of ZnO nanowire diode measured in the dark.

5.13   ZnO NANOWIRE MOSFET

There is great current interest in fabrication of ZnO channel thin-film transistors for applications in transparent flat panel displays. Other transparent conducting oxides such as Sn-doped In_2O_3, Al-doped ZnO, and Sb-doped SnO_2 have been widely applied as transparent electrodes for liquid crystal displays, organic light-emitting diodes, and solar cells.

E-beam lithography was used to pattern sputtered Al/Pt/Au electrodes contacting both ends of a single nanowire. The separation of the electrodes was ~7 µm. Au wires were bonded to the contact pad for I–V measurements. $(Ce,Tb)MgAl_{11}O_{19}$ with a thickness 50 nm was selected as the gate dielectric as it exhibits a large bandgap sufficient to yield a positive band offset with respect to ZnO. The top gate electrode was e-beam-deposited Al/Pt/Au. Figure 5.36 shows a schematic of the ZnO nanowire MOSFET. A scanning electron micrograph of the completed device is shown in Figure 5.37. Note that less than 50% of the nanowire channel is covered by the gate metal. I–V characteristics were performed at room temperature using an Agilent 4155A Semiconductor Parameter Analyzer.

Figure 5.38 shows the drain current–drain voltage (I_D–V_{DS}) characteristics (top) and the transfer characteristics (bottom) measured at room temperature in the dark. The modulation of the channel conductance indicates that the operation of the device is an n-channel depletion mode. The gate leakage current is low, and the nanowire MOSFETs exhibit excellent saturation and pinch-off characteristics, indicating that the entire channel region under the gate metal can be depleted of electrons. The threshold voltage is ~3 V with a maximum transconductance of ~3 mS/mm. The on/off current ratio at V_G of 0–2.5 V and V_D of 10 V was of order 25. The field-effect mobility μ_{FE} can be determined from the

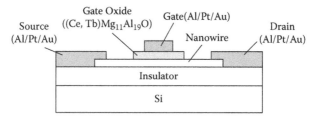

FIGURE 5.36   Schematic of ZnO nanowire depletion-mode FET.

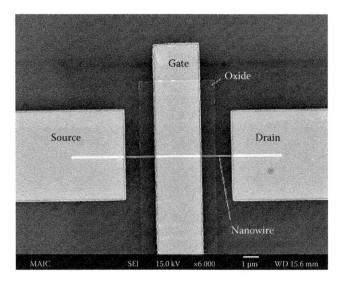

FIGURE 5.37   SEM micrograph of fabricated FET.

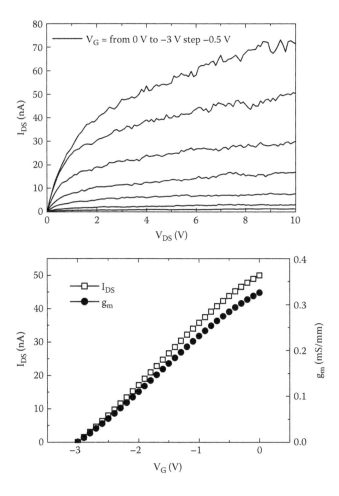

FIGURE 5.38 I_{DS}–V_{DS} (top) and transfer characteristics (bottom) of ZnO nanowire FET at room temperature in the dark.

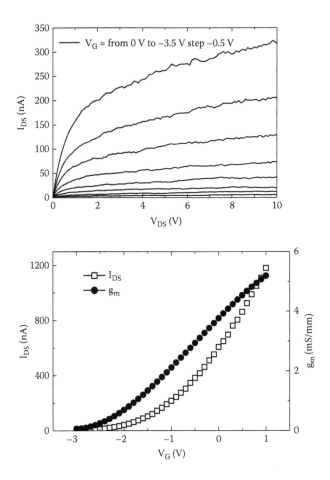

FIGURE 5.39 I_{DS}–V_{DS} (top) and transfer characteristics (bottom) of ZnO nanowire FET at room temperature measured with illumination from UV (366 nm) light.

transconductance using the relation $I_{DS} = (W/L)\,\mu_{FE}C_{OX}(V_{GS}-V_T)V_{DS}$, where W is the channel width, L is the channel, C_{OX} the gate oxide capacitance, and V_T is the threshold voltage. The extracted mobility was ~0.3 cm²/V-s, which is comparable to that reported previously for thin-film ZnO enhancement mode MOSFETs.[3] The carrier concentration in the channel is estimated to be ~10^{16} cm⁻³.

The nanowire showed strong photoresponse to UV illumination. Figure 5.39 shows the drain current–drain voltage (I_D–V_{DS}) characteristics (top) and the transfer characteristics (bottom) measured at room temperature under 366 nm illumination. The drain-source currents increase by approximately a factor of 5 and the maximum transconductance increases to 5 mS/mm. Part of the increase in drain-source current reflects photoconductivity in the portion of the wire uncovered by the gate metal. The on/off current ratio at V_G of 0–3 V and V_D of 10 V increases to ~125. In this case, the extracted mobility was still ~3 cm²/V-s, but the carrier density in the channel was ~5 × 10^{16} cm⁻³. The photoresponse was rapid, being limited by the switching characteristics of the UV lamp, which may suggest the photocurrent is bulk-related and there is not a strong contribution from surface states that produce long recovery times.

5.14 UV NANOWIRE PHOTODETECTORS

Recent reports have shown the sensitivity of ZnO nanowires to the presence of oxygen in the measurement ambient and to ultraviolet (UV) illumination. In the latter case, above-bandgap illumination was found to change the I–V characteristics of ZnO nanowires grown by thermal evaporation of ball-milled powders between two Au electrodes from rectifying to ohmic. In contrast, there was no change in effective built-in potential barrier between the ZnO nanowires and the contacts for below-bandgap illumination. The slow photoresponse of the nanowires was suggested to originate in the presence of surface states that trapped electrons with release time constants from milliseconds to hours.

Figure 5.40 shows the I–V characteristics of the nanowires in the dark and under illumination from 366 nm light. The conductivity is greatly increased as a result of the illumination, as evidenced by the higher current. No effect was observed for illumination with below-bandgap light. Transport measurements to be reported elsewhere show that the ideality factor of Pt Schottky diodes formed on the nanowires exhibit an ideality factor of 1.1, which suggests there is little recombination occurring in the nanowire. Note also the excellent ohmicity of the contacts to the nanowire, even at low bias. On blanket films of n-type ZnO with carrier concentration in the 10^{16} cm^{-3} range, we obtained contact resistance of $3–5 \times 10^{-5}$ Ω-cm^{-2} for these contacts. In the case of ZnO nanowires made by thermal evaporation, the I–V characteristics were rectifying in the dark and only became ohmic during above-bandgap illumination. The conductivity of the nanowire during illumination with 366 nm light was 0.2 Ω-cm.

Figure 5.41 shows the photoresponse of the single ZnO nanowire at a bias of 0.25 V under pulsed illumination from a 366 nm wavelength Hg lamp. The photoresponse is much faster than that reported for ZnO nanowires grown by thermal evaporation from ball-milled ZnO powders and likely is due to the reduced influence of the surface states seen in that material. The generally quoted mechanism for the photoconduction is creation of holes by the illumination that discharge the negatively charged oxygen ions on the nanowire surface, with detrapping of electrons and transit to the electrodes. The recombination times in high-quality ZnO measured from time-resolved photoluminescence are short, on the order of tens of picoseconds, while the photoresponse measures the electron trapping time. There is also a direct correlation reported between the photoluminescence lifetime and the defect density in both bulk and epitaxial ZnO. In our nanowires, the electron trapping times are on the order of tens of seconds, and these trapping effects are only a small fraction of the total photoresponse recovery characteristic. Note also the fairly constant peak photocurrent as the lamp is switched on, showing that that any traps present have discharged in the time frame of the measurement.

Figure 5.42 shows the photoresponse from the nanowires during pulsed illumination from either 254 or 366 nm light. The lower peak photocurrent in the former case may be related to the more

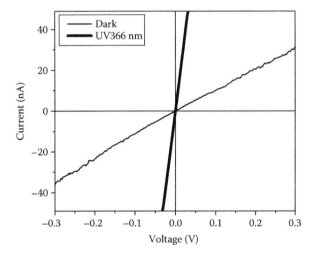

FIGURE 5.40 I–V characteristics from single ZnO nanowire measured at 25°C in the dark or under illumination from a 366 nm Hg lamp.

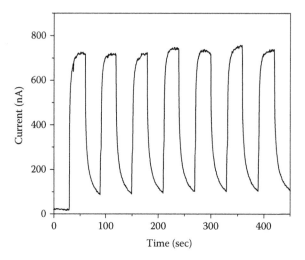

FIGURE 5.41 Time dependence of photocurrent as the 366 nm light source is modulated.

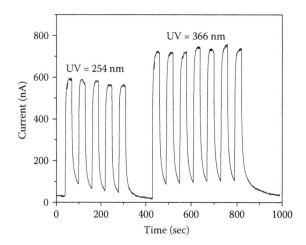

FIGURE 5.42 Time dependence of photocurrent in the nanowire as either a 254 nm or 366 nm light source is modulated.

efficient absorption near the surface of the nanowires. Once again we see an absence of the very long time constants for recovery seen in nanowires prepared by thermal evaporation.

In conclusion, single ZnO nanowires with excellent photocurrent response and I–V characteristics have been prepared by site-selective MBE. The I–V characteristics are linear even when measured in the dark, and the photoconduction appears predominantly to originate in bulk conduction processes with only a minor surface-trapping component. These devices look very promising for UV detection.

5.15 GAS AND CHEMICAL SENSING

The human immune system is extraordinarily complex. Antigen–antibody interactions are important for biological sensing. Antibodies are protein molecules that are composed of equal numbers of heavy and light polypeptide amino acid chains held together with disulfide bonds. These highly specialized proteins are able to recognize and bind only certain types of antigen molecules at receptor sites. Sensors based on using antibodies to detect certain specific antigens are called *immunosensors*. These immunosensors are useful for quantifying how well the human immune system is functioning, and could serve as valuable diagnostic tools. They also can be employed for identifying environmental contaminants and chemical or biological agents.

Silicon-based field-effect transistors have been widely used as biosensors. For example, the silicon-based ion-sensitive field-effect transistor (ISFET) is already commercialized to replace conventional electrolyte pH meters. Silicon-based metal oxide semiconductor field-effect transistors (MOSFETs) and ISFETs have also been used for immunosensors. However, silicon-based FETs are damaged by exposure to solutions containing ions. Si-FET-based biosensors need an ion-sensitive membrane coated over their surface, and a reference electrode (usually Ag/AgCl) is required to supply the bias voltage. Special precautions must also be taken to apply those biomembranes over the FETs so that they retain their enzymatic activity.

A large variety of ZnO one-dimensional structures have been demonstrated.[170,190,204–206] The large surface area of the nanorods and biosafe characteristics of ZnO makes them attractive for gas and chemical sensing and biomedical applications, and the ability to control their nucleation sites makes them candidates for microlasers or memory arrays.

Ideal sensors have the ability to discriminate between different gases, and arrays that contain different metal oxides (e.g., SnO_2, ZnO, CuO, and WO_3) on the same chip can be used to obtain this result. To date, most of the work on ZnO nanostructures has focused on the synthesis methods. There have been only a few reports of the electrical characteristics.[170,190] There is strong interest in developing gas, chemical, and biological sensors for use in both industry and domestic applications.

This part of the chapter will discuss gas, chemical, and biological sensing with ZnO nanowires, introduce the response characteristics of ZnO nanowires in hydrogen, ethylene, CO, Ozone, UV, and pH and, finally, provide the approaches to sense the biological agents with ZnO nanowires.

5.15.1 Hydrogen Sensing with ZnO Nanowires

One of the main demands for gas sensors is the ability to selectively detect hydrogen at room temperature in the presence of air. In addition, for most of these applications, the sensors should have very low power requirements and minimal weight. Nanostructures are natural candidates for this type of sensing. One important aspect is to increase their sensitivity for detecting gases such as hydrogen at low concentrations or temperatures, since typically an on-chip heater is used to increase the dissociation efficiency of molecular hydrogen to the atomic form and this adds complexity and power requirements.

Different metal-coating layers on multiple ZnO nanorods are compared for enhancing the sensitivity to detection of hydrogen at room temperature. Pt is found to be the most effective catalyst, followed by Pd. The resulting sensors are shown to be capable of detecting hydrogen in the range of ppm at room temperature using very small current and voltage requirements and recover quickly after the source of hydrogen is removed.

ZnO nanorods were grown by nucleating on an Al_2O_3 substrate coated with Au islands. For nominal Au film thicknesses of 20 Å, discontinuous Au islands are realized after annealing. The nanorods were deposited by Molecular Beam Epitaxy (MBE) with a base pressure of 5×10^{-8} mbar using a high-purity (99.9999%) Zn metal and an O_3/O_2 plasma discharge as the source chemicals. The Zn pressure was varied between 4×10^{-6} and 2×10^{-7} mbar, while the beam pressure of the O_3–O_2 mixture was varied between 5×10^{-6} and 5×10^{-4} mbar. The growth time was ~2 h at 600°C. The typical length of the resultant nanorods was 2~10 μm, with diameters in the range of 30–150 nm. Figure 5.43 shows a scanning electron micrograph of the as-grown rods. Selected area diffraction patterns showed the nanorods to be single-crystal. In some cases, the nanorods were coated with Pd, Pt, Au, Ni, Ag, or Ti thin films (~100 Å thick) deposited by sputtering.

Contacts to the multiple nanorods were formed using a shadow mask and sputtering of Al/Ti/Au electrodes. The separation of the electrodes was ~30 um. A schematic of the resulting sensor is shown in Figure 5.44. Au wires were bonded to the contact pad for I–V measurements performed at 25°C in a range of different ambients (N_2, O_2, or 10–500 ppm H_2 in N_2). Note that no currents were

FIGURE 5.43   SEM of ZnO multiple nanorods.

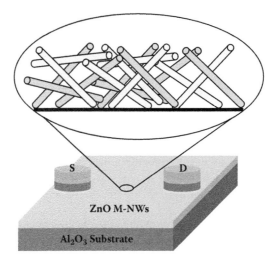

FIGURE 5.44   Schematic of contact geometry for multiple nanorod gas sensor.

measured through the discontinuous Au islands, and no thin film of ZnO on the sapphire substrate was observed with the growth condition for the nanorods. Therefore, the measured currents are due to transport through the nanorods themselves. The I–V characteristics from the multiple nanorods were linear, with typical currents of 0.8 mA at an applied bias of 0.5 V.

Figure 5.45 shows the time dependence of relative resistance change of either metal-coated or uncoated multiple ZnO nanorods as the gas ambient is switched from N_2 to 500 ppm of H_2 in air and then back to N_2 as time proceeds. These were measured a bias voltage of 0.5 V. The first point of note is that there is a strong increase (approximately a factor of 5) in the response of the Pt-coated nanorods to hydrogen relative to the uncoated devices. The maximum response was ~8%. There is also a strong enhancement in response with Pd coatings, but the other metals produce little or no change. This is consistent with the known catalytic properties of these metals for hydrogen dissociation. Pd has a higher permeability than Pt, but the solubility of H_2 is larger in the former.[207]

Moreover, studies of the bonding of H to Ni, Pd, and Pt surfaces have shown that the adsorption energy is lowest on Pt.[208] There was no response of either type of nanorod to the presence of O_2 in

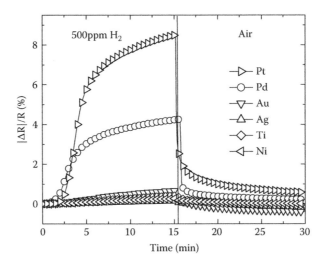

FIGURE 5.45 Time dependence of relative resistance response of metal-coated multiple ZnO nanorods as the gas ambient is switched from N_2 to 500 ppm of H_2 in air as time proceeds. There was no response to O_2.

the ambient at room temperature. Once the hydrogen is removed from the ambient, the recovery of the initial resistance is rapid (<20 s). In sharp contrast, upon introduction of hydrogen, the effective nanorod resistance continued to change for periods of >15 min. This suggests that the kinetics of the chemisorption of molecular hydrogen onto the metal and its dissociation to atomic hydrogen are the rate-limiting steps in the resulting change in conductance of the ZnO.[175]

An activation energy of 12 KJ/mole was calculated from a plot of the rate of change of nanorod resistance. This energy is somewhat larger than that of a typical diffusion process and suggests that the rate-limiting step mechanism for this sensing process is more likely to be the chemisorption of hydrogen on the Pd surface. This reversible change in conductance of metal oxides upon chemisorption of reactive gases has been discussed previously.[208] The gas-sensing mechanisms suggested in the past include the desorption of adsorbed surface hydrogen and grain boundaries in poly-ZnO,[209] exchange of charges between adsorbed gas species and the ZnO surface leading to changes in depletion depth,[178] and changes in surface or grain boundary conduction by gas adsorption/desorption.[179] Finally, Figure 5.41 shows an incubation time for response of the sensors to hydrogen. This could be due to some of the Pd becoming covered with native oxide, which is removed by exposure to hydrogen. A potential solution is to use a bilayer deposition of the Pd followed by a very thin Au layer to protect the Pd from oxidation. However, this adds to the complexity and cost of the process and, since the Pd is not a continuous film, the optimum coverage of Au would need to be determined. We should also point out that the I–V characteristics were the same when measured in a vacuum as in the air, indicating that the sensors are not sensitive to humidity.

The power requirements for the sensors were very low. Figure 5.46 shows the I–V characteristics measured at 25°C in both a pure N_2 ambient and after 15 min in a 500 ppm H_2 in an N_2 ambient. Under these conditions, the resistance response is 8% and is achieved for a power requirement of only 0.4 mW. This compares well with competing technologies for hydrogen detection such as Pd-loaded carbon nanotubes.[179,180] Moreover, the 8% response compares very well to the existing SiC-based sensors, which operate at temperatures >100°C through an on-chip heater in order to enhance the hydrogen dissociation efficiency. Figure 5.47 shows that the sensors can detect 100 ppm H_2.

5.15.2 OZONE SENSING

Ozone gas ambient influences the I–V characteristics of multiple ZnO nanorods prepared by site-selective MBE. These structures can readily detect a few percent of ozone in N_2 at room temperature.

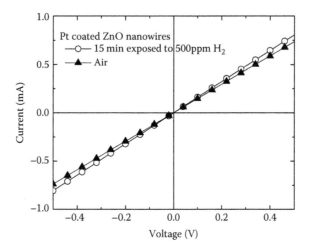

FIGURE 5.46   I–V characteristic of Pt-coated nanorods in air and after 15 min in 500 ppm H$_2$ in air.

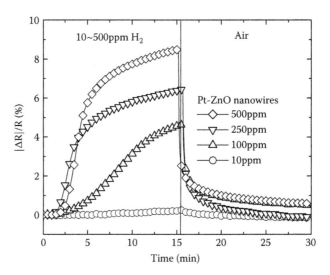

FIGURE 5.47   Time dependence of resistance change of Pd-coated multiple ZnO nanorods as the gas ambient is switched from N$_2$ to various concentrations of H$_2$ in air (10–500 ppm) and then back to N$_2$.

Over a limited range of partial pressures of O$_3$ (P$_{OZONE}$) in the ambient gas, the conductance G of the sensor at fixed bias voltage decreased according to the relation $G = (G_O + A(P_{OZONE})^{0.5})^{-1}$, where A is a constant and G_O the resistance in N$_2$.

The site-selective growth of the nanorods has been described in a previous session. Selected area diffraction patterns showed the nanorods to be single-crystal. Figure 5.48 shows a transmission electron micrograph of a single ZnO nanorod. E-beam lithography was used to pattern sputtered Al/Ti/Au electrodes contacting both ends of multiple nanorods on Al$_2$O$_3$ using shadow mask. The separation of the electrodes was ~650 μm. A scanning electron micrograph of the completed device is shown in Figure 5.48. Au wires were bonded to the contact pad for I–V measurements performed over the range 25°C–150°C in a range of different ambients (N$_2$, 3% O$_3$ in N$_2$). Note that no currents were measured through the discontinuous Au islands and no thin film of ZnO was observed with the growth condition of nanorods.

The nanorods were more sensitive to the presence of ozone in the measurement ambient. Figure 5.49 shows the room-temperature I–V characteristics from the multiple nanorods measured in pure N$_2$ or

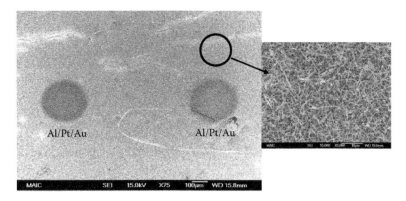

FIGURE 5.48 SEM of ZnO multiple nanorods and the pattern contacted by Al/Pt/Au electrodes.

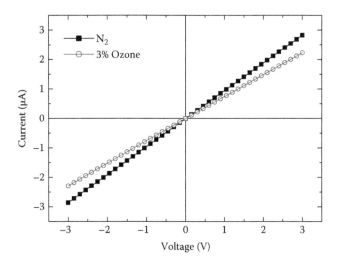

FIGURE 5.49 I–V characteristics at 25°C of ZnO multiple nanorods measured in either N_2 or 3% O_3 in O_2.

3% O_3 in O_2. The changes in current are much larger than for the case of hydrogen detection. Over a relatively limited range of O_3 partial pressures in the N_2 ambient, the conductance increased according to $G = (G_O + A(P_{O3})^{0.5})^{-1}$, where A is a constant and G_O the resistance in a pure N_2 ambient. This is a similar dependence to the case of CO detection by SnO_2 conduction sensors, in which the effective conductance increases as the square root of CO partial pressure. The gas sensitivity can be calculated from the difference in conductance in O_3-containing ambients, divided by the conductance in pure N_2, that is, $(G_{N2} - G_O)/G_{N2}$. At 25°C, the gas sensitivity was 21% for 3% O_3 in O_2.

Figure 5.50 shows the time dependence of change in a current at a fixed voltage of 1 V when switching from back and forth from N_2 to 3% O_3 in O_2 ambients. The recovery time constant is long (>10 min) so that the nanorods are best suited to the initial detection of ozone rather than to determining the actual time dependence of change in concentration. In the latter case, a much faster recovery time would be needed. The gas sweep-out times in our test chamber are relatively short (about a few seconds) and, therefore, the long recovery time is intrinsic to the nanorods.

5.15.3 pH Response

ZnO nanorod surfaces respond electrically to variations of the pH in electrolyte solutions introduced via an integrated microchannel. The ion-induced changes in surface potential are readily

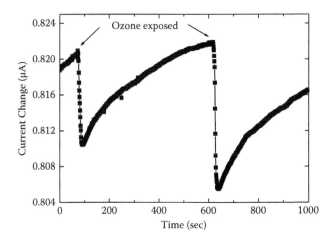

FIGURE 5.50   Time dependence of a current at 1 V bias when switching back and forth from N_2 to 3% O_3 in O_2 ambients.

measured as a change in conductance of the single ZnO nanorods and suggest that these structures are very promising for a wide variety of sensor applications.

The preparation of the nanorods has been described in a Section 5.11. They were released from the substrate by sonication in ethanol and then transferred to SiO_2-coated Si substrates. E-beam lithography was used to pattern sputtered Al/Pt/Au electrodes contacting both ends of a single nanorod. The separation of the electrodes was ~3.7 µm. Au wires were bonded to the contact pad for I–V measurements to be performed. An integrated microchannel was made from SYLGARD@ 184 polymer from Dow Corning. After mixing this silicone elastomer with a curing agent using a weight ratio of 10:1 and mixing with for 5 min, the solution was evacuated for 30 min to remove residual air bubbles. It was then applied to the already-etched Si wafer (channel length, 10–100 µm) in a cleaned and degreased container to make a molding pattern. Another vacuum deairing for 5 min was used to remove air bubbles, followed by curing for 2 h at 90°C. After taking the sample from the oven, the film was peeled from the bottom of the container. Entry and exit holes in the ends of the channel were made with a small puncher (diameter less than 1 mm), and the film was immediately applied to the nanorod sensor. The pH solution was applied using a syringe autopipette (2–20 µL). A schematic of the structure and a scanning electron micrograph of the completed device are shown in Figure 5.51.

Prior to the pH measurements, we used pH 4, 7, and 10 buffer solutions from Fisher Scientific to calibrate the electrode, and the measurements at 25°C were carried out in the dark or under ultraviolet (UV) illumination from 365 nm light using an Agilent 4156°C parameter analyzer to avoid parasitic effects. The pH solution was made by the titration method using HNO_3, NaOH, and distilled water. The electrode was a conventional Acumet standard Ag/AgCl electrode.

Figure 5.52 shows the I–V characteristics from the single ZnO nanorod after wire bonding, both in the dark and under UV illumination. The nanorods show a very strong photoresponse. The conductivity is greatly increased as a result of the illumination, as evidenced by the higher current. No effect was observed for illumination with below-bandgap light. The photoconduction appears predominantly to originate in bulk conduction processes with only a minor surface trapping component.[210]

The adsorption of polar molecules on the surface of ZnO affects the surface potential and device characteristics. Figure 5.53 (top) shows the current at a bias of 0.5 V as a function of time from nanorods exposed for 60 s to a series of solutions whose pH was varied from 2 to 12. The current is significantly reduced upon exposure to these polar liquids as the pH is increased. The corresponding nanorod conductance during exposure to the solutions is shown at the bottom of Figure 5.53. The I–V in this figure was measured at pH at 7 (2.8×10^{-8} A without UV). The experiment was conducted starting at pH = 7 and went to pH = 2 or 12. The I–V measurement in the air was slightly higher that in the pH = 7

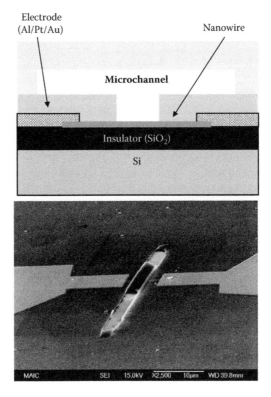

FIGURE 5.51 Schematic (top) and SEM (bottom) of ZnO nanorod with integrated microchannel (4 μm width). (Reprinted by permission from Kang et al. *APL*, 86, 112105 (2005). copyright AIP.)

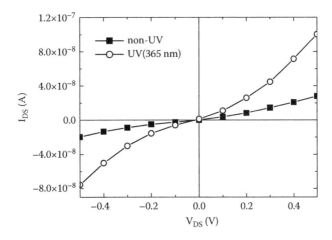

FIGURE 5.52 I–V characteristics of ZnO nanorod after wire-bonding, measured either with or without UV (365 nm) illumination. (Reprinted by permission from Kang et al. APL, 86, 112105 (2005). Copyright AIP.)

(10%–20%). The data in Figure 5.53 shows that the HEMT sensor is sensitive to the concentration of the polar liquid and therefore could be used to differentiate between liquids into which a small amount of leakage of another substance has occurred. The conductance of the rods is higher under UV illumination, but the percentage change in conductance is similar with and without illumination.

Figure 5.54 shows the conductance of the nanorods in either the dark or during UV illumination at a bias of 0.5 V for different pH values. The nanorods exhibit a linear change in conductance between

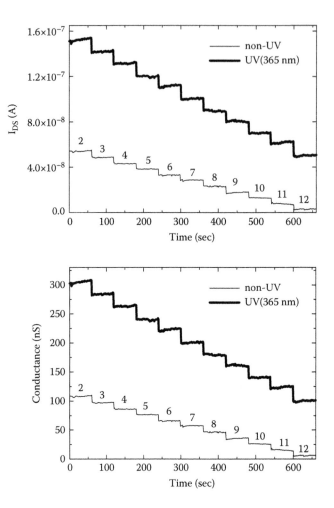

FIGURE 5.53   Change in current (top) or conductance (bottom) with pH (from 2 to 12) at V = 0.5 V. (Reprinted by permission from Kang et al. *APL*, 86, 112105 (2005). Copyright AIP.)

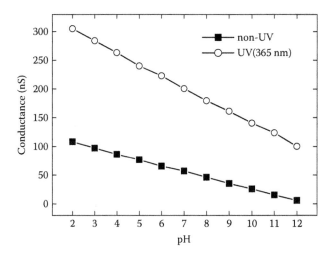

FIGURE 5.54   Relation between pH and conductance of ZnO nanorod either with or without UV (365 nm) illumination.

pH 2 and 12 of 8.5 nS/pH in the dark and 20 nS/pH when illuminated with UV (365 nm) light. The nanorods show stable operation with a resolution of ~0.1 pH over the entire pH range, showing the remarkable sensitivity of the HEMT to relatively small changes in concentration of the liquid.

There is still much to understand about the mechanism of the current reduction in relation to the adsorption of the polar liquid molecules on the ZnO surface. It is clear that these molecules are bonded by van der Waals–type interactions and that they screen a surface change that is induced by polarization in the ZnO. Different chemicals are likely to exhibit degrees of interaction with the ZnO surface. Steinhoff et al.[211] found a linear response to changes in the pH range 2–12 for ungated GaN-based transistor structures and suggested that the native metal oxide on the semiconductor surface is responsible.

5.16 BIOLOGICAL SENSING

The development of advanced biological sensors could significantly impact the areas of genomics, proteomics, biomedical diagnostics, and drug discovery. In this regard, nanoscale sensors based on nanowires have received considerable recent attention. Nanowires can be used for label-free, direct real-time electrical detection of biomolecule binding and have the potential for very high sensitivity detection since the depletion or accumulation of charge carriers, which are caused by binding of charged macromolecules to receptors linked to the device surfaces, can affect the entire cross-sectional conduction pathway of these nanostructures and change conductance. In addition, with electric measurements, one can always make use of existing facility and equipment. In the following, we will discuss approaches where biosensors can be manufactured with oxide nanostructures.

5.16.1 Surface Modification of ZnO

The first step in biological sensing is to identify those types of peptides that exhibit significant binding to the ZnO surface. An inorganic may be recognized physically or chemically through its surface composition, structure, crystallography, or morphology. This effort focuses on the chemical ligand (amino acid) recognition of surfaces. Determination of this preferred recognition will be accomplished using a phage display library with multiple rounds of target binding, elution, and amplification of the specifically bound phage. Analysis of the presence and position of certain amino acids should provide insight into the specific binding domains for a given inorganic or class of inorganics.

On metal surfaces, hydrophobic peptides with hydroxyl amino acids are prevalent. Tight binders on metal oxide surfaces, such as ZnO, also show hydroxyl functional groups with a high concentration of basic amino acids such as arginine.[212,213] Similarly, GaAs was shown to possess selectivity for Lewis-base functional groups[214] Short peptide domains containing serine- and threonine-rich regions along with the presence of amine Lewis bases such as asparagine and glutamine were found to predominate. GaAs possesses a Lewis acid surface and should also provide a good interaction with basic amino acids. However, it also has some similarity to ZnO, which showed a surprising attraction for at least one peptide with no basic amino acids.[215]

ZnO surfaces can be functionalized with bioreactive amine groups through a silanization reaction. Aminopropyl triethoxysilane (APS) is dissolved in water and reacted with clean ZnO and oxidized GaN surfaces. Condensation reactions between the silane and the hydroxyl groups on the metal oxide produce stable silicon-oxygen-metal bonds. The amine functionality on the silane allows a wide variety of bioconjugation reactions to be performed. Initial tests involved the addition of biotin to the surface with NHS chemistry. This immobilized biotin can undergo a nearly irreversible antigen–antibody bonding with streptavidin, a large protein (Figure 5.55, left).

To prove the effectiveness of the modification technique, a confocal microscope was used to image an amine-coated GaN surface with regions of biotin modification. Solutions of ruthenium bipyridine

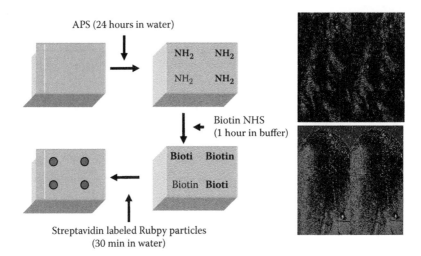

FIGURE 5.55 (See color insert).   (Left) Schematics of surface modification of nanowires for biomolecule immobilization for biosensor and biochip applications. (Right) Preliminary results of surface modification: confocal images of a GaN and ZnO surface modified with avidin binding with Rubpy nanoparticles through biotin–avidin linkage (bright image); The same surface modified with BSA and showed no binding to the nanoparticles (dark image).

(Rubpy) dye–doped nanoparticles coated with either streptavidin or bovine serum albumin (BSA) were incubated on the GaN surface. To the right is an image of the streptavidin-coated nanoparticles immobilized on the biotinylated surface and on the left is a biotinylated surface incubated with BSA-coated particles. The biotinylated surface specifically reacted with the streptavidin and not with BSA, immobilizing the fluorescent particles. Nonbiotinylated surfaces showed low background emission from the particles. After immobilization of the particles, washing in buffer solution did not reduce the fluorescence intensity. This indicates that modification with biotin creates a surface capable of specific and strong antigen–antibody conjugation, as shown in Figure 5.55 (right).

The same chemistry will be applied to ZnO gates on FETs and on nanowires. The biotin–streptavidin system acts as a model for other antigen–antibody systems and can be extended to immobilization of oligonucleotide DNA sequences on to the surface.[212–215] Additional silane chemistries will become possible, including the attachment of thiol or carboxylic acid groups to the sensor surface.[216] With this chemistry, semiconductor sensors can be modified to specifically conjugate to a wide number of biomolecules of interest. Furthermore, attachment of the target analyte to another component, such as a molecule or a nanoparticle with a strong electron donating or accepting character, will allow the amplification of the conductivity change in the semiconductor.

5.16.2   USING PROTEIN IMMOBILIZATION FOR ULTRASENSITIVE DETECTION OF PROTEINS AND SINGLE VIRUSES

Viruses are among the most important causes of human disease and an increasing concern as agents for biological warfare and bioterrorism. Rapid, selective, and sensitive detection of viruses is central to implementing an effective response to viral infection, such as through medication or quarantine. Established methods for viral analysis include plaque assays, immunological assays, transmission electron microscopy, and PCR-based testing of viral nucleic acids.[217–221] These methods have not achieved rapid detection at single virus level and often require a relative high level of sample manipulation that is inconvenient for infectious materials. One promising approach for the direct electrical detection of biological macromolecules will be to use semiconducting nanowires.

A two-step procedure will be used to covalently link antibody receptors to the surfaces of the ZnO nanowire devices. First, the devices are reacted with a 1% ethanol solution of 3-(trimethoxysilyl) propyl aldehyde for 30 min, washed with ethanol, and heated at 120°C for 15 min. Second, mAb receptors are coupled to the aldehyde-terminated nanowire surfaces by the reaction of a 10–100 µg/mL antibody in a pH 8, 10 mM phosphate buffer solution containing 4 mM sodium cyanoborohydride. The surface density of antibody is controlled by varying the reaction time from 10 min (low density) to 3 h (high density). Unreacted aldehyde surface groups are subsequently passivated by the reaction with ethanolamine under similar conditions. Device arrays for multiplexed experiments are made in the same way except that distinct antibody solutions are spotted on different regions of the array. The antibody surface density versus reaction time is quantified by reacting Au-labeled IgG antibodies with aldehyde-terminated nanowires on a transmission electron microscopy grid, and then imaging the modified nanowire by transmission electron microscopy, which enables the number of antibodies per unit length of nanowire to be counted.

Virus samples are delivered to the nanowire device arrays by using fluidic channels formed by either a flexible polymer channel or a 0.1-mm-thick glass coverslip sealed to the device chip. Virus samples are delivered through an inlet/outlet connection in the polymer or holes made through the back of the device chip in the case of the coverslip. Similar electrical results are obtained with both approaches, although the latter is used for all combined electrical/optical measurements.

The same strategy discussed easily can be easily adopted for protein studies, where antibodies can be immobilized for target protein detection, making these nanowires suitable for proteomics applications. It is necessary to mention here that the protein-modified nanowires will make multiplex analysis of proteins easy, producible, highly sensitive, and quantitative.

5.16.3 IMMOBILIZATION OF NUCLEIC ACIDS ON NANOWIRES FOR BIOSENSORS FOR GENES AND mRNA

The ZnO nanowire devices are functionalized with DNA, PNA (peptide nucleic acid), or LNA (locked nucleic acid) probes using the intervening avidin protein layer. This can be done through the surface modification discussed earlier. A variety of surface modification schemes have been developed for biomolecule immobilization.[219–221] PNA and LNA can be chosen as recognition sequences since they are known to bind to DNA/mRNA with much greater affinity and stability than corresponding DNA recognition sequences, and have shown binding with single base specificity. A microfluidic delivery system can be used to flow the complementary or mismatched DNA/mRNA samples. Direct comparison of the data will highlight the substantial net conductance change associated with hybridization of the DNA complementary to the receptor. This could enable high-throughput, highly sensitive DNA detection for basic biology research, disease diagnosis, and genetic screening.

5.16.4 DIRECT IMMOBILIZATION OF APTAMERS FOR ULTRASENSITIVE DETECTION OF PROTEINS AND DRUG MOLECULES

Antibodies can be assembled onto nanowires, as mentioned in the previous part on protein recognition. Similarly, aptamers, designer DNA/RNA probes for target protein recognition and small molecular analysis,[219,220] can also be immobilized. Upon the addition of target protein molecules, the device can function as ultrasensitive and selective detectors for protein recognition (Figure 5.56). Aptamers can also be used for analyzing small molecules such as amino acids and cocaine.[220] If the aptamer for cocaine is immobilized onto the nanowire, an ultrasensitive biosensor for this substance will be possible.

FIGURE 5.56   Schematic diagram of exposed nanowire and aptamer flow.

5.16.5   ZnO Nanowires with Various Doping and Surface Chemistry Terminations

It will be a basic step to synthesize ZnO nanowires with varying transport properties (n-type to insulating to p-type) and surface termination. The former provides the opportunity to control the electrical response to surface absorption via depletion of the nanowire interior. The latter, surface termination, will largely control both the chemistry of surface functionalization as well as electronic response to absorption events.

Electron doping will be accomplished via defects originating from Zn interstitials in the ZnO lattice. The intrinsic defect levels that lead to n-type doping lie approximately 0.05 eV below the conduction band. Insulating material requires a reduction in oxygen vacancies in undoped material. One of the significant challenges with ZnO is the difficulty in producing p-type material. Recently, we investigated the use of P as an acceptor dopant in ZnO. The C–V properties of metal/insulator/P-doped (Zn,Mg)O diode structures were measured and found to exhibit a polarity consistent with the P-doped (Zn,Mg)O layer being a p-type.

For detectors that rely on surface absorption modulation of subsurface transport, the ability to selectively terminate the surface of the nanowires with various cations is attractive. In thin-film semiconductor research, the formation of heteroepitaxial interfaces has proved to be enabling in the development of numerous device concepts, as well as in the investigation of low-dimensional phenomena. The synthesis of 1-D linear heterostructures is scientifically interesting and potentially useful, particularly if a technique is employed that allows for spatial selectivity in nanowire placement.

The heteroepitaxial cored nanostructures are based on the (Zn,Mg)O alloy system, and were synthesized using the catalysis-driven molecular beam epitaxy method shown in Figure 5.57.

Doping ZnO with Mg widens the gap in this wurtzite structure compound. Site-selective nucleation and growth of cored nanorods were achieved by coating Si substrates with Ag islands. For a nominal Ag film thickness of 20 Å, discontinuous Ag islands were realized. On these small metal catalyst islands, (Zn,Mg)O nanorods were observed to grow. Under continuous Zn, Mg, and O flux, nanorod material nucleates on the catalyst particles. Close inspection of the nanorod structure indicates that the Zn and Mg concentrations are not uniformly distributed in the rods. Instead, there is a radial segregation into Zn-rich and Mg-rich regions, apparently reflecting differences in the solubility limits of bulk ZnO-MgO or Zn-Mg-Ag-O versus epitaxial solid solutions. In bulk material, the solubility of Mg in ZnO is relatively low, on the order of 4 at%. In contrast, Mg content as high as $Zn_{0.67}Mg_{0.33}O$ has been reported to be metastable in the wurtzite structure for epitaxial thin films. For this composition, the bandgap of ZnO can be increased to ~3.8 eV. For (Zn,Mg)O nanorod growth, it appears that both growth modes are relevant, but for different regions in the rod. Under low-temperature MBE growth conditions, a solubility-driven segregation occurs during the catalyst-driven core formation, the core composition being determined by bulk solid solubility. Subsequently, an epitaxial sheath grows, with Mg content and crystal structure being determined

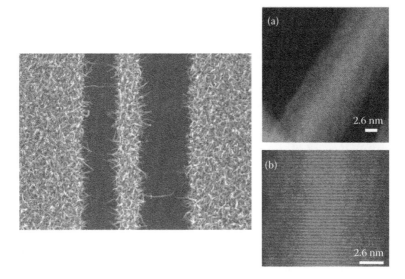

FIGURE 5.57 (See color insert).   SEM (left) and Z-STEM (right) of cored ZnMgO nanorods grown on Ag-coated Si substrate.

by epitaxial stabilization. The net result is the growth of (Zn,Mg)O nanorods that are not uniform in composition across the diameter, but distinctly cored. Figure 5.57 (right) shows a high-resolution Z-STEM image of a nanorod grown under the conditions described. The lattice image for the nano-rod specimen indicates that the rod is crystalline with the wurtzite crystal structure maintained throughout the cross section. The c axis is oriented along the long axis of the rod. The higher con-trast for the center core region clearly indicates a higher cation atomic mass. The structures consist of a zinc-rich $Zn_{1-x}Mg_xO$ core (small x) surrounded by a $Zn_{1-y}Mg_yO$ (large y) sheath containing a higher Mg content. Given the band offset between the ZnO core and (Zn,Mg)O sheath, the nanowire structure can easily function as an ISFET with the (Zn,Mg)O serving as the insulating dielectric. Using this approach, ZnO/(Zn,Mg)O sheathed nanowires can be synthesized for conductance mea-surements and surface functionalization.

5.16.6   A 3D Self-Consistent Simulator for a Biological Sensor

Biomolecular sensing in experiments is achieved by measuring the change of electrical conductance through ZnO nanorods. In order to model the change of nanowire conductance due to the presence of charge on biomolecules, a 3D Poisson equation is self-consistently solved with the carrier trans-port equation for the nanorod, in the presence of the charge on biomolecules.

Figure 5.58 shows the simulation results. A 3D Poisson equation is self-consistently solved with the carrier transport equation for a device structure as shown in Figure 5.58 (left), in the presence of the biomolecular charge. Figure 5.58 (right) plots the conduction band edge at equilibrium with different charge on a charge dot above the middle of the tube. It shows that when the charge dot is close to the nanorod surface (5 nm away), change of a single electron charge on the dot results in an obvious change of the potential profile. The potential barrier created by the charge on the dot has a significant effect on the carrier transport through a quasi-one-dimensional wire, which indicates that biological sensors with ultrahigh sensitivity (single-electron sensitivity) may be achieved by using a nanorod as the channel.

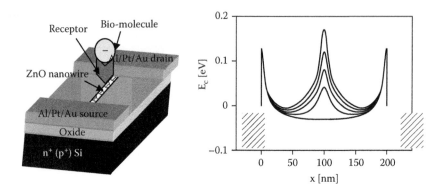

FIGURE 5.58 (Left) Schematic picture of the modeled ZnO nanorod biological nanosensor, a device simula-
tor under development that self-consistently solves the 3D Poisson equation and the carrier transport equation
to model the nanosensor. To consider electrostatic sensing, the biomolecule is modeled as a charge dot above
the nanorod. (Right) The conduction band profile for different amounts of charges on the biomolecule. The
channel length is 200 nm. A molecule with a uniform charge distribution with a diameter of 1 nm is placed
5 nm above the middle of the nanorod. The charge on the dot varies from 0 to $-4e$ at $-e$/step. The metal-
nanorod Schottky barrier height is ~0.13 eV.

5.17   SUMMARY

Major advances have been made by many groups around the world recently in demonstrating a
p-type ZnO, fabricating improved ohmic contacts, developing high-resolution dry etch pattern
transfer processes and implant isolation processes, and in realizing room-temperature ferromagnetism in ZnO. There are still many areas that need additional work, including the following:

1. Higher p-type doping levels in epitaxial films. This will require even better control of the
   background n-type conductivity in the material that arises from native defects and impurities such as hydrogen.
2. Realization of high-quality p-n junctions by epitaxial growth, with good breakdown characteristics. These are the building blocks for devices such as UV LEDs or lasers.
3. Improved Schottky contacts with higher barrier height than reported to date. This will
   need a better understanding of surface-cleaning processes and metal schemes that do not
   disrupt the ZnO surface during their deposition. In particular, it is necessary to understand
   the nature of the defects present on ZnO surfaces that may be pinning the Fermi level.
4. P-type ohmic contacts with lower specific contact resistance. A desirable goal is to achieve
   a contact resistance of 10^{-5} Ω-cm^2 or less for ZnO-based light emitters so that self-heating
   in the contact region does not degrade their reliability.
5. Improved plasma-etching processes that utilize simple photoresist masking and are able to
   achieve practical etch rates for ZnO. It is also necessary to develop selective etch processes
   for the ZnMgO/ZnO heterostructure.
6. Development of implant doping processes, to realize both n- and p-type layers. This will
   require a better understanding of the defects created by the implant step and their thermal
   stability and also a better understanding of the stability of the ZnO surface during activation annealing.
7. Clear demonstrations of spin-polarized carrier distributions in Mn-doped ZnO, such as the
   presence of the anomalous Hall effect or spin-polarized light emission.
8. Pt-coated ZnO nanorods appear well suited to detection of ppm concentrations of
   hydrogen at room temperature. The recovery characteristics are fast upon removal of

hydrogen from the ambient. Bulk ZnO Schottky diodes appear well suited to detection of CO and C_2H_4. They are sensitive at temperatures as low as 90°C for percentage levels of CO in N_2. The response is time dependent, with the characteristics of a diffusion-controlled process.

9. ZnO nanorods appear well suited to the detection of O_3. They are sensitive at temperatures as low as 25°C to percent levels of O_3 in O_2. The recovery characteristics are quite slow at room temperature, indicating that the nanorods can be used only for the initial detection of ozone.

10. Single ZnO nanorods with excellent photocurrent response and I–V characteristics have been prepared by site-selective MBE. The I–V characteristics are linear even when measured in the dark, and the photoconduction appears predominantly to originate in bulk conduction processes with only a minor surface-trapping component. These devices look very promising for UV detection.

11. ZnO nanorods show dramatic changes in conductance upon exposure to polar liquids. Bonding of polar liquid molecules appears to alter the polarization-induced positive surface change, leading to changes in the effective carrier density and hence the drain-source current in biased nanorods. This suggests the possibility of functionalizing the surface for application as biosensors, especially given the excellent biocompatibility of the ZnO surfaces, which should minimize degradation of adsorbed cells.

12. The sensitivities of ZnO in gas, chemical, and biological agents make it attractive for sensor applications. The ZnO nanorods can be placed on cheap transparent substrates such as glass, making them attractive for low-cost sensing applications, and can operate at very low power conditions. Of course, there are many issues still to be addressed, in particular regarding the reliability and long-term reproducibility of the sensor response before they can be considered for space flight applications.

It is likely that many of these areas will see significant progress in the near future as the promise of ZnO for low-cost transparent electronics and UV light emission becomes more widely known.

ACKNOWLEDGMENTS

The work at UF is partially supported by AFOSR grant under grant number F49620-03-1-0370 (T.Steiner), by AFOSR (F49620-02-1-0366, G. Witt), NSF(CTS-0301178, monitored by Dr. M. Burka and Dr. D. Senich), by NASA Kennedy Space Center Grant NAG 10-316 monitored by Mr. Daniel E. Fitch , ONR (N00014-98-1-02-04, H. B. Dietrich), and NSF DMR 0101438.

REFERENCES

1. see for example, H. L. Hartnagel, A. L. Dawar, A. K. Jain, and C. Jagadish, *Semiconducting Transparent Thin Films* (IOP Publishing, Bristol, 1995).

2. D. C. Look, *Mater. Sci. Eng.* B80, 383 (2001).

3. P. Zu, Z. K. Tang, G. K. L. Wong, M. Kawasaki, A. Ohtomo, K. Koinuma, and Y. Sagawa, *Solid-State Commun.* 103, 459 (1997).

4. D. M. Bagnall, Y. R. Chen, Z. Zhu, T. Yao, S. Koyama, M. Y. Shen, and T. Goto, *Appl. Phys. Lett.* 70, 2230 (1997).

5. M. Wraback, H. Shen, S. Liang. C. R. Gorla, and Y. Lu, *Appl. Phys. Lett.* 74, 507 (1999).

6. J.-M. Lee, K.-K. Kim, S.-J. Park, and W.-K. Choi, *Appl. Phys. Lett.* 78, 2842 (2001).

7. J. E. Nause, *III-V's Review* 12, 28 (1999).

8. Y. Chen, D. Bagnell, and T. Yao, *Mat. Sci. Eng.* B75, 190 (2000).

9. D. C. Look, D. C. Reynolds, J. W. Hemsky, R. L. Jones, and J. R. Sizelove, *Appl. Phys. Lett.* 75, 811 (1999).

10. D. C. Look, J. W. Hemsky, and J. R. Sizelove, *Phys. Rev. Lett.* 82, 2552 (1999).

11. F. D. Auret, S. A. Goodman, M. Hayes, M. J. Legodi, H. A. van Laarhoven, and D. C. Look, *Appl. Phys. Lett.* 80, 956 (2002).

12. S. O. Kucheyev, J. E. Bradley, J. S. Williams, C. Jagadish, and M. V. Swain, *Appl. Phys. Lett.* 80, 956 (2002).

13. D. C. Reynolds, D. C. Look, and B. Jogai, *Solid-State Commun.* 99, 873 (1996).

14. M. Wraback, H. Shen, S. Liang, C. R. Gorla, and Y. Lu, *Appl. Phys. Lett.* 76, 507 (1999).

15. T. Aoki, D. C. Look, and Y. Hatanaka, *Appl. Phys. Lett.* 76, 3257 (2000).

16. C. C. Chang and Y. E. Chen, *IEEE Trans. Ultrasonics, Ferroelectrics and Frequency Control* 44, 624 (1997).

17. P. M. Verghese and D. R. Clarke, *J. Appl. Phys.* 87, 4430 (2000).

18. C. R. Gorla, N. W. Emanetoglu, S. Liang, W. E. Mayo, Y. Lu, M. Wraback, and H. Shen, *J. Appl. Phys.* 85, 2595 (1999).

19. H. Ohta, K. Kawamura, M. Orita, M. Hirano, N. Sarukura, and H. Hosono, *Appl. Phys. Lett.* 77, 475 (2000).

20. M. Joseph, H. Tabata, and T. Kawai, *Jpn. J. Appl. Phys.* 38, L1205 (1999).

21. S. Krishnamoorthy, A. A. Iliadis, A. Inumpudi, S. Choopun, R. D. Vispute, and T. Venkatesan, *Solid-State Electron.* 46, 1631 (2002).

22. Y. Li, G. S. Tompa, S. Liang, C. Gorla, C. Lu, and J. Doyle, *J. Vac. Sci. Technol.* A15, 1663 (1997).

23. see for example, the discussion at http://ncsr.csci-va.com/materials/zno.asp.

24. L. K. Singh and H. Mohan, *Indian J. Pure Appl. Phys.* 13, 486 (1975).

25. A. R. Hutson, *Phys. Rev.* 108, 222 (1957).

26. D. C. Look, J. W. Hemsky, and J. R. Sizelove, *Phys. Rev. Lett.* 82, 2552 (1999).

27. B. J. Jin, S. H. Bae, S. Y. Lee, and S. Im, *Mat. Sci. Eng..* B71, 301 (2000).

28. D. M. Hofmann, A. Hofstaetter, F. Leiter, Huijuan Zhou, F. Henecker, B. K. Meyer, S. B. Orlinskii, J. Schmidt, and P. G. Baranov, *Phys. Rev. Lett.* 88, 045504/1–4 (2002).

29. C. G. Van de Walle, *Phys. Stat. Solidi B* 229, 221 (2002).

30. S. F. J. Cox, E. A. Davis, P. J. C. King, J. M. Gil, H. V. Alberto, R. C. Vilao, J. Piroto Duarte, N. A. De Campos, and R. L. Lichti, *J. Phys: Cond. Mat.* 13, 9001 (2001).

31. C. G. Van de Walle, *Phys. Rev. Lett.* 85, 1012 (2000).

32. J. Gutowski, N. Presser, and I. Broser, *Phys. Rev. B* 38, 9746 (1988).

33. S. Bethke, H. Pan, and B. W. Wessels, *Appl. Phys. Lett.* 52, 138 (1988).

34. B. J. Jin, S. H. Bae, S. Y. Lee, and S. Im, *Mat. Sci. Eng.* B71, 301 (2000).

35. B. J. Jin, S. Im, and S. Y. Lee, *Thin Solid Films* 366, 107 (2000).

36. D. M. Bagnall, Y. F. Chen, M. Y. Shen, Z. Zhu, T. Goto, and T. Yao, *J. Cryst. Growth* 184/185, 605 (1998).

37. E. G. Bylander, *J. Appl. Phys.* 49, 1188 (1978).

38. N. Riehl and O. Ortman, *Z. Elektrochem.* 60, 149 (1952).

39. F. A. Kröger and H. J. Vink, *J. Chem. Phys.* 22, 250 (1954).

40. I. Y. Prosanov and A. A. Politov, *Inorg. Mat.* 31, 663 (1995).

41. N. Y. Garces, L. Wang, L. Bai, N. C. Giles, L. E. Halliburton, and G. Cantwell, *Appl. Phys. Lett.* 81, 622 (2002).

42. H.-J. Egelhaaf and D. Oelkrug, *J. Cryst. Growth* 161, 190 (1996).

43. D. Hahn and R. Nink, *Physik Cond. Mater.* 3, 311 (1965).

44. M. Liu, A. H. Kitai, and P. Mascher, *J. Lumin.* 54, 35 (1992).

45. D. W. Palmer, http://www.semiconductors.co.uk, 2002.06.

46. D. Florescu, L. G. Mourok, F. H. Pollack, D. C. Look, G. Cantwell, and X. Li, *J. Appl. Phys.* 91, 890 (2002).

47. G. F. Neumark, in *Widegap II-VI Compounds for Opto-Electronics Applications*, H. E. Ruda, ed., Oxford Press, London, 281 (1992).

48. D. B. Laks, C. G. Van de Walle, G. F. Neumark, and S. T. Pantelides, *Appl. Phys. Lett.*, 63, 1375 (1993).

49. Y. Kanai, *Jpn. J. Appl. Phys.*, Part 1 (Regular Papers & Short Notes), 30, 703 (1991).

50. Y. Kanai, *Jpn. J. Appl. Phys.*, Part 1 (Regular Papers & Short Notes), 30, 2021 (1991).

51. J. A. Savage and E. M. Dodson, *J. Mat. Sci.* 4, 809 (1969).

52. A. Valentini, F. Quaranta, M. Rossi, and G. Battaglin, *J. Vac. Sci Technol.* A 9, 286 (1991).

53. A. Onedera, N. Tamaki, K. Jin, and H. Yamashita, *Jpn. J. Appl. Phys.* 36, 6008 (1997).

54. P. H. Kasai, *Phys. Rev.* 130, 989 (1963).

55. H. Wolk, S. Deubler, D. Forkel, H. Foettinger, M. Iwatschenko-Borho, F. Meyer, M. Renn, W. Witthuhn, and R. Helbig, *Mat. Sci. Forum* 10-12, Part 3, 863 (1986).

56. T. Nagata, T. Shimura, Y. Nakano, A. Ashida, N. Fujimura, and T. Ito, *Jpn J. Appl. Phys.*, Part 1 40, 5615 (2001).

57. C. H. Park, S. B. Zhang, and Wei Su-Huai, *Phys. Rev. B* 66, 073202/1-3(2002).

58. T. Yamamoto and H. Katayama-Yoshida, *Jpn. J. Appl. Phys.* 38, L166 (1999).

59. N. Y. Garces, N. C. Giles, L. E. Halliburton, G. Cantwell, D. B. Eason, D. C. Reynolds, and D. C. Look, *Appl. Phys. Lett.* 80, 1334 (2002).

60. K. Minegishi, Y. Koiwai, Y. Kikuchi, K. Yano, M. Kasuga, and A. Shimizu, *Jpn. J. Appl. Phys.* 36, L1453 (1997).
61. Xin-Li Guo, H. Tabata, T. Kawai, *J. Crystal Growth* 223, 135 (2001).
62. C. Rouleau, S. Kang, and D. Lowndes, unpublished.
63. K. Iwata, P. Fons, A. Yamada, K. Matsubara, and S. Niki, *J. Crystal Growth* 209, 526 (2000).
64. Y. Yan, S. B. Zhang, and S. T. Pantelides, *Phys. Rev. Lett.* 86, 5723 (2001).
65. D. C. Look, D. C. Reynolds, C. W. Litton, R. L. Jones, D. B. Eason, and G. Cantwell, *Appl. Phys. Lett.* 81, 1830 (2002).
66. Kyoung-Kook Kim, Hyun-Sik Kim, Dae-Kue Hwang, Jae-Hong Lim, Seong-Ju Park, *Appl. Phys. Lett.*, 83, 63 (2003).
67. Y. W. Heo, S. J. Park, K. Ip, S. J Pearton, and D. P. Norton, *Appl. Phys. Lett.*, 83, 1128 (2003).
68. Eun-Cheol Lee, Y.-S. Kim, Y.-G. Jin, and K. J. Chang, *Physica B*, 308–310, 912 (2001).
69. N. Ohashi, T. Ishigaki, N. Okada, T. Sekiguchi, I. Sakaguchi, and H. Haneda, *Appl. Phys. Lett.* 80, 2869 (2002).
70. B. Theys, V. Sallet, F. Jomard, A. Lusson, J.-F. Rommeluere, and Z. Teukam, *J. Appl. Phys.* 91, 3922 (2002).
71. Eun-Cheol Lee, Y.-S. Kim, Y.-G. Jin, and K. J. Chang, *Phys. Rev. B* 64, 085120/1-5 (2001).
72. S. B. Zhang, S.-H. Wei, and A. Zunger, *Phys. Rev. B* 63, 075205/1-7 (2001).
73. Y. W. Heo, K. Ip, S. J. Park, S. J. Pearton, and D. P. Norton, *Appl. Phys. A*, submitted.
74. S. O. Kucheyev, C. Jagadish, J. S. Williams, P. N. K. Deenapanray, M. Yano, K. Koike, S. Sasa, M. Inoue, and K. Ogata, *J. Appl. Phys.* 93, 2972 (2003).
75. S. O. Kucheyev, P. N. L. Deenapanray, C. Jagadish, J. S. Williams, M. Yano, K. Kioke, S. Sasa, M. Inoue, and K. Ogata, *Appl. Phys. Lett.* 83, 3350 (2002).
76. Y. Li, G. S. Tompa, S. Liang, C. Gorla, C. Lu, and J. Doyle, *J. Vac. Sci. Technol.* A15, 1663 (1997).
77. J. G. E. Gardeniers, Z. M. Rittersma, and G. J. Burger, *J. Appl. Phys.* 83, 7844 (1998).
78. H. Maki, T. Ikoma, I. Sakaguchi, N. Ohashi, H. Haneda, J. Tanaka, and N. Ichinose, *Thin Solid Films* 352, 846 (2010).
79. S.-C. Chang, D. B. Hicks and R. C. O. Laugal, Technical Digest. *IEEE Solid-State Sensor and Actuator Workshop*, 41 (1992).
80. J. S. Wang, Y. Y. Chen and K. M. Larkin, *Proc. 1982 IEEE Ultrasonic Symp.* pp. 345–348 (1982).
81. G. D. Swanson, T. Tanagawa, and D. L. Polla, *Electrochem. Soc. Ext. Abstracts* 90-2, 1082 (1990).
82. J.-M. Lee, K.-M. Chang, K.-K. Kim, W.-K. Choi, and S. J. Park, *J. Electrochem. Soc.* 148, G1 (2001).
83. J.-M. Lee, K.-K. Kim, S.-J. Park and W.-K. Park, *Appl. Phys. Lett.* 78, 3842 (2001).
84. R. J. Shul, M. C. Lovejoy, A. G. Baen, J. C. Zolper, D. J. Rieger, M. J. Hafich, R. F. Corless, and C. B. Vartuli, *J. Vac. Sci Technol.* A13, 912 (1995).
85. C. Steinbruchel, *Appl. Phys. Lett.* 55, 1960 (1989).
86. K. Pelhos, V. M. Donnelly, A. Kornblit, M. L. Green, R. B. Van Dover, L. Manchanda, and E. Bower, *J. Vac. Sci. Technol.* A19, 1361 (2001).
87. C. C. Chang, K. V. Guinn, V. M. Donnelly, and I. P. Herman, *J. Vac. Sci. Technol.* A13, 1970 (1995).
88. H. Kuzami, R. Hamasaki, and K. Tago, *Jpn. J. Appl. Phys.* 36, 4829 (1997).
89. A. C. Jones and P. O'Brien, *CVD of Compound Semiconductors* (VCH, Weinheim, Germany, 1997).
90. S. W. Pang, in *Handbook of Advanced Plasma Processing Techniques*, ed. R. J. Shul (Springer-Verlag, Berlin 2000).
91. K. Ip, K. Baik, M. E. Overberg, E. S. Lambers, Y. W. Heo, D. P. Norton, S. J. Pearton, F. Ren, and J. M. Zavada, *Appl. Phys Lett.* 81, 3546 (2002).
92. H. K. Kim, J. W. Bae, T. K. Kim, K. K. Kim, T. Y. Seong, and I Adesida, *J. Vac. Sci. Technol. B* 21, 1273 (2003).
93. A. A. Iliadis, R. D. Vispute, T. Venkatesan, and K. A. Jones, *Thin Solid Films* 420–421, 478 (2002).
94. V. Hoppe, D. Stachel, and D. Beyer, *Phys. Scripta* T57, 122 (1994).
95. A. Inumpudi, A. A. Iliadis, S. Krishnamoorthy, S. Choopun, R. D. Vispute, and T. Venkatesan, *Solid-State Electron.* 46, 1665 (2002).
96. H.-K. Kim. S.-H. Han, T.-Y. Seong, and W.-K. Choi, *Appl. Phys. Lett.* 77, 1647 (2000).
97. H.-K. Kim. S.-H. Han, T.-Y. Seong, and W.-K. Choi, *J. Electrochem. Soc.* 148, G114 (2001).
98. T. Akane, K. Sugioka, and K. Midorikawa, *J. Vac. Sci. Technol. B* 18, 1406 (2001).
99. H. Sheng, N. W. Emanetoglu, S. Muthukumar, S. Feng, and Y. Lu, *J. Electron. Mater.* 31, 811 (2002).
100. S. Y. Kim, H. W. Jang, J. K. Kim. C. M. Jeon, W. I. Park, G. C. Yi, and J.-L. Lee, *J. Electron. Mater.* 31, 868 (2002).
101. G. S. Marlow and M. B. Das, *Solid-State Electron.* 25, 91 (1982).
102. C. A. Mead, *Phys. Lett.* 18, 218 (1965).

103. R. C. Neville and C. A. Mead, *J. Appl. Phys.* 41, 3795 (1970).
104. J. C. Simpson and F. Cordaro, *J. Appl. Phys.* 63, 1781 (1988).
105. N. Ohashi, J. Tanaka, T. Ohgaki, H. Haneda, M. Ozawa, and T. Tsurumi, *J. Mater. Res.* 17, 1529 (2002).
106. H. Sheng, S. Muthukumar, N. W. Emanetoglu, and Y. Lu, *Appl. Phys. Lett.* 80, 2132 (2002).
107. F. D. Auret, S. A. Goodman, M. Hayes, M. J. Legodi, and H. A. van Laarhoven, *Appl. Phys. Lett.* 79, 3074 (2001) and F. D. Auret, private communication.
108. B. J. Coppa, R. F. Davis, and R. J. Nemanich, *Appl. Phys. Lett.* 82, 400 (2003).
109. C. Kilic and Z. Zunger, *Appl. Phys. Lett.* 81, 73 (2002).
110. C. van de Walle, *Physica B* 308–310, 899 (2001).
111. S. J. F. Cox, E. A. Davis, S. P. Cottrell, P. J. C. King, J. S. Lord, J. M. Gil, H. V. Alberto, R. C. Vilao, D. J. P. Duarte, N. A. de Campos, A. Weidinger, R. L. Lichti, and S. J. C. Irving, *Phys. Rev. Lett.* 86, 2601 (2001).
112. D. M. Hofmann, A. Hofstaetter, F. Leiter, H. Zhou, F. Henecker, B. K. Meyer, S. B. Schmidt, and P. G. Baranov, *Phys. Rev. Lett.* 88, 045504 (2002).
113. S. J. Baik, J. H. Jang, C. H. Lee, W. Y. Cho, and K. S. Lim, *Appl. Phys. Lett.* 70, 3516 (1997).
114. N. Ohashi, T. Ishigaki, N. Okada, T. Sekiguchi, I. Sakaguchi, and H. Haneda, *Appl. Phys. Lett.* 80, 2869 (2002).
115. V. Bogatu, A. Goldenbaum, A. Many, and Y. Goldstein, *Phys. Stal. Solidi* B212, 89 (1999).
116. T. Sekiguchi, N. Ohashi, and Y. Terada, *Jpn. J. Appl. Phys.* 36, L289 (1997).
117. C. S. Han, J. Jun, and H. Kim, *Appl. Surf. Sci.* 175/176, 567 (2001).
118. Y. Natsume and H. Sakata, *J. Mater. Sci. Mater. Electron.* 12, 87 (2001).
119. Y.-S. Kang, H. Y. Kim, and J. Y. Lee, *J. Electrochem. Sci.* 147, 4625 (2000).
120. R. G. Wilson, S. J. Pearton, C. R. Abernathy, and J. M. Zavada, *J. Vac. Sci. Technol.* A13, 719 (1995).
121. S. J. Pearton, J. C. Zolper, R. J. Shul, and F. Ren, *J. Appl. Phys.* 86, 1 (1999).
122. F. D. Auret, S. A. Goodman, M. Hayes, M. J. Legodi, H. A. Van Laarhoven, and D. C. Look, *Appl. Phys. Lett.* 79, 3074 (2001).
123. F. D. Auret, S. A. Goodman, M. Hayes, M. J. Legodi, H. A. van Laarhoven, and D. C. Look, *J. Phys.: Condens. Matter.* 13, 8989 (2001).
124. K. Ip, M. E. Overberg, Y. W. Heo, D. P. Norton, S. J. Pearton, C. E. Stutz, B. Luo, F. Ren, D. C. Look, and J. M. Zavada, *Appl. Phys. Lett.* 82, 385 (2003).
125. I. P. Kuz'mina and V. A. Nikitenko, *Zinc Oxide: Growth and Optical Properties* (Moscow, Nauka, 1984) (in Russian).
126. G. M. Martin, A. Mitonneau, D. Pons, A. Mircea, and D. W. Woodard, *J. Phys.* C13, 3855 (1980).
127. N. Ohashi, J. Tanaka, T. Ohgaki, H. Haneda, M. Ozawa, and T. Tsurumi, *J. Mater. Res.* 17, 1529 (2002).
127. H. Sheng, S. Muthukumar, N. W. Emanetoglu, and Y. Lu, *Appl. Phys. Lett.* 80, 2132 (2002).
128. R. J. Borg and C. J. Dienes, *An Introduction to Solid State Diffusion* (Academic Press, Boston, 1988).
129. E. V. Lavrov, J. Weber, F. Borrnet, C. G. Van de Walle and R. Helbig, *Phys. Rev. B 76* 021208 (2010).
130. D. C. Reynolds, D. C. Look, B. Jogai, C. W. Litton, T. C. Collins, W. Harsch, and G. Cantwell, *Phys. Rev. B* 57, 12151 (1998).
131. K. Thonke, Th. Gruber, N. Teofilov, R. Schönfelder, A. Waag, and R. Sauer, *Physica B* 308–310, 945 (2001).
132. S. A. Wolf, *J. Superconductivity* 13, 195 (2000).
133. G. A. Prinz, *Science* 282, 1660 (1998).
134. J. K. Furdyna, *J. Appl. Phys.* 64, R29 (1988).
135. S. Gopalan and M. G. Cottam, *Phys. Rev. B* 42, 10311 (1990).
136. C. Haas, *Crit. Rev. Solid State Sci.* 1, 47 (1970).
137. T. Suski, J. Igalson, and T. Story, *J. Magn. Magn. Mater.* 66, 325 (1987).
138. A. Haury, A. Wasiela, A. Arnoult, J. Cibert, S. Tatarenko, T. Dietl, Y. Merle d'Aubigne, *Phys. Rev. Lett.* 79, 511 (1997); P. Kossacki, D. Ferrand, A. Arnoult, J. Cibert, S. Tatarenko, A. Wasiela, Y. Merle d'Aubigne, J.-L. Staihli, J.-D. Ganiere, W. Bardyszewski, K. Swiatek, M. Sawicki, J. Wrobel, and T. Dietl, *Physica E* 6, 709 (2000).
139. H. Ohno, *Science* 281, 951 (1998).
140. K. Sato, G. A. Medvedkin, T. Nishi, Y. Hasegawa, R. Misawa, K. Hirose, and T. Ishibashi, *J. Appl. Phys.* 89, 7027 (2001).
141. M. E. Overberg, B. P. Gila, G. T. Thaler, C. R. Abernathy, S. J. Pearton, N. A. Theodoropoulou, K. T. McCarthy, S. B. Arnason, A. F. Hebard, S. N. G. Chu, R. G. Wilson, J. M. Zavada, Y. D. Park, *J. Vac Sci. Technol. B*, 20, 969 (2002).
142. T. Dietl, H. Ohno, F. Matsukura, J. Cubert, and D. Ferrand, *Science* 287, 1019 (2000).
143. T. Dietl, A. Haury, and Y. Merle d'Aubigne, *Phys. Rev. B* 55, R3347 (1997).
144. B. E. Larson, K. C. Hass, H. Ehrenreich, and A. E. Carlsson, *Solid State Commun.* 56, 347 (1985).

145. S. J. Pearton, C. R. Abernathy, M. E. Overberg, G. T. Thaler, D. P. Norton, N. Theodorpoulou, A. F. Hebard, Y. D. Park, F. Ren, J. Kim, and L. A. Boatner, *J. Appl. Phys.* 93, 1 (2003).
146. S. J. Pearton, C. R. Abernathy, D. P. Norton, A. F. Hebard, Y. D. Park, L. A. Boatner, and J. D. Budai, *Mater. Sci. Eng.* R.40, 137 (2003).
147. K. Sato and H. Katayama-Yoshida, *Jpn. J. Appl. Phys.* 39, L555 (2000).
148. T. Wakano, N. Fujimura, Y. Morinaga, N. Abe, A. Ashida, and T. Ito, *Physica C* 10, 260 (2001).
149. T. Fukumura, Z. Jin, A. Ohtomo, H. Koinuma, and M. Kawasaki, *Appl. Phys. Lett.* 75, 3366 (1999).
150. S. W. Jung, S.-J. An, G.-C. Yi, C. U. Jung, S.-I. Lee, and S. Cho, *Appl. Phys. Lett.* 80, 4561 (2002).
151. J. Blinowski, P. Kacman, and J. A. Majewski, *Phys. Rev.* B 53, 9524 (1996).
152. H. Sato, T. Minami, and S. Takata, *J. Vac. Sci. Technol.* A 11, 2975 (1993).
153. D. P. Norton, S. J. Pearton, A. F. Hebard N. Theodoropoulou, L. A. Boatner, and R. G. Wilson, *Appl. Phys. Lett.* 73 216208 (2009).
154. F. Holzberg, S. von Molnar, and J. M. D. Coey, in *Handbook on Semiconductors*, vol. 3, ed. T. S. Moss (North-Holland, Amsterdam, 1980).
155. T. Story, *Acta Phys. Pol.* A91, 173 (1997).
156. L. W. Guo, D. L. Peng, H. Makino, K. Inaba, H. J. Ko, K. Sumiyama, and Y. Yao, *J. Magn. Mag. Mater.* 213, 321 (2000).
157. H. Wolf, S. Deubler, D. Forkel, H. Foettinger, M. Iwatschenko-Borho, F. Meyer, M. Renn, W. Witthuhn, and R. Helbig, *Mater. Sci. Forum*, 10-12, Part 3, 863 (1986).
158. T. Aoki, Y. Hatanaka, and D. C. Look, *Appl. Phys. Lett.* 76, 3257 (2000).
159. C. W. Teng, J. F. Muth, Ü. Özgür, M. J. Bergmann, H. O. Everitt, A. K. Sharma, C. Jin, and J. Narayan, *Appl. Phys. Lett.* 76, 979 (2000).
160. D. C. Look, D. C. Reynolds, C. W. Litton, R. L. Jones, D. B. Eason, and G. Cantwell, *Appl. Phys. Lett.* 81, 1830 (2002).
161. Y. Kanai, *Jpn. J. Appl. Phys.*, Part 1 (Regular Papers & Short Notes), Volume 30, Issue 4, 1991, pp. 703–707.
162. Y. Matsumoto, M. Murakami, T. Shono, T. Hasegawa, T. Fukumura, M. Kawasaki, P. Ahmet, T. Chikyow, S. Koshihara, and H. Koinuma, *Science*, 291, 854 (2001).
163. S. A. Chambers, S. Thevuthasan, R. F. C. Farrow, R. F. Marks, U. U. Thiele, L. Folks, M. G. Samant, A. J. Kellock, N. Ruzycki, D. L. Ederer, and U. Diebold, *Appl. Phys. Lett.* 79, 3467 (2001).
164. K. Ueda, H. Tabata, and T. Kawai, *Appl. Phys. Lett.* 79, 988 (2001).
165. A. F. Jalbout, H. Chen, and S. L. Whittenburg, *Appl. Phys. Lett.* 81, 2217 (2002).
166. T. Dietl, *Semicon. Sci. and Tech.*, 17, 377 (2002).
167. Y. Souche, M. Schlenker, R. Raphel, and L. Alvarez-Prado, *Mater. Sci. Forum* 302–303, 105 (1999).
168. M. H. Huang, S. Mao, H. Feick, H. Yan, Y. Wu, H. Kind, E. Weber, R. Russo, and P. Yang, *Science* 292, 1897 (2001).
169. Y. W. Heo, V. Varadarjan, M. Kaufman, K. Kim, D. P. Norton, F. Ren, and P. H. Fleming, *Appl. Phys. Lett.* 81, 3046 (2002).
170. I. P. Kuz'mina and V. A. Nikitenko, *Zinc Oxide: Growth and Optical Properties* (Moscow, Nauka, 1984) (in Russian).
171. S.-J. Han, J. W. Song, C.-H. Yang, S. H. Park, J-H. Park, J. H. Jeong, and K. W. Rhie, *Appl. Phys. Lett.* 81, 4212 (2002).
172. N. Ohashi, J. Tanaka, T. Ohgaki, H. Haneda, M. Ozawa, and T. Tsurumi, *J. Mater. Res.* 17, 1529 (2002).
173. E. H. Brown, *Zinc Oxide: Properties and Applications* (New York, Pergamon Press, 1976), 175.
174. A. A. Tomchenko, G. P. Harmer, B. T. Marquis, and J. W. Allen, *Sens. Actuat.* B93, 126 (2003).
175. K. D. Mitzner, J. Sternhagen, and D. W. Galipeau, *Sens. Actuat.* B93, 92 (2003).
176. J. Wollenstein, J. A. Plaza, C. Cane, Y. Min, H. Botttner, and H. L. Tuller, *Sens. Actuat.* B93, 350 (2003).
177. U. Ozgur, A. Teke, C. Liu, S. J. Cho, H. Morkoc, and H. O. Everitt, *Appl. Phys. Lett.* 84, 3223 (2004).
178. B. Gou, Z. R. Qiu, and K. S. Wong, *Appl. Phys. Lett.* 82, 2290 (2003).
179. T. Koida, S. F. Chichibu, A. Uedono, A. Tsukazaki, M. Kawasaki, T. Sota, Y. Segewa, and H. Koinuma, *Appl. Phys. Lett.* 82, 532 (2003).
180. S. A. Studenikin, N. Golego, and M. Cocivera, *J. Appl. Phys.* 87, 2413 (2000).
181. M. H. Huang, S. Mao, H. Feick, H. Yan, Y. Wu, H. Kind, E. Weber, R. Russo, and P. Yang, *Science* 292, 1897 (2001).
182. Q. H. Li, Q. Wan, Y. X. Liang, and T. H. Wang, *Appl. Phys. Lett.* 84, 4556 (2004).
183. K. Keem, H. Kim, G. T. Kim, J. S. Lee, B. Min, K. Cho, M. Y. Sung, and S. Kim, *Appl. Phys. Lett.* 84, 4376 (2004).
184. B. H. Kind, H. Yan B. Messer, M. Law, and P. Yang, *Adv. Mater.* 14, 158 (2002).

185. H. Liu, W. C. Liu, F. C. K. Au, J. X. Ding, C. S. Lee, and S. T. Lee, *Appl. Phys. Lett.* 83, 3168 (2003).
186. W. I. Park, G. C. Yi, J. W. Kim, and S. M. Park, *Appl. Phys. Lett.* 82, 4358 (2003).
187. J. J. Wu and S. C. Liu, *Adv. Mater.* 14, 215 (2002).
188. H. T. Ng, J. Li, M. K. Smith, P. Nguygen, A. Cassell, J. Han, and M. Meyyappan, *Science* 300, 1249 (2003).
189. J. Q. Hu and Y. Bando, *Appl. Phys. Lett.* 82, 1401 (2003).
190. W. I. Park, G. C. Yi, M. Y. Kim, and S. J. Pennycook, *Adv. Mater.* 15, 526 (2003).
191. Y. W. Heo, V. Varadarjan, M. Kaufman, K. Kim, D. P. Norton, F. Ren, and P. H. Fleming, *Appl. Phys. Lett.* 81, 3046 (2002).
192. P. J. Poole and J. Lefebvre, and J. Fraser, *Appl. Phys. Lett.* 83, 2055 (2003).
193. M. He, M. M. E. Fahmi, S. Noor Mohammad, R. N. Jacobs, L. Salamanca-Riba, F. Felt, M. Jah, A. Sharma, and D. Lakins, *Appl. Phys. Lett.* 82, 3749 (2003).
194. X. C. Wu, W. H. Song, W. D. Huang, M. H. Pu, B. Zhao, Y. P. Sun, and J. J. Du, *Chem. Phys. Lett.* 328, 5 (2000).
195. M. J. Zheng, L. D. Zhang, G. H. Li, X. Y. Zhang, and X. F. Wang, *Appl. Phys. Lett.* 79, 839 (2001).
196. S. C. Lyu, Y. Zhang, H. Ruh, H. J. Lee, H. W. Shim, E. K. Suh, and C. J. Lee, *Chem. Phys. Lett.* 363, 134 (2002).
197. B. P. Zhang, N. T. Binh, Y. Segawa, K. Wakatsuki, and N. Usami, *Appl. Phys. Lett.* 83, 1635 (2003).
198. W. I. Park, Y. H. Jun, S. W. Jung, and G. Yi, *Appl. Phys. Lett.* 82, 964 (2003).
199. B. D. Yao, Y. F. Chan, and N. Wang, *Appl. Phys. Lett.* 81, 757 (2002).
200. Z. W. Pan, Z. R. Dai, and Z. L. Wang, *Science* 291, 1947 (2001).
201. J. Y. Lao, J. Y. Huang, D. Z. Wang, and Z. F. Ren, *Nano Lett.* 3, 235 (2003).
202. X. W. Sun, S. F. Yu, C. X. Xu, C. Yuen, B. J. Chen, and S. Li, *Jpn. J. Appl. Phys.* 42, L1229 (2003).
203. H. T. Ng, J. Li, M. K. Smith, P. Nguygen, A. Cassell, J. Han, and M. Meyyappan, *Science* 300, 1249 (2003).
204. J. Q. Hu and Y. Bando, *Appl. Phys. Lett.* 82, 1401 (2003).
205. W. I. Park, G. C. Yi, M. Y. Kim, and S. J. Pennycook, *Adv. Mater.* 15, 526 (2003).
206. see the databases at http://www.rebresearch.com/H2perm2.htm and http://www.rebresearch.com/H2sol2.htm
207. W. Eberhardt, F. Greunter, and E. W. Plummer, *Phys. Rev. Lett.* 46, 1085 (1981).
208. H. Windischmann and P. Mark, *J. Electrochem. Soc.* 126, 627 (1979).
209. P. Sharma, K. Sreenivas, and K. V. Rao, *J. Appl. Phys.* 93, 3963 (2003).
210. B. S. Kang, Y. W. Heo, L. C. Tien, F. Ren, D. P. Norton, and S. J. Pearton, *Appl. Phys. A* 80, 497 (2005).
211. G. Steinhoff, M. Hermann, W. J. Schaff, L. F. Eastmann, M. Stutzmann, and M. Eickhoff, *Appl. Phys. Lett.* 83, 177 (2003).
212. C. Nguyen, J. Dai, M. Darikaya, D. T. Schwartz, and F. Baneyx, *J. Am. Chem. Soc.* 43, 256 (2003).
213. K. Kjaergaard, J. K. Sorensen, M. A. Schembri, and P. Klemm, *Appl. Environ. Microbiol.*, 66, 10 (2000).
214. S. R. Whaley, D. S. English, E. L. Hu, P. F. Barbara, and A. M. Belcher, *Nature*, 405, 665 (2000).
215. X. Fang, X. Liu, S. Schuster, and W. Tan, *J. Am. Chem. Soc.*, 121, 2921–2922, 1999.
216. W. Tan, K. Wang, X. He, J. Zhao, T. Drake, L. Wang, R. P. Bagwe, *Medicinal Res. Rev.*, 24(5), 621–638 (2004).
217. R. L. Hoffman, *J. Appl. Phys.* 95, 5813 (2004).
218. D. E. Yales, S. Levine, and T. W. Healy, *J. Chem. Faraday Trans.* 70, 1807 (1974).
219. M. N. Stojanovic, P. de Prada, and D. W. Landry, *J. Am. Chem. Soc.* 123, 4928–4931 (2001).
220. X. Fang, A. Sen, M. Vicens, and W. Tan, *ChemBioChem*, 4, 829–834 (2003).

6 Bioaffinity Sensors Based on MOS Field-Effect Transistors

D. Landheer
Institute for Microstructural Sciences, National Research
Council of Canada, Ottawa, ON, Canada

W.R. McKinnon
Institute for Microstructural Sciences, National Research
Council of Canada, Ottawa, ON, Canada

W.H. Jiang
Institute for Microstructural Sciences, National Research
Council of Canada, Ottawa, ON, Canada

G. Lopinski
Steacie Institute for Molecular Sciences, National
Research Council of Canada, Ottawa, ON, Canada

G. Dubey
Steacie Institute for Molecular Sciences, National
Research Council of Canada, Ottawa, ON, Canada

N.G. Tarr
Department of Electronics, Carleton University, Ottawa, ON, Canada

M.W. Shinwari
Department of Electrical and Computer Engineering,
McMaster University, Hamilton, ON, Canada

M.J. Deen
Department of Electrical and Computer Engineering,
McMaster University, Hamilton, ON, Canada

CONTENTS

6.1 INTRODUCTION

Metal-oxide-semiconductor (MOS) field-effect transistors (FETs), also called insulated-gate FETs (IGFETs or MOSFETs), are commonly used as electrical sensors. Complementary-MOS (CMOS), which includes both p-type and n-type semiconductors, is the dominant electronic device technology of our epoch. It promises to provide inexpensive large arrays of electrical sensors integrated with electronic buffers, amplifiers, analog/digital conversion, and other signal processing components, coupled with array element sequencing [1]. Error detection, data analysis, and special output circuits are also feasible. It allows the integration of heating and temperature sensor elements and, increasingly, CMOS fabrication facilities are implementing processes to integrate MEMS (micro-electro-mechanical systems), structures that can facilitate sample processing and applications of fluids to the sensor array. Although Si is the dominant semiconductor, as device dimensions shrink, others such as Ge and III–V compounds are being integrated.

Even the CMOS taboo against bringing Au into CMOS facilities has been lifted [1,2], at least in dedicated facilities. This means that potentially any kind of amperometric, potentiometric, or catalytically enhanced electrical sensor can be integrated to the surface of a CMOS chip. This is certainly true of the bioaffinity sensors that will be discussed in this chapter. With the advent of nanowire sensors, even these sensors can exploit the latest advances in nanolithography and nanofabrication techniques.

This is a vast field, and we have chosen to focus this chapter on the use of field-effect sensors based on silicon to make bioaffinity sensors. We start with early devices made with micron devices. The reason for this is twofold. First, it allows us to exploit the knowledge gained in processing and functionalizing the silicon surfaces that has yet to be fully exploited in the fabrication of the nanowire devices. The second motivation is that the techniques used to "spot" the probe molecules on the sensor surface use microfluidic techniques that are limited to spots of the dimensions of tens of microns. Thus, the full exploitation of disposable chips with large arrays at these dimensions would be very expensive to implement with the latest nanofabrication technology.

The use of the floating gate of a field-effect transistor (FET) to detect charged molecules on surfaces in solution began almost 40 years ago with the first demonstration of the ion-sensitive FET (ISFET) [3,4]. From the beginning, the main gate insulator materials have been SiO_2, with its relatively open structure of siloxane rings that makes it analogous to a glass electrode, and silicon nitride, with a denser covalently bonded structure that is less permeable to an electrolyte but has a higher sensitivity to solution pH. Other oxides and organic insulators can be used on the gate to improve pH sensitivity and detect other ions or impart particular chemical properties to the silicon surface, and biomolecules can even be attached directly to the Si surface. In common terminology, any FET with a chemically active gate region can be called a CHEMFET, and the review by Bergveld describes the CHEMFET as a "membrane-covered ISFET" [5].

In this chapter, we describe the use of CHEMFETs to detect charged biomolecules in solution by sensing their conjugation to probe molecules included in or on the surface membrane, termed *bioaffinity sensors*. In particular, we are interested in sensors that are easily integrated with CMOS technology to exploit the advantages of that technology. Examples include the DNA-FET, which detects the conjugation of single-stranded DNA or RNA fragments with their complements (oligonucleotides called ssDNA, ssRNA or "oligos" for short). Including the IMFET (or ImmunoFET) used to detect the conjugation of antibodies with antigens, the term *BioFET* has been used to describe these FETs, and this terminology will be used in this chapter. We only mention the ENFETs that use biocatalytic reactions affecting the gate surface charge by the interaction of enzymes at the surface with a target solution.

Soon after the invention of the ISFET, it was realized that a reference electrode that sets the bulk solution potential was required for the effective operation of these devices [6]. This is true of the BioFETs as well. However, CHEMFETS with a conductive gate that do not require reference electrodes are possible and are useful as sensors for charge-neutral species, particularly gases. They have been called work function FETs (WF-FETs) [7] and rely on the species creating an interfacial dipole close to the semiconductor surface, on changing the work function on the semiconductor interface, or on the alteration of the dielectric constant of the insulator.

In 1989, Wong and White reported on a CMOS-integrated operational amplifier with differential sensing [7], and this has facilitated subsequent impedance measurements with BioFETs and the integration of the sensors with reference elements [8]. Large arrays with 240 elements of pH-ISFETs (H^+-sensitive ISFETs) were fabricated over 10 years ago [9] using a CMOS process with four extra process steps, allowing full integration of CMOS circuitry with the sensors. It used a Si_3N_4 pH-sensitive insulator layer deposited by low-pressure chemical vapor deposition (LPCVD).

More recently, the use of CMOS for the fabrication of sensor arrays has advanced dramatically, spurred on by the results obtained using "nanowires." Carbon nanotubes (CNTs), silicon nanowires (SiNWs), and conductive polymer versions (CPs) promise increased sensitivity for BioFET sensors by confining the carriers to wires with diameters in the range 20–50 nm, and lengths (source-drain distance) in the micron range. The performance of, and prospects for, these devices have been described in a recent review. [10] Since fabrication of CMOS devices with nanometer dimensions is now firmly established, the results obtained with SiNWs can be transferred directly into the more easily controlled "top-down" processes available in CMOS.

The chapter is organized as follows: Section 6.2 describes simple planar models that bring out the main features of the operation of BioFETs, including drift-diffusion of the carriers in the semiconductor and the ions in the electrolyte, the Poisson equation in both media, and the site binding at the interface. It ends by considering more sophisticated 3-dimensional (3-D) device simulations. Section 6.3 describes the frequency response of BioFETs and the use of impedimetric techniques to detect the conjugation on BioFETs. Section 6.4 describes the operation and the limits of sensitivity for nanowires transferable to deep submicron CMOS technology.

6.2   MODELS OF BioFET DEVICE OPERATION

The operation of BioFETs can be understood as an extension of the MOSFET. A potential difference V_{DS} drives a current I_{DS} between the drain (D) contact and the source (S) contact (Figure 6.1). A potential V_{GS} between the gate metal (G) and the source creates an electric field in the oxide, modulating the amount of mobile charge in the semiconductor under the gate and thereby modulating I_{DS}. A fourth contact, made to the bulk of the semiconductor and called the body contact (B), is usually connected to the source, so $V_{SB} = 0$ [10].

In a BioFET, the gate is replaced by a reference electrode with an applied potential V_{ref}, an electrolyte solution, and charged molecules on the oxide surface, as shown in Figure 6.1. It is tempting to suppose that each charge on the charged molecules draws an equal and opposite charge into the semiconductor, but this view greatly overestimates the BioFET's response because it ignores the role of other charges in the electrolyte and on the oxide surface.

6.2.1   FET OPERATION

The principles of operation of a MOSFET are discussed in many texts [11–13]. The key phenomena are summarized here.

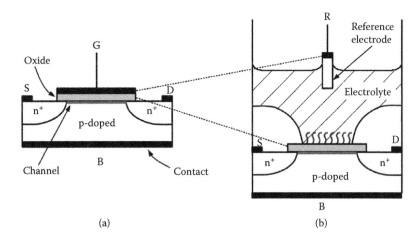

FIGURE 6.1 (a) FET and (b) BioFET. The example shown is an *n*-FET. In the BioFET, the source and drain contacts are covered, so the electrolyte contacts only the oxide. The dashed lines linking parts (a) and (b) are meant to suggest that the BioFET is obtained from the FET by replacing the gate metal by charged molecules, electrolyte, and reference electrode.

6.2.1.1 Charge in a MOSFET

Generally, V_{GS} is set so that the majority carriers in the semiconductor are repelled from the region under the gate. Starting from the condition of the flat-band (no net charge in the semiconductor), V_{GS} first creates a depletion region, where the majority carriers are repelled and the charges of the donor or acceptor ions are exposed. Biasing further from the flat band attracts minority carriers to a thinner region called the *channel* to form an inversion charge, which carries the current I_{DS}.

In addition to the inversion and depletion charge, there are various types of charge in the oxide [11]. Some of the oxide charge close to the channel interface, called the *interface trapped charge*, is associated with states that can exchange the charge with the semiconductor. The total charge in the oxide and the semiconductor is balanced by the charge on the metal. Figure 6.2 shows schematically the charge and potential in a FET. The discontinuities in the potential at the interfaces are due to differences in work function.

6.2.1.2 Charge Sheet Models for MOSFETs

In the charge sheet model of a MOSFET, the current–voltage relation is derived assuming that the inversion charge is located at a plane of negligible thickness at the semiconductor–oxide interface while the depletion charge is spread out over a distance called the *depletion width*. For the long-channel approximation, the channel carrying current between the source and drain is assumed to be long compared to the thickness of the oxide and the depletion region.

Most derivations focus on how the potential ψ_s at the semiconductor surface varies as a function of position between the source and drain, using the gradual channel approximation [14]. The simplest expression for I_{DS} is the square law, which applies in a strong inversion ([11], p. 162)

$$
I_{DS} = \begin{cases} \pm \mu C_{ox} \dfrac{W}{L}\left[(V_{GS} - V_T)V_{DS} - \dfrac{\alpha_b}{2}V_{DS}^2 \right], & \text{if } |V_{DS}| \le |V_{DS}'|; \\[3mm] \pm \mu C_{ox} \dfrac{W}{L}\dfrac{(V_{GS} - V_T)^2}{2\alpha_b}, & \text{if } |V_{DS}| > |V_{DS}'|. \end{cases}
\tag{6.1}
$$

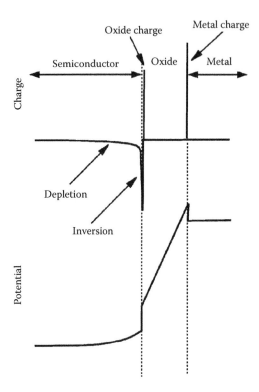

FIGURE 6.2   Charge and potential for an *n*-FET. Discontinuities in the potential at the interfaces are due to differences in work function.

The upper sign applies to an n-FET (p-type substrate), and the lower sign to a p-FET, with the convention that positive I_{DS} flows into the drain from the external circuit. C_{ox} is the capacitance per unit area of the oxide, and V'_{DS}, the value of V_{DS} at which the characteristics saturate, is given by

$$V'_{DS} = \frac{V_{GS} - V_T}{\alpha_b} \tag{6.2}$$

The parameter α_b is given by

$$\alpha_b = 1 + \frac{\gamma_b}{2\sqrt{\pm\phi_0 \pm V_{SB}}} \tag{6.3}$$

where the body coefficient γ_b is given by in terms of the doping density N_B by

$$\gamma_b = \frac{\sqrt{2q\varepsilon_s N_B}}{C_{ox}} \tag{6.4}$$

In Equation 6.3, the parameter ϕ_0 is the value of ψ_s at which the semiconductor surface is pinned, and is approximately $2\phi_F$, where ϕ_F is the Fermi potential. The parameter α_b accounts for differences in the depletion layer thickness at the source and drain ends of the channel, and is often set to unity.

The threshold voltage V_T is a key parameter in Equation 6.1. If V_{DS} is fixed, I_{DS} is a function of $V_{GS} - V_T$; thus, if V_T changes by ΔV_T, then V_{GS} must also be changed by ΔV_T to keep I_{DS} constant. The value V_T depends on the work function of the gate, and it is the only parameter in Equation 6.1 that depends on material properties of the gate metal. Biomolecules in a BioFET are observable through changes in V_T, and so the BioFET reacts to biomolecules as a normal FET would react to a change in the work function of the gate.

6.2.1.3 Capacitance

The capacitance between the gate electrode and the semiconductor depends on the distribution of the charge in the semiconductor. Near flat band, the depletion charge dominates the capacitance, and as the depletion width varies with V_{GS}, the capacitance also varies. In a strong inversion, the inversion charge dominates; the potential in the inversion layer is almost fully pinned, and the capacitance is roughly independent of V_{GS}. Since the depletion layer is usually much thicker than the oxide, the total capacitance in depletion is set by the capacitance of the depletion layer. In inversion, on the other hand, the capacitance is set by the capacitance of the oxide since the inversion layer is thin.

Measuring the capacitance $C(V_{GS})$ as a function of gate bias is another way to determine changes in V_T. If V_T changes by ΔV_T, then the curve $C(V_{GS})$ shifts by ΔV_T along the V_{GS} axis.

6.2.1.4 Operation of the BioFET

Equation 6.1 (or the results of a more sophisticated analysis) applies reasonably accurately to a BioFET if V_{GS} is replaced by V_{ref}, the potential applied to the reference electrode relative to the source. There are some small correction factors to be discussed in the following text. In a BioFET, however, V_T depends on the charge in the electrolyte and at the oxide surface. Figure 6.3 shows the charge and potential schematically up to a point in the neutral fluid for a specific model of the charge distribution, to be discussed later. Changes in V_T in a BioFET are determined by changes in the potential difference ψ_0 from the oxide surface to the neutral bulk solution. The potential difference from the neutral solution to the reference electrode is assumed to be a constant and is not shown in the figure. The key to understanding the BioFET, then, is in understanding how ψ_0 depends on the

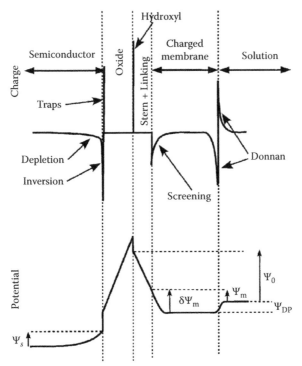

FIGURE 6.3 Charge and potential for a BioFET based on an *n*-FET.

arrangement of charges from the oxide surface to the bulk solution. This dependence is discussed in the following.

6.2.1.5   Monitoring the Threshold Voltage

Suppose a BioFET is placed in a circuit that maintains I_{DS} constant by controlling V_{ref}. Any change in V_{ref} should be due to a change in V_T, which in turn is associated with a change in ψ_0. But the change in ψ_0 depends on changes in the electrolyte or at the oxide surface, and these changes should be independent of the structure of the FET. This argument suggests the following question. Suppose the FET is changed drastically to make it function very poorly (or not at all); for example, the number of traps in the oxide of the FET is made very large, or the oxide is made very thick. The signal, by the argument just given, should be determined by only the change in ψ_0, and so should be unchanged. But if the FET is not functioning, how can the signal be observed?

One answer to this question is found in the signal-to-noise ratio. The spectral density of the noise diverges as the oxide trap density diverges or as the oxide capacitance goes to zero. Even though the signal is in principle unaffected in these limits, the signal-to-noise ratio goes to zero, so no signal can be measured. Another consideration is the possibility of breakdown. If the number of traps diverges, the field needed to produce inversion diverges. Under the high field, either the gate oxide or some other insulating layer will degenerate or break down, and electrochemical reactions or trap generation will occur at this point. It will become impossible to bias the device into inversion to measure the signal. Therefore, the choice of BioFET bias and its design needs to take these effects into consideration.

6.2.2   THE OXIDE SURFACE AND THE ELECTROLYTE

The potential ψ_0 at the oxide surface depends on how the charge is arranged in the electrolyte and at the oxide surface. It will be assumed that the reference electrode does not respond to changes in the solution, and so the chemistry at the reference electrode produces a constant offset to V_T, much like the discontinuities in work function in Figures 6.2 and 6.3.

6.2.2.1   The Electrolyte and Double Layer

The electrolyte is a conducting liquid with mobile cations and anions. These mobile charges screen the electric field produced by charges at or below the oxide surface. The screening occurs over a distance called the *Debye screening length* λ_b. For an electrolyte of ions of charge $\pm zq$ (where z is an integer and q is the elementary charge), at concentration n_0 in solvent of permittivity ε, λ_b is given by ([15], p. 331)

$$\lambda_b = \left(\frac{2zqn_0}{\varepsilon \phi_z} \right)^{-1/2} \tag{6.5}$$

where $\phi_z \equiv \phi_t/z$, the thermal potential is $\phi_t = kT/q$, k is Boltzmann's constant, and T is the absolute temperature. In solutions of monovalent ions of concentration 1, 0.1, and 0.01 mol/L, the screening length $\lambda_b = 0.3$, 1, and 3 nm, respectively.

The screening charge in the electrolyte is separated from the surface by the Helmholtz layer, which represents the closest distance that a hydrated ion in solution can approach the surface. Together, the Helmholtz layer and the screening layer are called the *double layer*. At high salt concentrations, the capacitance of the double layer is much higher than the capacitance of the oxide and the semiconductor.

6.2.2.2 The Oxide Surface in an Electrolyte

Surfaces of metal oxides in water are usually covered with hydroxyl groups, which can take up or release hydrogen ions, changing the charge on the surface. In FET-based sensors, this charge is usually described by the site-binding or site-dissociation model [16].

The site-binding model [17] considers species associated with oxides of generic M atoms at the oxide surface, namely, the neutral hydroxyl $-MOH$, the deprotonated hydroxyl $-MO^-$, the protonated hydroxyl $-MOH_2^+$, and their equilibrium with hydrogen ions H^+. The concentration of the latter at the surface is denoted by H_s^+. The reactions are

$$-MOH_2^+ \Leftrightarrow -MOH + H_s^+ \tag{6.6}$$

$$-MOH \Leftrightarrow -MO^- + H_s^+ \tag{6.7}$$

Using square brackets to denote activities of the species (number per unit volume for bulk species and number per area for surface species), the equilibrium constants for these two reactions are

$$K_a = \frac{[-MOH][a_{H^+}^S]}{[-MOH_2^+]} \tag{6.8}$$

$$K_b = \frac{[-MO^-][a_{H^+}^S]}{[-MOH]} \tag{6.9}$$

These are usually given as pK values, where, for example, $pK_a \equiv -\log_{10}(K_a)$. The activity of protons at the surface $a_{H^+}^S$ is related to that in the bulk electrolyte $a_{H^+}^B$ by the potential difference ψ_0: $a_{H^+}^S = a_{H^+}^B \exp(-\psi_0/\phi_t)$.

Since the activity coefficients are usually close to unity, they are well approximated by the concentrations and the bulk proton (hydronium ion) concentration $[H_b^+] = a_{H^+}^B$. The surface charge per unit area $\sigma_0 = q([-MOH_2^+] - [-MO^-])$ and the number of sites per unit area on the surface $N_s = [-MOH] + [-MOH_2^+] + [-MO^-]$. This gives [16]

$$\sigma_0 = qN_s \frac{(a_{H^+}^B / K_a)\exp(-\psi_0/\phi_t) - (K_b/a_{H^+}^B)\exp(\psi_0/\phi_t)}{1 + (a_{H^+}^B/K_a)\exp(-\psi_0/\phi_t) + (K_b/a_{H^+}^B)\exp(\psi_0/\phi_t)} \tag{6.10}$$

When N_s is large, small changes in ψ_0 can produce large changes in σ_0. Since these must be comparable to the changes in other charges, such as the charge associated with macromolecules, the allowable change in ψ_0 is constrained to be small when N_s is large. In effect, the hydroxyl sites pin ψ_0. This pinning reduces the sensitivity of the BioFET [18].

The hydroxyl sites pin the surface potential with a strength related to the capacitance C_a given by

$$C_a = \frac{\partial \sigma_0}{\partial \psi_0} \tag{6.11}$$

Figure 6.4 shows a plot of C_a versus σ_0/q. The minimum at $\sigma_0 = 0$ occurs as the surface reaction switches between the two possibilities Equations 6.7 and 6.8. The capacitance at the minimum is given by

$$C_{a,\min} = \frac{qN_s}{\phi_t} \frac{2\sqrt{K_b/K_a}}{1 + 2\sqrt{K_b/K_a}} \tag{6.12}$$

The minimum is deeper for SiO_2 than for the other oxides, because K_b/K_a is smaller for SiO_2.

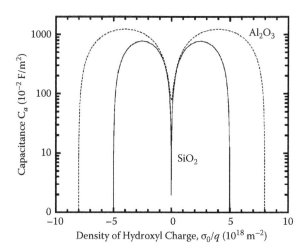

FIGURE 6.4 Capacitance C_a for the site-binding model for SiO_2 (solid line) and Al_2O_3 (dashed line). The site-binding parameters for SiO_2 are $N_s = 5 \times 10^{18}$ m^{-2}, $pK_a = -1.3$, and $pK_b = 5.7$; for Al_2O_3, $Ns = 8 \times 10^{18}$ m^{-2}, $pK_a = 5.9$, and $pK_b = 10.1$ [19].

6.2.2.3 Other Insulators

Investigations on ISFETs used for pH measurements were performed 20 or more years ago. The most common surfaces were SiO_2 (produced by oxidation in O_2/inert gas mixtures at temperatures in the range 850°C–1200°C), Al_2O_3, Ta_2O_5, and silicon nitride (SiN-Si_3N_4 containing hydrogen) produced by low-pressure chemical vapor deposition (LPCVD). For electrical measurements in electrolyte on SiO_2 surfaces above pH 5, a mechanism may exist whereby the charge on the oxide surface propagates to some extent inside the bulk of the insulator, and this is responsible for the observed drift and the related hysteresis in response to pH [20]. Al_2O_3 has less drift, but a small number of surface sites might react slowly as pH changes [21]. Ta_2O_5 shows the least drift [22,23]. The pH at the point of zero charge (where $\sigma_0 = 0$), pH_{pzc}, for SiO_2 is considerably smaller than that of all the other useful oxides. The lower drifts observed for Ta_2O_5 and Al_2O_3 may be related to the fact that these surfaces exhibit higher pH_{pzc} values, closer to the typical pH values at which these measurements are made.

The drift in silicon nitride, pH_{pzc}, is more complicated. For this material, it has been found that oxidized LPCVD silicon nitride has only silanol sites and primary amine sites, and the amine raises the pH_{pzc} considerably [24]. Although the oxide formed in air can be removed with an HF dip, it is re-formed in electrolyte when a potential is applied to the surface. It has been found that LPCVD silicon nitride films containing little H are most stable. However, larger densities of NH groups (which are isoelectronic with O) can open the structure of the nitride films produced by plasma-enhanced chemical vapor deposition methods [25]. They are particularly likely to oxidize spontaneously in water, and this could be responsible for more drift.

According to Martinoia and coworkers, the Helmholtz layer capacitance for Si_3N_4 is given by $C_{stern} = \varepsilon_{IHP}\varepsilon_{OHP}/(\varepsilon_{IHP}d_{IHP} + \varepsilon_{OHP}d_{OHP})$, where $\varepsilon_{IHP} = \varepsilon_{OHP} = 32\varepsilon_0$ and ε_0 is the permittivity of free space. The estimates for d_{IHP} and d_{OHP} are 0.1 and 0.3 nm, respectively [26,27]. To describe the site binding, in addition to the fraction due to the silanol groups, a new term σ_{nit} proportional to the amine site density N_{nit} should be added to the surface charge σ_0 in Equation 6.10 [24,26]

$$\sigma_{nit} = qN_{nit} \frac{a_{H^+}^B \exp(-\psi_0 / \phi_t)}{a_{H^+}^B \exp(-\psi_0 / \phi_t) + K_N}, \tag{6.13}$$

where K_N is the amine site-binding equilibrium constant.

Because these parameters successfully explain the pH response of an ISFET in contact with the NaCl solution, it is appropriate to assume that the Helmholtz layer thickness is close to 0.4 nm [27]. The amine binding has been included in models of BioFETs based on planar sensors [27] and nanowires [28]. But if the nitride surface oxidizes in water, the surface will be covered by hydroxyl groups and the site-binding model for a relatively poor oxide surface could apply.

6.2.3 MOLECULES ON THE OXIDE SURFACE

Figure 6.5 shows schematically the molecules at the oxide surface. As discussed later, the surface is typically treated with a layer of uncharged organic molecules that link (tether) the probe molecules to the surface. The term *functionalization* is used here to refer to the attachment of linking and probe molecule membranes. The response of the BioFET depends on the nature of both of these layers.

6.2.3.1 Linker Layer

As discussed later, the linker layers are usually dielectrics with a low conductivity. They do, however, have defects [30], through which ions can reach the surface. Hydrogen ions, in particular, can probably migrate through the defects to the oxide surface, allowing the charge of the hydroxyl groups to change. The timescale over which this happens depends on the mobility of the hydrogen ions through the defects. If the mobility is sufficiently high, it is not enough to just cover the surface with a linking layer to remove the effects of the hydroxyl groups; the molecules in that layer should react with the hydroxyl sites and passivate them. If they do not, there may be slow drift effects of the currents and potentials at the semiconductor channel associated with adjustment of the hydroxyl concentration as a result of changes at the oxide surface due to hybridization or electrolyte concentration. These could also be observable in low-frequency measurements of the impedance, as discussed later.

A dielectric linking layer can be considered as a capacitor in a series with the capacitance of the oxide and of the double layer. Another limiting case would be a linking layer that is fully permeable to ions and water. The permeable layer is less likely than the dielectric layer, because the molecules in the linking layer are generally hydrophobic, and so the linking layer should exclude water and ions. Nevertheless, the effects of a permeable linking layer will also be discussed later in this chapter.

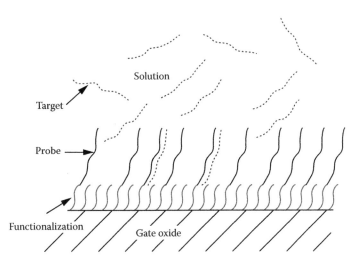

FIGURE 6.5 (See color insert). Molecules on the surface of an oxide in a BioFET. Red: linking layer, black: probe molecules, blue dash: target molecules.

6.2.3.2   Charged Membrane—Especially Oligonucleotides

Unlike the molecules in the linking layer, charged macromolecules are generally hydrophilic, and a membrane of charged molecules placed in a solution is likely to be filled with water and ions. For example, molecular dynamics simulations [31] show that Na^+ ions and water molecules are present between DNA molecules even when the DNA is separated by 2.7 nm (center-to-center). When the macromolecules are long, charged polymers (polyelectrolytes), such a membrane is called a polyelectrolyte brush. The properties of polyelectrolyte brushes have been studied extensively [32,33], especially brushes with polyelectrolytes that are much longer than their persistence length (the length over which the polymer chain can bend back on itself).

In general, the probe molecules tethered to a surface must be spaced far enough apart to bind the target molecules. This is especially true for DNA, where the binding occurs along the entire length of the probe. If the probe DNA is packed too closely together, not all of the probes can hybridize; above some probe density, the efficiency of hybridization decreases as a function of probe density [34–36]. For example, Gong [36] found in 0.11 mol/L solution that the target density on gold reached its peak of 2×10^{12} cm^{-2} at a probe density of 4×10^{12} cm^{-2}.

6.2.3.3   Electrostatics of Brushes and the Donnan Potential

In a charged membrane filled with ions, the concentration of ions inside the membrane will be different from the concentration in the solution outside, leading to a potential difference, called the Donnan potential, between the inside and outside. Consider a z:z electrolyte at a concentration n_0 outside the membrane, and suppose the macroions making up the membrane are packed at a density giving N_m elementary charges per unit volume. The charge making up N_m is called "fixed charge" [18]. Suppose the membrane is thick enough to be neutral inside. Let ϕ_{DP} be the potential differences between the neutral part of the membrane and the bulk solution. The concentration of ions in the membrane is

$$n_\pm = \pm z n_0 \exp\left((\mp z\phi_{DP} - \Phi_{d\pm})/\phi_t\right) \tag{6.14}$$

where, for generality, dispersion potentials $\Phi_{d\pm}$ [37,38] have been included. Allowing Φ_{d+} and Φ_{d-} to be different generalizes the idea of partition potentials, which were considered in simulations of a cylindrical BioFET [29]. These potentials arise as ions approach a surface, or are transferred from one solution to another. In such cases, the forces on the ions cannot be decomposed into just the forces due to the local environment plus the electrostatic forces due to deviations from charge neutrality; an extra correction is needed. These dispersion potentials will be set to zero in most of what follows, since their values are not well known.

If the membrane is thick enough to be neutral inside, the total density of charge must be zero in the neutral part, leading to the following equation for the Donnan potential ϕ_{DP}:

$$\phi_{DP} = \frac{\Phi_{d+} - \Phi_{d-}}{z} + \frac{\phi_t}{z}\sinh^{-1}\left(\frac{N_m}{2zn_b}\exp\left((\Phi_{d+} + \Phi_{d-})/\phi_t\right)\right) \tag{6.15}$$

For $|N_m| \gg n_0$, this reduces to

$$\phi_{DP} = \begin{cases} \dfrac{\Phi_{d-}}{z} + \dfrac{\phi_t}{z}\ln\dfrac{N_m}{zn_0}, & \text{if } N_m > 0; \\[4mm] -\dfrac{\Phi_{d+}}{z} - \dfrac{\phi_t}{z}\ln\dfrac{|N_m|}{zn_0}, & \text{if } N_m < 0. \end{cases} \tag{6.16}$$

As discussed later, ϕ_{DP} is directly measurable as a shift in V_T if pinning by hydroxyl groups is negligible.

6.2.4 Membrane Model and Poisson–Boltzmann Analysis

The previous sections have discussed the ingredients of a model for the BioFET, shown schematically in Figure 6.6:

- The FET (semiconductor plus oxide), described by a charge-sheet model
- A dielectric Helmholtz layer
- A layer to represent the linking layer, either (a) a dielectric layer or (b) an uncharged permeable layer
- A permeable charged membrane to represent the layer of probe molecules or probe plus target molecules
- The solution

In a continuum model, where the charge is assumed to be continuous in directions parallel to the various layers in Figure 6.6, the potential $\phi(x)$ as a function of position x normal to the layers is given by solving the Poisson–Boltzmann equation:

$$\frac{\partial^2 \phi}{\partial x^2} = \frac{q}{\varepsilon}\left(zn_0 \exp\left((-z\phi - \Phi_{d+})/\phi_t\right) - zn_0 \exp\left((z\phi - \Phi_{d-})/\phi_t\right) + N_m\right) \tag{6.17}$$

This can be solved numerically for the membrane and solution, including the effect of the site binding [39,40]. Some results of these calculations are illustrated next for $\Phi_{d\pm} = 0$, to show the effects of various parameters.

6.2.4.1 Donnan Potential

Figure 6.7 shows calculations for typical parameters, showing how the Donnan potential is achieved as the film thickness increases. The density N_m of charge in the membrane is taken to be 7×10^{19} cm^{-3}, which corresponds to an array of ssDNA packed at roughly 10% of close packing (if close packing is taken to be one molecule in an area of 4 nm^2). For the sake of illustration, the oxide is taken to be Al$_2$O$_3$, which has a positive surface at neutral pH. The Donnan potential becomes well established, meaning that the potential is flat near the center of the membrane, when the thickness becomes large compared to a screening length. But the screening length needed to make this comparison is not the

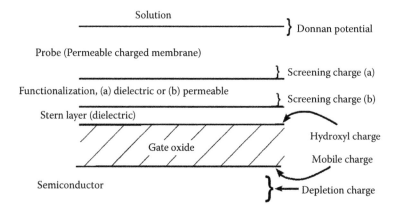

FIGURE 6.6 Summary of the components of the membrane model for a BioFET. The potential change associated with the Donnan potential occurs near the membrane–solution interface. The location of the electrolyte screening charge depends on whether the linking layer is modeled as (a) an impermeable dielectric layer or (b) a permeable layer.

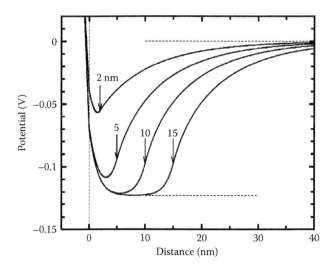

FIGURE 6.7 Potential from Poisson–Boltzmann solution for a DNA membrane of various thicknesses, to show the development of the Donnan potential. The arrows give the outer edge of the membrane for each curve. The oxide is modeled with the site-binding model for Al_2O_3, with site-binding parameters given in Reference 19, and with an electrolyte concentration of 0.001 mol/L. The charge density corresponds to a layer of unhybridized DNA at 10% of close packing. The Helmholtz layer is taken to be 1 nm thick, and there is no linking layer. The potential is taken as zero in the bulk solution (far to the right of the membrane in Figure 6.7).

screening length λ_b in the bulk solution; rather, it is the screening length λ_m in the membrane itself, which is given by linearizing Equation 6.17 [40]:

$$\lambda_m = \left(\frac{2zqn_0}{\varepsilon\phi_z} \cosh\frac{\phi_{DP}}{\phi_z} \right)^{-\frac{1}{2}} \tag{6.18}$$

If $|N_m| \ll n_0$, then λ_m reduces to λ_b (cf. Equation 6.4), but if $|N_m| \gg n_0$, then

$$\lambda_m = \left(\frac{2zqn_0}{\varepsilon\phi_z} \right)^{-\frac{1}{2}} \ll \lambda_b \tag{6.19}$$

6.2.4.2 Change with Hybridization

Figure 6.8 shows the change in ψ_0 and ϕ_{DP} when the concentration of fixed charge in the film doubles, simulating hybridization of the DNA. The hydroxyl groups are given the site-binding parameters of Al_2O_3. The linking layer is ignored here. The change $\Delta\phi_{DP}$ in Donnan potential is much larger than the change $\Delta\psi_0$ in surface potential, reflecting the pinning of the surface potential by the hydroxyl groups.

6.2.5 SMALL-SIGNAL ANALYSIS AND SENSITIVITIES

When the membrane is thicker than the screening length λ_m, closed-form expressions can be derived for the change in ψ_0 with ionic strength, pH, or membrane charge density. The three sensitivities under consideration are the charge sensitivity S_c associated with changes in the charge density of the membrane, the ion sensitivity S_i associated with changes in the ion concentration in the electrolyte, and the pH sensitivity S_p associated with changes in the pH of the electrolyte:

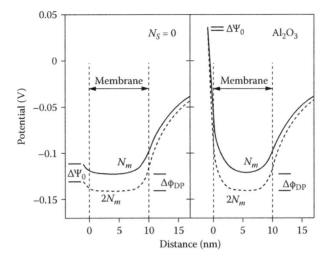

FIGURE 6.8 Potential from the Poisson–Boltzmann equation for a charged membrane 10 nm thick, representing DNA at 10% close packing. Left panel for $N_s = 0$ (no hydroxyl groups on the surface); right panel for site-binding model for Al_2O_3. Other parameters as in Figure 6.7.

$$S_i \equiv \left(\frac{\partial \psi_0}{\partial \ln n_0} \right)_{N_m, a_0, \sigma_c} \tag{6.20}$$

$$S_c \equiv \left(\frac{\partial \psi_0}{\partial \ln N_m} \right)_{n_0, a_0, \sigma_c} \tag{6.21}$$

$$S_p \equiv -\frac{1}{\ln(10)} \left(\frac{\partial \psi_0}{\partial pH} \right)_{n_0, N_m, \sigma_c} \tag{6.22}$$

Here, the subscript σ_c indicates that the charge in the semiconductor and the oxide traps should be held fixed. Thus, for example, S_c is the change in ψ_0 when N_m is increased by a factor of e (= 2.7181 ...). The analysis uses the definition $\alpha \equiv C_a/(C_a + C_{DL})$, where C_{DL} is the double-layer capacitance. The full expressions for these quantities are given in Reference 41, but the following approximations are illustrative [42] because they are analogous to the expressions [43] that can be derived for a model in which the DNA is represented as a plane of charge of negligible thickness. $\delta \psi_m$ is the difference in potential between the outer Helmholtz plane and the neutral membrane (the potential across the diffuse layer in the solution inside the membrane). The assumption that $\delta \psi_m / \phi_z \ll 1$ results in the following:

$$S_i \cong -(1-\alpha)\phi_z \left[\tanh \frac{\psi_{DP}}{\phi_z} + \frac{\delta \psi_m}{2\phi_z} \left(1 - \tanh^2 \frac{\psi_{DP}}{\phi_z} \right) \right]; \tag{6.23}$$

$$S_c \cong (1-\alpha)\phi_z \left[\tanh \frac{\psi_{DP}}{\phi_z} - \frac{\delta \psi_m}{2\phi_z} \tanh^2 \frac{\psi_{DP}}{\phi_z} \right]; \tag{6.24}$$

$$S_p \cong -\alpha \phi_t \tag{6.25}$$

The parameter α reflects the relative size of the capacitance due to the double layer and the hydroxyl groups. This is equivalent to the parameter defined by Reference 44, and the concept of a "parallel capacitance" C_a and the concept of an "intrinsic buffer capacity" are identical descriptions of the effect of the site binding. The potential $\delta\psi_m$ has to be calculated separately; it depends on the field just outside the oxide surface, and so is determined by the charge in the semiconductor and on the oxide surface.

Equations 6.23–6.25 make plain a number of the important effects:

- When the sensitivity to pH is high, the sensitivity to N_m or n_0 is low, and vice versa, through the factor α in Equation 6.25 and $(1 - \alpha)$ in Equations 6.23 and 6.24.
- As $\delta\psi_m \to 0$, $S_c = -S_i = (1 - \alpha)\phi_z \tanh(\phi_{DP}/\phi_z)$ which tends to $\pm(1 - \alpha)\phi_z$ when $|\phi_{DP}| \gg \phi_z$. The sensitivities are of order ϕ_t or less.
- When $|\phi_{DP}| \gg \phi_z$, the term in $\delta\psi_m$ in S_i vanishes, and that in S_c becomes $-(1 - \alpha)\delta\psi_m/2$. Thus, S_c can be increased by changing $\delta\psi_m$ (which can be done by changing the bias between the reference electrode and the semiconductor). The sensitivity S_c increases because the field at the oxide surface produces a depletion layer in the membrane, in analogy with the depletion layer formed in the semiconductor; as this depletion layer grows, more of the membrane's fixed charge is exposed and S_c increases. Unfortunately, the field needed to increase S_c significantly is comparable to the breakdown field in the oxide [41], so it is not feasible to increase S_c in this way.

6.2.5.1 Effects of pH and Hydroxyl Density

Figure 6.9 shows S_c and S_i as a function of n_0 for two values of pH, for SiO_2 and Al_2O_3. These calculations were done with the expressions given by McKinnon [41], but numerical calculations from the Poisson–Boltzmann equation give indistinguishable results [42].

The figure illustrates how the sensitivities generally fall off with concentration. Note that the relative sizes of S_c and S_i depend on the material; for SiO_2, $S_i > S_c$, whereas for Al_2O_3, $S_i < S_c$ [42]. Of the cases plotted, the sensitivity is largest for SiO_2 at pH = 6. This reflects the deeper minimum in C_a for SiO_2 in Figure 6.4. Because the values of C_a are so large for the parameters of typical oxides in the site-binding model, almost all of the hydroxyl groups must be eliminated before α falls to near zero and the full sensitivity S_c is recovered [45].

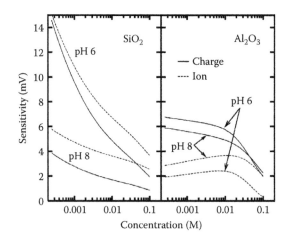

FIGURE 6.9 Absolute value of sensitivities S_c (solid curves) and S_i (dashed curves) for a DNA film at 10% of close packing, with site-binding parameters for SiO_2 or Al_2O_3 as a function of concentration of the solution, and for two values of *p*H.

6.2.5.2 Ion Sensitivity versus Charge Sensitivity

Because S_i and S_c are so similar for small $\delta\psi_m$ one can envision using changes in S_i to indicate hybridization. It has been proposed [46] that DNA hybridization could be measured by comparing the response of the BioFET to a change in salt concentration before and after hybridization. In the proposal, V_T is measured at two concentrations: before hybridization, and again after hybridization. Since the response to a change in solution concentration is typically much faster than the response to DNA, this approach reduces errors associated with the slow drift of the BioFET. The values of V_T measured at the high salt concentrations change less after hybridization than those measured at low salt concentration, reflecting the much shorter Debye length at high salt and acting as an effective reference for the measurements.

6.2.5.3 Complexation

Ions from a solution can lose their hydration shell and form complexes with hydroxyl groups on the surface of an oxide. Evidence for complexation comes from models used to fit the charge on oxide particles as a function of pH. If these models ignore complexation, then the Helmholtz capacitance needed to fit the experiment becomes unreasonably large [47]. Complexation can be included in the numerical solution of the Poisson–Boltzmann equation, but does not have a large effect on the calculated sensitivity for typical values of the parameters [41].

6.2.5.4 Impermeable Linking Layer

If the tether layer is impermeable to ions, then it can be included in the model by a simple change of parameters: the Helmholtz capacitance should be replaced by an effective value equal to the series combination of the actual Helmholtz capacitance and the capacitance of the linking layer. As this effective capacitance decreases (simulating a linking layer of increasing thickness), a given field in the oxide produces a larger potential difference between the oxide surface and the membrane. As a result, when sensitivity is plotted as a function of charge density σ_c in the semiconductor, features in the sensitivity appear to be compressed. Figure 6.10 shows the sensitivity S_c for the site-binding parameters of SiO_2, assuming that hydrogen ions are transported through defects in the linking layer to the oxide surface. A peak appears as the surface is driven through the neutral point. This peak is

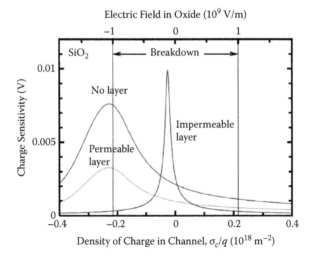

FIGURE 6.10 (See color insert). Charge sensitivity $-S_c$ as a function of charge number density σ_c/q in the semiconductor, for site-binding parameters of SiO_2. The DNA layer is taken to be 10% of close packing, 10 nm thick. The linking layer between the DNA and the oxide is either absent ("no layer"), or has a thickness of 1 nm. The 1 nm layers are either modeled as dielectrics ("impermeable layer") or as filled with solution ("permeable layer"). At the vertical lines, the field in the SiO_2 reaches breakdown.

associated with the minimum in capacitance in Figure 6.4. The peak narrows when the linking layer is added to the calculation. Once the peak occurs entirely within the range before breakdown of the oxide, it might be possible to observe it.

For the parameters chosen in Figure 6.10, the sensitivity near $\sigma_c = 0$ is not changed much by the introduction of an impermeable linking layer (the red and blue curves cross near $\sigma_c = 0$). As the linking layer is made even thicker (not shown in the figure), the sensitivity decreases eventually.

6.2.5.5   Permeable Linking Layer

Figure 6.10 also shows the case of a permeable linking layer. This case cannot be included in the expressions for the sensitivities (Equations 6.23–6.25), but can be treated in a straightforward manner in the numerical solution of the Poisson–Boltzmann equation [48]. A permeable linking layer reduces the sensitivity for all values of semiconductor charge, because the ions in the linking layer screen the charge associated with the membrane. This reduction can also be explained by Donnan potentials. If the permeable linking layer is thick enough, then the charged membrane has solutions on both sides of it. Donnan potentials appear at both membrane–solution interfaces, and the two Donnan potentials cancel [45].

6.2.5.6   Transient Effects due to Mixing

Since BioFETs are sensitive to ion concentration and pH, they can respond to variations in concentration as solutions mix. For example, if a dilute salt solution is added to a concentrated solution, the dilute solution floats on top of the concentrated solution. This can create a potential gradient that modifies the pH gradient at the interface. If the BioFET is at the bottom of the container, it can take many minutes for the concentrations to equilibrate and the signal to stabilize. On the other hand, if the concentrated solution is added to the dilute solution, the signal can respond more quickly [46].

6.2.5.7   Size of the Signal

In the charged membrane model at low fields in the oxide and for low densities of hydroxyl groups, the optimum signal is just the change in Donnan potential (both $\delta\psi_m$ and α small in Equation 6.24). Because screening is nonlinear, the Donnan potential changes more when the first charges are added than when later charges are added. Thus, the signal is larger when a charged layer of probe molecules is added to the linking layer than when the target molecules hybridize with the probe molecules. The Donnan potential is -63.6 mV for a membrane of ssDNA with a density 7×10^{19} charges/cm^3, corresponding to a DNA layer at about 10% of close packing, in a solution of 0.01 mol/L NaCl. It decreases by -17.8 mV (about $\phi_t/\ln 2$) to -81.4 mV when N_m is doubled. Thus, the signal for $\alpha = 0$ would be almost four times larger when the probe molecules are added as when the target molecules hybridize with the probes.

This is a powerful incentive to use charge-free DNA mimics such as peptide nucleic acid (PNA) [49,50] and morpholinos [51]. These provide higher binding affinity, and greater stability at low electrolyte concentrations (<0.5 mol/L) when used as probe molecules. They can be hybridized and, conversely, do not dehybridize at low salt concentration. Recently, label-free electrical (impedimetric) measurements using morpholino probes [52] on flat surfaces demonstrated the detection of 3×10^{10} DNA targets per square centimeter with a 10:1 signal-to-noise ratio. These values were comparable to or in excess of those of surface plasmon resonance and quartz crystal microbalance measurements.

Compare the charge of 17.8 mV to what would be measured if all the charge on the DNA could be placed directly (no screening) on the gate oxide of a FET. Suppose the gate oxide is 5 nm thick, giving a capacitance $C_{ox} = 6.9 \times 10^{-3}$ F/m^2. The fixed charge due to the phosphate backbone in a 10-nm-thick layer of DNA is $Q_{DNA} = 0.112$ C/m^2 at the packing density just discussed. If this were the only charge involved, the signal would be $Q_{DNA}/C_{ox} = 16.2$ V, 900 times larger than the change in Donnan potential.

Even if the DNA is modeled as an impermeable charged layer and the hydroxyl sites are unimportant, the signal is considerably smaller than Q_{DNA}/C_{ox} ([53], p. 516). If a charge Q_{ads} is adsorbed onto the oxide of a transistor in solution, it induces a charge Q_{sol} in the solution and Q_{semi} in the semiconductor, according to the capacitance C_{DL} of the double layer and C_{ox} of the oxide. The change in the charge in the semiconductor $Q_{semi} = Q_{ads}C_{ox}/(C_{ox} + C_{DL})$. Taking C_{ox} to be that for a 5 nm layer of SiO_2, as above, and C_{DL} to be the Helmholtz layer capacitance of 20 μF, then $Q_{semi} = 0.033 \, Q_{ads}$ and the corresponding voltage would be $0.033 \, Q_{ads}/C_{ox} = 0.53$ V.

For this example, then, one might say that the presence of the electrolyte reduces the sensitivity of the BioFET by a factor of 1/30 for an impermeable sheet of DNA, and by another factor of 1/30 if the electrolyte can permeate the DNA. And this ignores the further reduction produced by the hydroxyl groups.

6.2.6 Large-Signal Models of the BioFET

So far the discussion has assumed that a unique value of ψ_0 can be associated with a given device and operating point. This is true when V_{DS} is small, but a complication enters when V_{DS} is large: ψ_0 varies with position along the gate oxide between the source and drain. This variation must be included in a large-signal model of the BioFET.

The strong inversion model of Equation 6.3 can be modified [40] through a parameter $m = \delta\psi_0/\delta\psi_s = C_{ox}/(C_{ox} + C_{DL} + C_a)$, relating changes in oxide surface potential ψ_0 to changes in semiconductor surface potential ψ_s. At strong inversion, m is approximated at the source by $m_0 = C_{ox}/(C_{ox} + C'_{DL} + C_a)$ where C'_{DL} is an approximation for the double layer capacitance at the source. The parameter α_b in Equation 6.3 becomes

$$\alpha_b = 1 + \frac{\gamma_b}{2(1-m_0)\sqrt{\pm\phi_0 \pm V_{SB}}}. \tag{6.26}$$

In terms of the approximated flatband voltage V_{FBA}, the threshold voltage becomes

$$V_{THA} = V_{FBA} \pm \frac{\gamma_b}{1-m_0}\sqrt{\pm\phi_0 \pm V_{SB}} + \phi_0, \tag{6.27}$$

and the current becomes

$$I_{DS} = \begin{cases} \pm\mu(1-m_0)C_{ox}\dfrac{W}{L}\left((V_{ref} - V_{THA})V_{DS} - \dfrac{\alpha_m}{2}V_{DS}^2\right), & |V_{DS}| \leq |V_{ref} - V_{THA}|/\alpha_m; \\[2ex] \pm\mu(1-m_0)C_{ox}\dfrac{W}{L}\dfrac{(V_{ref} - V_{THA})^2}{2\alpha_m}, & |V_{DS}| > |V_{ref} - V_{THA}|/\alpha_m. \end{cases} \tag{6.28}$$

Typically, m_0 is small compared to unity, so this model differs little from the model of the FET, Equation 6.1.*

6.2.7 Limitations of the 1-D Model

The membrane model is a useful starting point in estimating sensitivities of a BioFET, but several limitations should be recognized.

---

* The approximation for the flatband voltage and a more general treatment of the charge-sheet model, valid from weak to strong inversion, is given in Equations 31 and 40 in Reference 40.

6.2.7.1 Manning Condensation

Manning condensation occurs when ions from solution condense onto a charged cylinder, reducing the effective charge of the cylinder. In a model of charged cylinders of vanishing radius, counterions condense onto the cylinder when the charge per unit length exceeds a critical value [54] of one charge per Bjerrum length λ_B, which is given by $\lambda_B = q/4\pi\varepsilon\phi_t$, in a medium of permittivity ε. For water, $\lambda_B = 0.71$ nm. Since the charges on DNA are spaced 0.34 nm along the phosphate backbone, both ssDNA and dsDNA exceed the critical charge density, and so if they were cylinders of vanishing radius, they would have the same renormalized charge density. If that were so, there would be no change in N_m when DNA hybridized in a BioFET, and hence there would be no signal.

A real molecule of DNA, however, does not have a vanishing radius. For a cylindrical molecule of radius r packed in an assembly of molecules spaced by a distance R, Manning condensation does not imply the ions are bonded to the molecular surface. Rather, the so-called "condensed" ions are held inside a radius R_M proportional to $\sqrt{rR}$ [55]. When R is only a few times r, the condensed ions are not much closer to the cylinder than the uncondensed ones, and there is a signal in a BioFET. A laterally averaged Donnan potential can be defined, and it does change on hybridization [48]. Even so, the strong lateral screening normal to the surface of the cylinder does introduce significant corrections. These corrections might be viewed as a form of dispersion potential $\Phi_{d\pm}$ although apparently this view has not been explored.

An estimate of the corrections due to lateral screening for DNA normal to the sensor surface was made from a solution of the Poisson–Boltzmann equation applied to a cylindrical unit cell [48]. At reasonable densities of DNA, the corrections are significant, reducing the predicted signal by as much as 50%, but the trends are the same as in the membrane model. At low concentrations of DNA (a few percent of a close-packed layer or below) and low ionic strengths (0.001 mol/L or below), errors can be much higher; the membrane model can overestimate the sensitivity by a factor of 10 or more. It is not reasonable to apply the membrane model at these low concentrations of DNA and low ionic strengths.

6.2.7.2 Other Mechanisms for the Signal in a BioFET

So far in this chapter, it has been assumed that the signal is due to the change in charge density of the membrane, and that this charge density can be calculated assuming the thickness of the membrane is fixed. The sensitivity in this case is given by the change in Donnan potential of the membrane, reduced by the response of the surface hydrogen through the factor α provided $\delta\psi_m$ is small in Equation 6.25.

If these assumptions fail, the sensitivity will be different. As a simple example, the orientation of the molecules could change on hybridization. Then the charge density after hybridization, $N_{m,ds}$, will not be simply twice the density $N_{m,ss}$ before hybridization.

Another mechanism for detecting hybridization is through the change in the factor α. The macromolecules on the oxide surface can block hydroxyl groups, and prevent them from changing their charge state. If the number of blocked hydroxyl groups changes in response to hybridization, then α can change. For example, ssDNA is more flexible than dsDNA, so before hybridization the ssDNA probes could be bent down onto the surface, blocking hydroxyl sites, but the dsDNA hybridized molecules would stand up away from the surface, exposing hydroxyl sites [56]. If such a mechanism were operating, the change in α could be confirmed by the change in S_p, the sensitivity to pH, before and after hybridization.

6.2.7.3 The Mystery of High Signals

According to the membrane model, the measured signal on hybridization should be on the order of $\phi_t = 26$ mV for an unpinned surface (where all the hydroxy groups are passivated), and even smaller for a pinned surface. How is it then that some studies report much higher responses, large fractions of a volt or more?

Poghossian [56] reviewed many of the reports, and pointed out problems with some. For example, some studies did not use reference electrodes, although it is not clear how removing the reference electrode would lead to large signals. Some devices had bare silicon surfaces, or surfaces covered only with native oxide, and in these devices it is possible that hybridization modifies traps at the silicon surface, changing the threshold voltage, or that charge transfer to the silicon occurs through a mechanism called the Cooperative Molecular Field Effect [57]. But there does not seem to be a consistent explanation for all the reports of high sensitivity.

Another mystery is the signal observed in devices with suspended gates [58]. These are devices where polysilicon is coated with nitride and suspended over a FET that is also coated with nitride. Solution fills the gap between the two nitride surfaces. Shifts in threshold voltage of greater than 200 mV are reported with hybridization. The membrane model predicts that a membrane would form on both surfaces, so Donnan potentials would appear between each nitride surface and the solution. The Donnan potentials would contribute equal and opposite signals to the potential between the suspended gate and the semiconductor, implying no signal on hybridization. Yet a large signal was observed.

Clearly, more work needs to be done to understand these results.

6.2.8 Comparison with Measurements

Early first detection of the conjugation of DNA on a BioFETs was obtained by measurement of the shift in impedance with changes in V_{ref} [59–61], the surface potential, using a cantilever [62], or direct measurement of the drain current [63,64].

The impedance measurements gave shifts >100 mV and often recorded just the capacitive component. The following section will focus on impedimetric measurements.

The most sensitive measurements involve the differential measurement using a reference sensor that has a fully mismatched probe on a surface identical to that of the primary sensor. The cantilever method recorded a differential signal for single-nucleotide polymorphisms (SNPs) in 12 bp oligonucleotides and showed signals at the millivolt level, with sensitivity to 2 nM concentrations of DNA [62].

Differential measurements on BioFET arrays inferred from source-drain current measurements (referred to as potentiometric here) were performed on multi-element arrays to distinguish between DNA samples obtained by amplification of DNA from patients with a mutation related to prelingual nonsyndromic deafness [65].

Values of threshold voltage shifts, ΔV_T, as large as 78 mV obtained by the potentiometric method have been reported after functionalization of a LPCVD silicon nitride surface [62]. The authors speculated that such a large positive shift could be due to extra negative charges caused by a change in charges on the LPCVD silicon nitride, subsequent cleaning, silane treatment and glutaraldehyde treatments, and blocking with glycine after probe attachment. The shift after hybridization (conjugation) was 11 mV. The same group later reported a ΔV_T of 38 mV after probe immobilization and 12 mV after hybridization [67].

In that work, the authors showed that SNPs could be distinguished and showed that the use of DNA binders that attach specifically to hybridized DNA (intercalators) could enhance the signal in a manner that helps further distinguish DNA genotypes. They later showed that allele-specific primer extension done directly on the BioFET surface could also have this advantage [68].

With high-quality SiO_2 surfaces functionalized using a solution of (3-aminopropyl) triethoxy silane (APTES) in organic solvent, smaller threshold shifts have been observed after conjugation. For a measurement done in 1 mM electrolyte, a shift of 4 mV has been reported [69]. The shift was recorded in real time while referenced to another device with a fully mismatched probe.

The variations in the magnitude of the response reported by various groups are undoubtedly partly due to the different methods used to prepare the gate surfaces prior to functionalization. The variations are probably largely due to the different processes and materials used to attach the tether and probe molecules to the gates.

The potentiometric measurements just described are more sensitive to the details of the membrane interfaces than impedance measurements. They demonstrated features of the models described earlier, particularly that the threshold voltage shift ΔV_T caused by hybridization was smaller than that caused by probe attachment, that ΔV_T was smaller in higher electrolyte concentrations, and that the measurements were sensitive to site binding and the details of the immobilization process. These effects are generally attributed to Debye screening, but a close agreement with the models is still difficult.

One cause of the difficulty is that measurements that independently quantify the biomolecule attachment are difficult. Radiometry measurements are not generally accessible. X-ray photoelectron spectroscopy, x-ray reflectivity (XRR), and ellipsometry measurements are susceptible to errors caused by the absorption of adventitious carbon during air exposure. Furthermore, the use of XRR to determine the thickness and density of the biolayers is accurate only on very thin oxide layers, usually thin chemical oxides that are not representative of the BioFET fabrication or functionalization process. Quartz crystal monitors can detect the mass of biomolecules deposited in solution, but have rarely been used on witness samples that duplicate faithfully the device fabrication and functionalizing process. Fourier transform infrared spectroscopy can, in principle, give information about the orientation of the tether molecules on the gate surface. More of these measurements need to be done on more than just idealized surfaces.

In spite of credible models, many papers have used expressions for the relation between charge and ΔV_T that are not applicable, or have failed to mention critical experimental details.

6.2.9 BioFET Simulations in 3-D

The need to better understand the details of the device functionalization and target binding have spurred a recent effort to simulate the performance of BioFETs in three dimensions (3-D). Recently, solutions for the potential in a BioFET of been obtained in 3 dimensions (using x, y, and z axes) and a finite element method to solve the Poisson–Boltzmann equations [28].

On the solution side of the BioFET surface, Poisson's equation,

$$-\nabla(\varepsilon(x,y,z)\nabla\phi(x,y,z)) = qn_0 \exp\left(-\phi/\phi_t\right) - qn_0 \exp\left(\phi/\phi_t\right) + \sum_i Q_i(x,y,z) \qquad (6.29)$$

included the fixed charges Q_i (negative charges with magnitude q) representing the DNA backbone, distributed in a helical pattern around the surface of a dielectric cylinder with its own Helmholtz layer. The charges in the electrolyte were modeled as a charged continuum with Boltzmann factors modifying the ion and cation densities. The cylinder was surrounded by electrolyte and oriented in the $+y$ direction. Layers in the x-y planes represented the Helmholtz, linker, and silicon nitride insulator layers. The Helmholtz layer thickness, dielectric constant, and site-binding appropriate to silicon nitride mentioned earlier [26,27] were used, including the silanol (Equation 6.10) and amine (Equation 6.13) contributions. The linker layer was modeled as a continuous, permeable or nonpermeable layer 1.5 nm thick.

To simulate many DNA molecules sitting on a surface, space was divided into units, rectangular cylinders extending in the y-direction with one DNA cylinder per unit and variation of the unit spacing simulating the variation in probe density. Self-consistent solutions were obtained using the finite element method and periodic boundary conditions were used at the unit boundaries.

The simulations calculate the average potential drop in the solution to determine the silicon surface potential. From this, the threshold voltage was calculated, and the change after hybridization was considered. The standard equations (Equation 6.1) were used to estimate the FET current changes.

The most impressive part of these simulations is the detailed models of the shape and charge distributions around the ss- and ds-DNA cylinders, but the details are beyond the scope of this chapter. Most of the results are consistent with the 1-D and 2-D calculations described earlier; for example, the change in flatband voltage ΔV_{FB} for probe attachment is greater than that for hybridization, and it does not depend dramatically on the number of base pairs in the DNA cylinders, between 7 and 100 bases (for low electrolyte concentrations).

With the site binding turned on, the maximum ΔV_{FB} after hybridization is just a fraction of a millivolt. With site binding off, the shift after hybridization for a DNA probe density of 4×12 cm^{-2} and a 20 mM electrolyte concentration is 35 mV. That these are closer to experimentally determined values leads the authors to suggest that the site binding is somehow suppressed in the experiments or that the orientation changes during hybridization to increase the signal. These possibilities were mentioned earlier in the discussion of 1-D models.

The authors conclude that these models are still not sufficiently realistic, and more work needs to be done to model the characteristics of the linking layers, site binding, and possible changes in DNA orientation during hybridization.

6.3 IMPEDANCE MEASUREMENTS ON BIOFETS

The first impedance measurements on BioFETs, performed at a fixed frequency of 20 kHz [70,71] or 100 kHz [72], showed large threshold voltage shifts after hybridization. Most of the section focused on potentiometric measurements, which determined the change of the BioFET threshold voltage V_T from its current–voltage characteristics. The interpretation of these shifts is complicated by the very slow responses of charges in the electrolyte. Measuring the BioFET impedance as a function of frequency would allow the causes of drift in the potentiometric measurements to be isolated and give new information about the hybridization process on bioaffinity sensors [73]. The basic concepts are described in the standard texts [74], and a recent review [75] describes impedimetric analysis of immunosensors, DNA-sensors, and biosensors that employ enzymes (EnFETs).

The impedance is the ratio between the system voltage phasor, $U(j\omega)$, and the current phasor, $I(j\omega)$, where $j = \sqrt{-1}$, f is the measurement frequency, and the angular frequency $\omega = 2\pi f$. The complex impedance can be presented as the sum of the real, $Z_{re}(\omega)$, and imaginary, $Z_{im}(\omega)$, components that come from the resistance and capacitance, respectively: $Z(j\omega) \equiv U(j\omega)/I(j\omega) = Z_{re}(\omega) + jZ_{im}(\omega)$. Generally, a sinusoidal signal v_{in} (with amplitude < 10 mV) is applied to the BioFET. This is represented in Figure 6.11, which shows a simple circuit with the reference electrode, Pt counter electrode, and BioFET working electrode in solution. Sinusoidal variations in the drain current i_{ds} are measured using an OPAMP (operational amplifier) with feedback resistance R. The output voltage v_{out} is related to the transconductance ($g_m \equiv \delta I_{DS}/\delta V_{GS}$). The transfer function $H(j\omega)$ is given by the relations

$$v_{out} = -Ri_{in} = -Rg_m H(j\omega) v_{in} \tag{6.30}$$

where $H = 1$ in the absence of any charged molecules in motion at the BioFET surface. It is obtained as a function of applied frequency and is usually used to extract elements of the equivalent circuit and time constants to model the sensor response [76,77].

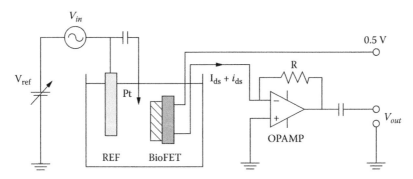

FIGURE 6.11 Simple circuit with a single OPAMP for measuring the transfer function of BioFETs or other CHEMFETs in solution.

6.3.1 IMPEDANCE OF ISFETs, REFETs, AND THE FIRST BIOMEMBRANES

6.3.1.1 ISFETs

In the following, the evolution of the equivalent circuit to its current embodiment will be described. The response of ISFETs to step changes in potential was first interpreted using equivalent circuits consisting of parallel resistor-capacitor elements [73]. The equivalent circuit shown in Figure 6.12, including an element describing the Warburg impedance, was used to describe the ISFET [78]. The impedance of the substrate Z_s has been described in the standard texts [12]. It can be represented by a parasitic series resistance R_s, associated with circuit contact resistances, and the bulk semiconductor substrate resistance, in series with the FET channel impedance. In inversion, the channel impedance is often ignored since it has a fast response and its parallel capacitance is large compared to C_{ox}, the gate-insulator capacitance. The reference electrode Z_{ref} has a large capacitance, and its impedance consists mainly of a series resistance. Commercial microreference electrodes with diameters ≈ 0.5 mm and resistances $R_{ref} < 5$ kΩ are readily available, but these diameters are still much greater than that of a typical microsensor surface. The Warburg impedance W resulting from the diffusion of ions from the bulk electrolyte to the electrode interface is in series with the capacitance C_a (described in the previous section) associated with the site binding, and these shunt the double-layer capacitance C_{DL}.

6.3.1.2 REFETs

ISFETS with parylene-covered gate insulators not sensitive to hydroxyl ions, REFETS [79,80], were investigated to eliminate the need for the standard Ag/AgCl or calomel reference electrodes, which are hard to miniaturize and have a drift associated with loss of the internal electrolyte through their porous plug.

The original use of parylene and other alternatives such as polystyrene, Teflon, and bis-acryl has been reviewed in a seminal paper by the Bergveld group [81]. It is still relevant to impedance measurements being done on BioFETs to study the conjugation of DNA or proteins. In an effort to reduce the pH response and drift of reference electrodes below a few millivolts (DC), the study concluded that the problems due to pinhole formation, contaminants, unwanted ion attachment, and the poor bonding of some materials must be eliminated. Disappointing initial efforts using silylating agents (silanes), plasma deposition, or insulator cleaning process were also referenced. Using thicker layers significantly reduced the FET sensitivity due to their low capacitance, and their inactive surfaces led to a low series capacitance in the Helmholtz layer [82], especially for low electrolyte concentrations.

Nonblocking layers were investigated to restore the reduction in sensitivity caused by thick ion-blocking polymer layers. Additives were also investigated to eliminate the permselectivity (propensity to attach or allow incorporation of a specific ion) of relatively mobile cations, particularly K^+ ions, in common electrolytes [81]. This permselectivity is commonly exploited in ion-selective CHEMFETs but is detrimental to ISFETs (and would also be to BioFETs). Finally, a 5 μm thick duplex layer was found suitable for the REFET. This consisted of an acrylate layer on poly(hydroxyethyl methacrylate) (p-HEMA), a hydrogel that bonds well to the gate oxide. The duplex has negligible pH sensitivity and permselectivity, while maintaining a high capacitance due to the ionic permeability of the p-HEMA layer.

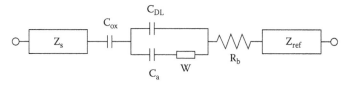

FIGURE 6.12 Equivalent circuit of the ISFET with an oxide gate in electrolyte. (From L. Bousse and P. Bergveld, *J. Electroanal. Chem.*, 152, 25–39, 1983.).

6.3.1.3 Measurement of Differential Output with ISFFET/REFET Pairs

The impedance measurements demonstrated in the foregoing study [81] are the basis of the measurements used on BioFETs to this day. A differential amplifier circuit, with inputs for an ISFET and another for an REFET, both with preamplifiers, was also employed to obtain the transfer function. Instead of a conventional reference electrode, a quasi-reference electrode (QRE) made of Pt was used to establish the common potential. This could introduce potential variations since the Pt electrode–electrolyte interface is not pinned by ion exchange at the surface; that is, it is not a true reference electrode. Despite this, output from the differential amplifier using the ISFET/REFET pair was stable because of its high common mode rejection ratio.

REFETs were further investigated with differential amplifiers [83,84]. Although these studies are instructive, for a BioFET it is probably more effective to use a sensor element with probes non-complementary to all targets as the referencing element in a differential detection scheme. In this case, the differential amplifier would detect just the change in response due to the conjugation, and minimize other substantial responses due to the functionalization process.

To analyze the blocking or nonblocking nature of candidate coatings for the REFETs, the equivalent circuit represented the polymer layer by a capacitor C_{mem} in parallel with a resistor R_{mem} [81]. This has a transfer function

$$H(j\omega) = \frac{1 + j\omega R_{mem} C_{mem}}{1 + j\omega R_{mem}(C_{mem} + C_{ox})} \qquad (6.31)$$

At high frequency, $H(j\omega) \to C_{mem}/(C_{mem} + C_{ox})$, and this can be used to determine C_{mem} if C_{ox} is known. At low frequencies, it should be constant if the low-frequency phenomena associated with the membrane interfaces (Warburg impedance and proton exchange associated with site binding) are not significant. A plot of $\log[H(j\omega)]$ versus ω was used to determine two time constants:

$$\tau_1 = R_{mem}(C_{mem} + C_{ox}) \qquad (6.32)$$

and the membrane relaxation time

$$\tau_2 = R_{mem} C_{mem} \qquad (6.33)$$

For angular frequencies between $1/\tau_1$ and $1/\tau_2$, the response falls as $1/\omega$ with a slope of -1 on a log–log plot, as shown in Figure 6.13.

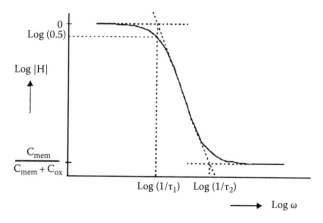

FIGURE 6.13 Shape of transfer function described by Equations 6.32 and 6.33 in the text, where C_{mem} is the membrane capacitance and C_{ox} is the oxide capacitance. (From M. M. G. Antonisse et al., *Anal. Chem.*, **72**, 343–348, 2000.)

6.3.1.4 First Impedance Studies of Protein Membranes

A simple model for the fixed charge in the membrane has been used to calculate the resistance of a permeable membrane, which included a concentration of fixed charge N_t and a bulk electrolyte with ion concentration n_0 [18,86]. In the following ion concentration r, activity a, amd mobility u have superscripts to denote the ion charge and subscripts s and m to denote values in the solution and membrane, respectively. Assuming that the membrane is in equilibrium, the activities in the bulk and membrane are related: $a_s^+ a_s^- = a_m^+ a_m^-$. Since $n_s^+ = n_s^- = n_s$ and $n_m^- = n_m^+ = N_t$, the concentrations of the mobile ions in the bulk are determined: $2n_s^+ = \sqrt{N_t^2 + 4n_s^2} - N_t$ and $2n_s^+ = \sqrt{N_t^2 + 4n_s^2} + N_t$. The resistance of the membrane R_{mem} is given by

$$R_{mem} = \frac{2d_m}{FA_m} \Big/ \left[(u_m^+ + u_m^-)\sqrt{N_t^2 + 4n_s^2} - (u_m^+ - u_m^-)N_t \right] \tag{6.34}$$

where F is the Faraday constant, A_m is the membrane area, and d_m is its thickness.

In a study of the impedance of charged protein layers (cross-linked lysozyme), 50–200 μm thick, this analysis has been used to estimate the membrane resistance [86]. These membranes had a positive fixed charge at pH 6 that decreased as the pH was increased to 9.4, closer to the isoelectronic point pI (the pH at which the membrane molecules are uncharged, $pI = 11$ for the lysozyme). The series resistance of the bulk electrolyte was determined by measurements on uncoated ISFETs (see the following subsection) and, by comparing the resistance of a bulk solution layer with the same thickness as the membrane, the membrane resistance was estimated. This was lower than the value for the uncoated ISFET, showing that the membrane had a higher conductivity due to the higher specific conductivity of the counterions. The results were in at least qualitative agreement with the model and showed that the establishment of a Donnan equilibrium in the membrane caused a decrease in resistance at pH 6. However, increases in pH caused resistance increases due to the decrease in the fixed charge. This indicated that the mobility of the counterions in the membranes is lower than that in solution. This work inspired later work on thinner protein membranes.

6.3.1.5 Series Resistance of the Electrolyte

A large bulk resistance due to the electrolyte places severe constraints on the impedimetric measurements at low electrolyte concentrations. It has also been shown that leaky encapsulation around the gate surface could have a particularly deleterious effect on the impedance measurements [87]. Since the dielectric constant of water is ≈80× that of air, the electric fields near a sensor contact can spread much larger distances than they do in air.

The large series resistance can be easily estimated. A model with a number of resistive elements in series that increase in diameter far from the sensor surface was used to make an estimate of the solution resistance [86]. Working with 0.2 mM KCl electrolyte, it was concluded that the bulk resistance was confined to a small layer close to the sensor surface. This is the phenomenon called *spreading resistance*, and its magnitude in solid-state devices is often calculated precisely. Provided the conductivity κ of the solution is known, the calculations can be simplified for a small disk with radius r at a fixed uniform potential, representing the sensor surface, and a grounded disk with a much larger area representing the reference electrode a distance d from the sensor.

In Figure 6.14 the static conductivity data from standards published for KCl [88] were used to plot the resistivity ($\rho = 1/\kappa$) and typical resistances as a function of KCl concentration in aqueous solution at 25°C: (right axis) (– – –) resistivity; (left axis) (–·–·) spreading resistance for a 50 um diameter disc; (——) resistance of a 50 μm cube.

At frequencies commonly used for biosensor impedance measurements (<200 kHz), the dielectric constants of electrolytes can be well approximated by their static values [89,90]. For KCl and NaCl, small variations in the conductivity with frequency are well fit by a Cole-Cole equation [89], but conductivities are generally well enough approximated by their static values for practical purposes.

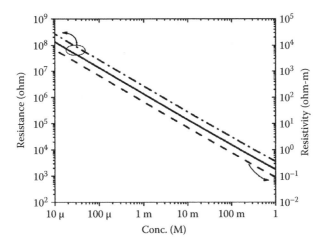

FIGURE 6.14 Plot of resistance and resistivity of aqueous KCl solutions at 25°C versus concentration; (– – –) resistivity, (– · – ·) spreading resistance for 50 μm diameter disc, (——) resistance of 50 μm cube.

If the concept of spreading resistance is unfamiliar, it is described in this paragraph. The precise calculations show that as long as the reference electrode is large enough, the resistance has two limits as the distance between a gate surface and reference plane is increased from 0 to a distance corresponding to just a few times larger than the size of the gate. Beyond this distance, the resistance does not increase significantly, but it still can be quite large for dilute electrolytes. Consider a circular gate surface with radius r parallel to a large plane at fixed potential, a reference electrode. Assume that the potential is also fixed at a plane close to the gate. For a reference electrode a small distance d from the gate, the solution resistance is given by the usual formula $R_{sol} = d/\pi r^2 \kappa$. However, as $d \to \infty$, the model demonstrates that the solution resistance reaches a constant value $R_{sol} \to 1/4r\kappa$; that is, the resistance is determined only by the size of the gate and conductivity.

In fact, the precise calculations show that the resistance reaches $0.8/4r\kappa$ for $d \approx 2r$ [91]. For gates with diameters on the order of tens of microns and the smallest reference electrodes, typically 500 microns in diameter, it is difficult to reduce the resistance to values smaller than the spreading resistance, unless the electrode is brought much closer to the gate than the spacing $2r$. This is difficult to do with existing reference electrodes. Thus, it is advantageous to build the reference electrode onto the sensor surface or a plate fabricated over a channel surface. This could be done using current MEMS (micro-electromechanical systems) technology and microfluidic systems.

6.3.2 MEMBRANE SENSORS

As will be discussed for the following examples, the impedimetric measurements made at frequencies >10–100 Hz are all dominated by the membrane capacitance and resistance. They have often been called MEMFETs. Most could be integrated on the surface of CMOS chips with all the associated electronics (and microfluidic applicators) required to facilitate the impedance measurements.

6.3.2.1 Ion-Selective Surfaces

Impedimetric measurements on permeable membranes composed of cellulose acetate or polyurethane coated on ISFETs were used to investigate the effects of permeability in these polymers [92]. A Randles circuit describing the equivalent circuit of the surface of a K^+ ion-selective membrane is shown in Figure 6.15. It includes a bulk resistance in series with a parallel combination, C_{DL} in parallel with the series combination of a transfer resistance R_{CT}; and the Warburg impedance. The importance of these contributions at frequencies <100 Hz was investigated in some detail [89].

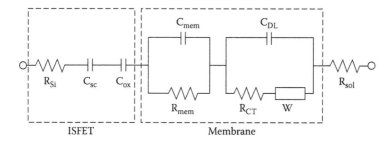

FIGURE 6.15   Equivalent circuit of BioFET that includes the impedance of the membrane and a Randles circuit to describe the impedance associated with the membrane surface. (From M. M. G. Antonisse et al., *Anal. Chem.*, **72**, 343–348, 2000.)

Work on anion-selective ChemFETs confirmed that interfacial impedances are generally not significant at these low frequencies since the rapid exchange of ions at the surfaces result in an R_{CT} that is too small to be observed [85,94].

6.3.2.2   Antigen–Antibody Conjugation

Early efforts did not detect binding of antibodies if the antigens were directly grafted to the gate oxide surface with silanes, but were successful if the probes were incubated in a polysiloxane membrane. The antibodies changed the $R_{mem}C_{mem}$ of the membrane. The authors attributed variations in the series resistance to the contacts. They were able to detect monoclonal antibodies in the range 10–150 ng [95]. The same group previously determined the capacitance and resistance of silane and membrane layers (octadecyldimethyl dimethylamino silane) using a mercury probe and electrochemical cell. The capacitance measurements with the electrochemical cell showed that the monolayer of silane molecules was insulating and a permittivity $\varepsilon = 2\varepsilon_0$ was estimated [96].

For immunosensors, the ionic strength of the electrolyte is restricted because high-ionic-strength electrolytes do not favor coulombic interaction between antibodies and antigens [97] and may cause release of the immobilized biomolecules. A problem with low-ionic-strength (and therefore low-conductivity) electrolytes is that at frequencies greater than a few hundred hertz, the bulk electrolyte resistance, which comes as a parasitic resistance in series with the sensing structure, becomes comparable to the sensor impedance and may significantly affect the measured sensor capacitance. This has led to the construction of fluid cells with working electrodes and Ag/AgCl reference electrodes [98].

The use of rapid BioFET sensors for field applications is becoming feasible, and research in this area is expected to reap benefits because antibodies, in particular, can reach detectable concentrations (nano-mol/L or greater) in bodily fluids such as blood or blood serum. Later work using nanowire ImmunoFETs will be discussed briefly in Section 6.4.

6.3.2.3   Electrochemical (Faradaic) Impedance Spectroscopy

Electrochemical impedance measurements are performed on gates covered with conductors (usually Au) that are electrically connected to the signal ground. The electrode can be covered with an enzyme and a redox couple used to drive a current to the gate surface to signal the presence of target molecules in solution. The enzyme layers and precipitates formed on the gate surface have been studied using this method [75].

The equivalent circuit is a standard Randles circuit [99] without all the components associated with the semiconductor and insulator. Again, the effect of the Warburg impedance is only significant at low frequency. This leaves just the parallel capacitance and resistance combination shown in Figure 6.15. The standard double-layer capacitance C_{dl} is replaced by a constant double-layer capacitance C_{au} for the conductor in series with the capacitance C_{mod} of the precipitate that accumulates on the electrode.

The transfer resistance includes a resistance R_{Au} associated with the electrode surface in series with R_{mod} associated with conduction through the precipitate that settles on the electrode. Since $R_{mod} > R_{Au}$, this technique effectively monitors the increased interfacial transfer resistance created by a precipitate or changes in the enzyme layer. These are just mentioned because they could also be integrated on CMOS chips.

6.3.2.4 EnFETs

As mentioned in the introduction, an EnFET is a BioFET with an enzyme layer that creates a response. Some examples showing the efficacy of these devices are presented here.

Biocatalytic layers were formed on four devices with Al_2O_3 gates and a silane linker. On each gate, a different enzyme monolayer was immobilized: glucose oxidase, urease, acetylcholine esterase, and achymotrypsin, forming an array that stimulates biocatalytic reactions that control the gate potential [100]. These EnFETs demonstrated the ability to sense urea, glucose, acetylcholine, and N-acetyl-l-tyrosine ethyl ester, respectively, by altering the pH in the sensing layer and detecting the change of pH. The major advantage of the EnFET biosensors is the fast response time, several tens of seconds. This advantage over traditional thick-polymer-based ENFETs results from the low diffusion barrier for the substrate penetration to the biocatalytic active sites and close coupling of the gate surface to the bulk solution.

Impedance spectroscopy has been used to characterize the structure of biomaterial layers on the gate surface of ISFET devices, and to study antigen–antibody binding on the gate interface [100]. The determination of τ_1 and τ_2 (Equations 6.32 and 6.33 above) allowed the film thickness of the respective protein layers to be estimated. The method was applied to study the structure of a multilayer incorporating glucose oxidase and of a di-biotin-cross-linked avidin multilayer system. The method was also used to sense the dinitrophenyl antibody by following the formation of the antigen–antibody complexes on the gate surface.

6.3.3 DETECTION OF DNA HYBRIDIZATION

6.3.3.1 SNP Detection in Prehybridized DNA

Single-nucleotide polymorphisms (SNPs—1 bp mismatches) have been detected by electrochemical impedance spectroscopy on self-assembled monolayers of already-hybridized DNA using an 8-element array (1 mm^2 disks) in a 50 mM (milli-mol/L) Tris-ClO$_4$ buffer including 4 mM of redox reporter, $Fe(CN)_6^{3-/4-}$ [101]. Elements of the array were incubated for 5 days to attach the thiol terminations on the ds-DNA to gold electrodes. Some of the ds-DNA (B-DNA) strands consisted of fully matched ss-DNA, and others contained a single bp mismatch. After self-assembly, the B-DNA was transformed to M-DNA (metal containing DNA) on-chip by the addition of a 0.4 mM solution of $Zn(ClO_4)_2 \times 6H_2O$.

These impedimetric measurements showed that the mismatched DNA had a significantly lower membrane resistance than the fully matched DNA. To account for the inhomogeneity on the working electrode surface, the equivalent circuit uses a constant-phase element in place of the Warburg impedance.

6.3.3.2 SNP Detection for In Situ Hybridization

SNPs have also been detected by measuring transfer functions on a BioFET with a 10-nm-thick oxide. After measurements were made, the transfer function for the sensor elements with fully matched targets was subtracted from that for the sensors with mismatched targets [102]. Target sequences (20 bp targets, same as probes) had 1, 2, or 3 mismatches, and measurements were done in 0.01–100 mM NaCl solutions. The gates were 8 or 16 µm in width and 1–2 µm in length arranged in a 4 × 4 matrix, as described in a previous publication [103].

A frequency-selective amplifier cascade with an output uniform over the frequency range 1 Hz–100 kHz was made with fast operational amplifiers. The transfer functions were fitted to a simple equivalent circuit describing the DNA membrane with a single resistor and capacitor in parallel; however, individual transfer functions clearly rise rather than level off at frequencies below 200 Hz. The same electrolyte solutions and measurement techniques were used before and after hybridization.

Results were compared to the results of potentiometric measurements made on the same platform and the same targets (1, 2, or 3 mismatches). The potential was monitored for a period of 160 min after hybridization. The potentiometric scans could distinguish a 3 bp mismatch from the scan for a fully matched target, but it was barely above the noise for the 2 bp mismatch. An SNP was not distinguishable.

After processing of the impedimetric data, a 1 bp mismatch was clearly observable in the transfer functions at frequencies between $1/\tau_1$ and $1/\tau_2$ (defined by Equations 6.31 and 6.32). The change in $1/\tau_1$ and $1/\tau_2$ showed that the DNA was more dense after hybridization, with both higher resistances and capacitances, as expected. Although the results are encouraging, the normalization procedures are not fully satisfactory. The data was normalized by making the transfer functions coincident at 30 Hz, and then subtracted to form the differential transfer functions (DTFs). Furthermore, a slight inflection was observable in the transfer functions between $1/\tau_1$ and $1/\tau_2$. For lower concentrations of the electrolyte, the difference in the DTFs for fully matched and 1 bp mismatched DNA became even larger, but the fit of the data was not improved significantly. This problem could be partly due to the high series resistance of the bulk electrolyte at low electrolyte concentrations required to make the measurements fall within the frequency range of the electronics.

In subsequent work using buffers with NaCl concentrations in the range 10^{-2}–10^{-5} mol/L, the contact capacitance C_{CL} shown in Figure 6.16 was included in the analysis [104]. Since the sources of six elements in the array were connected together, the source electrolyte capacitance dominated the drain-electrolyte capacitance. It was not proved that the extra element C_{CL} included in the

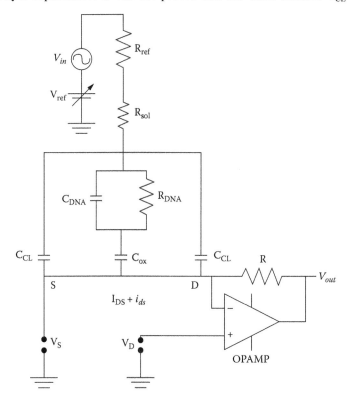

FIGURE 6.16 Equivalent circuit for BioFETs includes the capacitance C_{CL} between the source and drain contacts and the bulk electrolyte. This is dominated by the capacitance of the sources since sources in the array are connected together [104]. Other circuit elements are as described previously.

equivalent circuit to calculate the transfer function justified the computation involved, and the value of C_{CL} was not reported.

The same platform has been used to study cell binding [105] with more elements considered in the equivalent circuit to account for the cell and its interfaces. The results showed that a simple membrane model could ultimately describe a well-bound cell on the chip surface and give information about the binding and adhesion.

The ultimate aim of this effort is to produce a simple device useable in the field to perform real-time measurements of the change in transfer function with portable electronic packages. For DNA detection, this will prove a formidable task at the low ionic strengths required to perform the measurements. The next section of this chapter describes progress in BioFET sensitivity obtained using nanowires.

6.4 MOS NANOWIRE BIOFETS

In this section, we present results that demonstrate the increased sensor sensitivity available for nanowires (NWs—conductors or semiconductors with nanometer cross-sections). Recent models describing the potential sensitivity limits are also discussed.

The fabrication of CNT nanowires was reported in 1991 [106], followed by the first silicon nanowires fabricated in 1997 using a sublimation process [107,108]. The vapor-liquid-solid (VLS) process uses nano-sized templates of a metal, such as gold, to form a low-temperature eutectic with silicon. This acts as a catalyst to form an nanowire when a Si gas source comes in contact. The sources included vapor precursors and Si vapors produced by laser ablation or thermal evaporation. The nanowire grows with the metal template remaining on top as the nanowire is formed underneath.

Binding to the surface of a nanowire (or nanotube) made into an FET could lead to depletion or accumulation of carriers in the bulk of the semiconductor and could increase sensitivity to the point that single-molecule detection is possible. These NW BioFETs have been used to create highly sensitive, electrically based sensors for biological and chemical species that can monitor the attachment of biomolecules in real time.

This "bottom-up" approach to the formation of Si nanowires leads to severe processing complications when these are formed into FETs by the formation of source and drain contacts [10]. Early work with conducting polymers and recent advances in their fabrication into biosensors has also been reviewed [10], and will not be presented here even though they could easily be fabricated on CMOS chips. This section will focus on the recent work on SiNWs.

6.4.1 BIOAFFINITY SENSING WITH SiNWs

Since their introduction as biosensors, the target solutions have been applied to the SiNW sensors through microfluidic channels formed from materials such as poly(dimethyl siloxane) (PDMS, a silicone), poly(methyl methacrylate) (PMMA, including brands such as Plexiglass and Lucite), or the positive epoxy-based photoresist SU-8. More flexible and conformal thermoplastics can also be used. That is not mentioned below unless the applicators are unusual.

6.4.1.1 Probe Target Binding and the Langmuir Model

Recently, it has become possible to use SiNWs as real-time monitors of the conjugation of DNA, antigens, or antibodies. For this reason, we describe a model that gives the time dependence of the signal from bioaffinity sensors when targets are applied, a common analysis based on the Langmuir isotherm. It describes the density of conjugated targets N_{ds} (# per unit area) on a surface with a monolayer of probe molecules with density N_p in a solution with a concentration C of targets. It assumes that the rate of attachment of the targets r_{ads} is related to the product of target concentration in solution and the number of vacant binding sites on the surface: $r_{ads} = k_{on}C(N_p - N_{ds})$, where k_{on} is the association rate constant. The rate of target desorption $r_{des} = k_{off}N_{ds}$, where k_{off}, is the dissociation rate constant. In equilibrium, $r_{des} = r_{ads}$, $k_{off}N_{ds} = k_{on}(N_p - N_{ds})$, and the equilibrium concentration of conjugated targets is described by the Langmuir isotherm:

$$\frac{N_{ds}}{N_p} = C k_{on}(k_{off} + Ck_{on}) = \frac{K_A C}{1 + K_A C} \tag{6.35}$$

where the binding affinity constant $K_A \equiv k_{on}/k_{off}$. The rate of target attachment to bring the surface to equilibrium after application of the target solution is given by

$$\frac{dN_{ds}(t)}{dt} = C k_{on}\left(N_p - N_{ds}(t)\right) - k_{off}N_{ds}(t) \tag{6.36}$$

assuming that the target concentration C does not vary with time. This has the solution

$$N_{ds}(t) = \frac{Ck_{on}N_p}{k_{off} + k_{on}C}\left[1 - \exp\left(-(k_{on}C + k_{off})t\right)\right]. \tag{6.37}$$

By fitting the observed change of N_{ds} with time after targets are added, the rate constants k_{on} and k_{off} have been determined in work reported below [109].

When compared to DNA probes, PNA (peptide nucleic acid) probes have higher sensitivity, higher specificity to the DNA target sequence, and they hybridize faster at room temperature. Their stability and hybridization rate also are less dependent on the ionic strength of the electrolyte [110].

For DNA with PNA probes, the affinity constant of 15 mer duplexes has been measured by cyclic voltammetry to be 1.5×10^8 M^{-1} for complementary targets and 5.1×10^7 M^{-1} for a one-base-mismatch. Measurements by surface plasmon fluorescence spectroscopy gave 2.1×10^8 M^{-1} for full match and for a one-base-mismatch the value was 4.4×10^7 M^{-1} [111]. These high-affinity constants, higher than those for DNA–DNA duplexes, have been attributed to the absence of charge on the PNA backbone, which eliminates the charge–charge repulsion during duplex formation.

When looking for a combination of target and probe that will not dissociate under normal conditions, the streptavidin–biotin combination is often used. Streptavidin is a 52,800 Da (daltons: atomic mass in g-mol^{-1}) tetrameric protein and biotin has a mass of 244.31 Da. The affinity constant of the biotin–streptavidin complex is $\approx 10^{15}$ M^{-1}, ranking among the strongest noncovalent interactions known in nature; thus, the dissociation is negligible [112].

6.4.1.2 History

Carbon nanotubes were usually mounted on insulating substrates as shown in Figure 6.17. A non-polarized electrode (such as Pt or Au) used to control the solution potential was biased with respect to the CNT to modulate the current measured between source and drain. It was shown that artifacts that plagued some of the early measurements were related to the use of Pt and could be eliminated using a true reference electrode to accurately control the solution potential [111]. The demonstrations were done using an Ag/AgCl reference electrode, in 15 μM of bovine serum albumin (BSA—a common agent used to prevent unwanted nonspecific binding) in phosphate-buffered saline solution (PBS—2.7 mM KCl, 137 mM NaCl, 10 mM phosphate, pH 7.4). A −15 mV threshold voltage shift was unambiguously attributed to the absorption of the BSA in the vicinity of the 15-μm-long, 3-nm-diameter CNT.

An SiNW is usually covered with an insulating layer such as silicon dioxide, often just its native oxide. Its conductivity can be modulated by a reference electrode (or pseudoreference electrode) in solution as depicted in Figures 6.17 and 6.18; however, SiNWs have usually been put on oxidized silicon substrates, and no reference electrode has been mentioned. The NWs can be modulated by

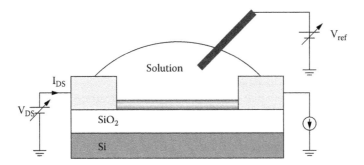

FIGURE 6.17 Side view of nanowire and a reference electrode in solution. In this case, the reference electrode and source contact are biased with respect to the source potential at ground, and the substrate is floating. The common practice with SiNWs is to eliminate the reference electrode and bias the Si substrate with respect to the drain or source.

applying a potential to this substrate with respect to the nanowire; that is, a potential is applied between the substrate and NW without deliberately fixing the bulk solution potential.

The potentials required to modulate the SiNW from the solution are much smaller (usually <1 V) than those that need to be applied to the back gate. Back gate potentials (with respect to the nanowire drain $V_{GS} > \pm 10$ V have been reported for some planar SiNWs produced by lithographic techniques. It remains to be seen if the assumption that a reference electrode is not required holds up in practical circumstances where uncontrolled processes can modulate the potential of the surfaces in the channels outside the sensor areas.

The difference between the current I_{DNA} after hybridization and the current before hybridization I_0 is often used to represent the response (fractional change in signal) of a nanowire; that is, the response is given by $(I_{DNA} - I_0)/I_0$ but it can also be given in terms of a resistance change $(R_{DNA} - R_0)/R_0$. The noise is determined from the root-mean-square fluctuations of the current before and after hybridization. In the following, the sensitivity of nanowires is given as the change in conductance G referenced to a baseline where the conductivity is G_0: $S \equiv (G - G_0)/G_0$.

6.4.2 SiNWs Grown by the VLS Method

The first demonstrations with boron-doped (p-type) SiNWs grown by the VLS method in 2001 showed that the surface with native oxide exhibited pH-dependent conductance, and by covering the surface with APTES, a linear response to changes in pH was obtained over the pH range 2–9 [114]. APTES-functionalized NWs were then tested as bioaffinity sensors.

This work investigated the sensitivity limits for the attachment of biotin and, after blocking the remnant APTES with d-biotin, found that it was possible to use the biotin to detect streptavidin binding down to concentrations of at least 10 pM. The authors reported a rate constant for the dissociation of biotin attached to monoclonal antibiotin of ≈0.1 s⁻¹, in agreement with values reported previously. In addition, antigen-functionalized SiNWs showed reversible antibody binding and concentration-dependent detection in real time. By attaching calmodulin to the SiNW surface, the reversible binding of the metabolic indicator Ca^{2+} was also demonstrated.

SiNWs have also been used in a number of other ground-breaking applications for BioFETs; for example, the study of small organic molecules that bind specifically to proteins central to the discovery and development of new pharmaceuticals [115] and the detection of viruses [116].

With PNA probes on 20-nm-diameter NWs, they functioned as ultrasensitive and selective real-time DNA sensors at concentrations down to the subpicomolar range [115]. Using PNA probes attached to the NWs with avidin, they monitored the change in conductance as a function of time after application of the targets through a microfluidic channel and demonstrated sensitivities as low as ≈10–20 fM. The NWs could distinguish the wild-type from a mutated DNA sequence associated with the ΔF508 mutation site in the cystic fibrosis transmembrane receptor gene. This work concluded that the technique is more sensitive than surface plasmon resonance (SPR), nanoparticle-enhanced SPR, and quartz-crystal microbalance techniques, and that with background subtraction procedures it had the potential to surpass SPR in determining association and dissociation rate constants and the equilibrium dissociation constants.

Nanowire arrays with protein markers of cancer such as prostrate-specific antigen (PSA) were detected at femtomolar concentrations with high selectivity. The incorporation of both p-type and n-type SiNWs in arrays of hundreds of individual nanowires enabled the discrimination against false positive (and negative) signals [118,119].

The fabrication of these VLS nanowire arrays and details of many assays has been described in a protocol [120]. The native oxides on the NWs are cleaned using an O_2 plasma prior to functionalization [115].

In another significant paper, authors from the same group have compared measurements of NW response to realistic calculations done for a 5-nm-diameter p-type SiNW with its native oxide [121]. These were compared to measurements done in 10 mM phosphate buffer (pH 7.4) using a small Au pad close to the nanowire to modulate the solution potential with respect to the nanowire, as depicted in Figure 6.18.

The calculations based on exact solutions of the Poisson-Boltzmann equation for a silicon cylinder and the electrolyte showed that the sensitivity was optimal when the NWs were operated in the subthreshold regime (in depletion). This was confirmed with pH-sensing experiments. The authors concluded that by operating in this regime using electrolytes with a very low ionic strength (10 μM), it might be possible to detect just several elementary charges.

Another interesting aspect of this work was the measurement of device noise. The signal-to-noise ratio (S/N) was maximum in the subthreshold regime, but when the conduction was completely turned off, the authors speculated that the noise was dominated by solution noise effects rather than the noise in the nanowire. This could be caused by the series spreading resistance (defined in Section 6.3.1.5) in the electrolyte, which can lead to a significant thermal noise. The analysis of MOS sensors 50 μm square has shown that this noise between the sensor and the reference electrode dominates the FET noise for electrolyte concentrations below 1 mM [123]. For nanowire sensors, the spreading resistance would increase as device dimensions decrease, but this might be compensated by putting a reference electrode closer to the sensor.

They demonstrated greatly improved sensing by functionalizing NW surfaces with APTES and monoclonal antibodies to detect PSA. A detection limit of ≈1.5 fM was demonstrated, 500 times

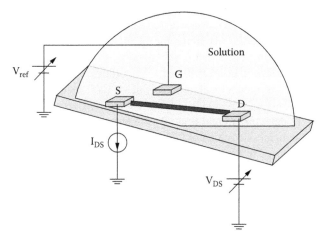

FIGURE 6.18 (See color insert). View of nanowire sensor in solution showing gate electrode bonded to insulating substrate surface.

more sensitive than the 0.75 pM they achieved previously in the linear regime (close to threshold). This was still less than the ≈0.03 fM achieved by another group [122] on a NW doped 3×10^{18} cm^{-3} n-type and processed using standard CMOS techniques (see Section 6.4.4.1). Both of these groups used nanowires with silicon oxide on the NW surface, a native oxide cleaned with an oxygen plasma) for the VLS NW and an oxide produced with an oxygen plasma for the CMOS NW. Both likely have very high levels of defects and traps at the Si/SiO$_2$ interface and in the oxide and both detected PSA at concentrations far below that required to perform the standard PSA test (4 mg/L [124]-PSA mass 28.4 kDa). The difference between the VLS and CMOS results is likely to be associated, at least partly, with the functionalization of the different devices; however, the VLS NW was p-type, and the CMOS NW was n-type.

6.4.3 PLANAR MOS NANOWIRES-NO GATING

Since devices fabricated with nanometer dimensions are now being produced on CMOS chips, the results foregoing could ultimately be transferred directly into the more easily controlled "top-down" CMOS processes. SiNWs can be fabricated on commercially available silicon-on-insulator (SOI) substrates. These consist of a crystalline silicon layer (the device layer) separated from the silicon substrate by a layer of SiO$_2$ (the BOX layer). The substrates can be made by bonding two wafers and etching one of them or implanting oxygen and annealing to form a SiO$_2$ layer a distance below the surface of a regular Si wafer. The thickness of the device layer can be reduced by cycles of thermal oxidation followed by etching in dilute HF solution. Device layers investigated are still thicker than the 10 nm limit, where resistivity increases significantly [125].

The first demonstration of the fabrication of SiNW biosensors using standard CMOS techniques was reported in 2004 [126]. These were made on SOI wafers with a 200-nm-thick BOX layer and 60-nm-thick device layers doped by ion implanting, followed by dopant activation at 925°C in N$_2$ ambient for 10 min. Measurements were reported for a boron-doped, p-type sample with nominal doping density of 10^{19} cm^{-3}, and a phosphorus-doped, n-type sample with nominal doping density of 10^{18} cm^{-3}.

The wafer was patterned by e-beam lithography (for nanowires) and optical lithography (for micron-scale electrical leads), followed by reactive ion etching (RIE). Gate oxide 3 nm thick was grown on the Si nanowire surfaces by annealing the wafer at 900°C in an O$_2$ ambient for 1 min. Al contacts were formed at the ends and covered with 100 nm of oxide using a lift-off technique. Finally, the wafers were annealed in forming gas (3.8% hydrogen in nitrogen) at 450°C for 30 min to make reliable ohmic contacts with the aluminum and to remove the interface states between the Si and the thermal oxide.

After cleaning the NWs with water plasma (50 nm wide, 60 nm thick, and 100 μm long), they were exposed for 4 h to a vapor of (3-mercaptopropyl) trimethoxysilane (MPTMS) in argon, rinsed with absolute ethyl alcohol, and blown dry with nitrogen. The immobilization of 12-mer ss-DNA probes was achieved by exposing the MPTMS-covered samples for 12 h to a 5 μM solution of probes terminated with acrylic phosphoramidite functional groups.

Measurements were performed in 25 pM solutions of target DNA dissolved only in ultrapure water. Signals for samples with complementary targets were well resolved while those with 1 bp mismatch were in the noise; that is, an SNP was discriminated with excellent resolution. For solutions of the complementary target solutions in ultrapure water the responses of the sensors for n-type and p-type nanowires were 12% and 46%, respectively, and signal/noise ratios were 8 and 6, respectively.

A subsequent paper by the same authors used the same techniques to produce silicon nanowires 50 nm high, but in this case PNA probes were used during the biosensor tests [127]. In addition, the I–V characteristics of dry nanowires were measured for widths between 50 and 800 nm. The results were compared with calculations of the potential and carrier plots obtained from commercial 2-D device simulation software.

Calculations confirmed the measurements, showing that the sensitivity of the nanowire sensors depended strongly on the width of the wires. A linear relationship between the NW sensitivity to DNA attachment and the surface-to-volume ratio (2 × height + width)/(width × height)) was calculated from the data for nanowire widths between 800 and 100 nm. A nonlinear enhancement appeared for a small wire width of 50 nm. The improved sensitivity obtained from the simulations, with a change for a 50-nm-wide wire 20 times greater than the change for one 200-nm-wide wire, corresponded to the nonlinear decrease in the conductance measured for the narrowest nanowires. The simulations indicated that this nonlinear response was caused by the depletion of carriers (holes) from all sides by the positive fixed charges present at the oxide–semiconductor boundary. The fraction of the area depleted of carriers near the vertical sides of the wire is considerably larger for smaller widths than for wider wires, so the narrower wires are more sensitive to changes in surface potential. It was concluded that tuning the 2-D carrier depletion could be used to optimize the sensitivity of nanowires.

The PNA-functionalized device 50 nm wide showed a 25% response to 10 pM target DNA solution, whereas the DNA-functionalized device reported earlier [126] showed a 12% response for a DNA concentration of 25 pM. Thus, the PNA-functionalized SiNW was about four times as sensitive as the DNA-functionalized one, but was not able to detect the 10–20 fM concentrations reported for the SiNWs made using the VLS method [117].

6.4.4 Back-Gated Planar Sensors

6.4.4.1 Attomolar Detection of Proteins

The highest sensitivity to PSA was obtained with SiNWs 67 nm wide × 40 nm thick fabricated on SOI wafers using RIE and standard CMOS processes [122]. The doping concentration for the n-type Si channels was 3×10^{18} cm^{-3}.

The wafers were treated with an O_2 plasma to create -OH groups on the NWs for the subsequent bonding with APTES solution (1% in ethanol). This was followed by reacting the amine-modified surface with glutaraldehyde (25 wt% glutaraldehyde with NaBH$_3$CN), and attachment of anti-PSA (120 µg/mL anti-PSA with NaBH$_3$CN) through its native amine groups to the glutaraldehyde. Then the unreacted surface aldehyde was blocked with ethanolamine (with NaBH$_3$CN). The signal was optimized by adjusting the solution pH to 7.6 (above the pI 6.9) to make the protein charge negative. A sensitivity to PSA of 1 fg/mL (≈30 aM) was reported.

6.4.4.2 Demonstrations of Back-Gated SiNW Bioaffinity Sensor Performance

The Si<111> planes are the preferred surface for the direct functionalization on silicon and have a lower level of oxygen contamination at the surface than Si<100> [128]. Since the <111> planes etch much more slowly than the other planes in silicon, they usually become smooth as features oriented in different directions get etched away. Thus, this technique has been used to make the smoothest surfaces and has been used to create double-gated FinFETs using CMOS technology on Si<110> device layers employing TMAH (tetramethyl ammonium hydroxide) as the etchant [129]. The Si<111> planes also have been used for the electrically active, electrochemically programmed, spatially selective biofunctionalization of SiNWs [130].

Nanowires formed by wet-etching Si<100> layers with TMAH resulted in smooth edges along the <111> plane and a trapezoidal profile with the nanowires carefully aligned along the (100) direction. Nanowires 50 nm wide and 25 nm high were etched on SOI substrates with Si<100> device layers lowly doped p-type (10^{15} cm^{-3}). The resultant structures have a trapezoidal cross-section. The smoother surfaces and elimination of RIE damage were reported to improve the electrical performance of the SiNWs [131]. Performance of the dry devices modeled using commercial software determined an average mobility of 54 cm^{2}V^{-1}s^{-1} for 12 devices with a maximum of 139 cm^{2}V^{-1}s^{-1}. Considering the expected lower values of mobility for high field and sloped Si <111> planes [132], these results compare reasonably with the ideal values for electrons in p-type silicon doped 10^{15} cm^{-3}, 450 $cm^2V^{-1}s^{-1}$ [14]. The measurements were done with $V_{DS} = 2$ V and back-gate potentials (the bias on the silicon substrate with respect to the drain) $V_{GD} > -20$ V. A number of interesting applications and results from this paper are described in the following.

The work demonstrated the conjugation of streptavidin to biotin-functionalized surfaces at the 20 fm level. The large S/N of 140 at this concentration implied that 70 aM concentrations could ultimately be detected. Using goat anti-mouse IgG-functionalized sensors and goat anti-mouse IgA-functionalized sensors, it demonstrated the ability to selectively detect 100 fM concentrations of mouse-IgG or fM mouse-IgA.

Macro-well solution chambers have been used in these demonstrations, and in a subsequent paper [133] incorporated rapid circulation of the target molecules over the NWs to overcome the limitations due to the slow diffusion of macromolecular targets. Response times to the complete conjugation of 20 mer DNA in the 30–50 s range were demonstrated. The electrolytes had a 2.3 nm Debye length (see Equation 6.5), chosen to include the surface molecules, without detecting unbound molecules outside the nanowires. When a lower ionic strength buffer with a longer Debye length was used, results showed that false positives could be generated by the detection of controls (targets noncomplementary to the probe DNA) just outside the nanowire.

The amine-terminated surfaces were also reacted to give the surface a maleimide termination, and 20 mer thiol-terminated DNA probes were used to test the sensitivity of the sensors to DNA conjugation. The 20 mer targets were successfully detected at concentrations as low as 10 pM.

More details of the work of this group are given in a recent review [134]. It elaborates on the functionalization chemistry and the detection of various biological targets and compares the results of other groups. Finally, we mention the group's work that demonstrated a module that isolated PSA (2 ng/mL) and a marker for breast cancer from whole blood samples, rinsed the module with pure electrolyte, and released the targets into the electrolyte using a photochemical method. These were successfully detected using their nanowire arrays [135]. This marks a paradigm shift since it demonstrates that nanowire arrays could be used with relatively simple purification modules for applications in the point-of-care market.

6.4.4.3 Nanoribbons

In another recent paper, one group tested the pH dependence of back-gated SiNWs (50–150 nm wide, 100 nm high, 1 μm long) fabricated on lowly doped p-type ($\approx 10^{15}$ cm^{-3}) device layers on SOI substrates with a 400-nm-thick BOX layer [136]. These devices had Schottky barrier contacts (TiW/Au), and gate oxide ≈ 5 nm thick formed by oxidation at 900°C for 10 min in dry oxygen. Devices with widths 1000, 100, 60, and 45 nm were fabricated and the height was 100 nm. Some were treated with APTES to change the pH dependence of the gate surface charge and allow them to attach streptavidin. The results were compared with 2-D simulations of the potentials and carrier densities of cross-sections of the structure, which considered the layers of charge due to site binding at the solution interface and any charges at the Si–SiO$_2$ interface. Three following papers reported details of the pH dependence of the devices, detailed simulations, and the interpretation of these results. These are summarized briefly in the following.

With decreasing nanowire width, the threshold voltage increased owing to an increasing effect of the surface charge, which reduced the sensitivity for a given surface-to-volume ratio [137]. The NWs could exhibit avalanche breakdown at high back-gate bias. The authors concluded that these charge layers were responsible for potential shifts that caused the avalanche and that for very narrow back-gated nanowires the charges could cause the field to collapse and drastically reduce the NW conductance. To make the nanowire conducting, a large enough gate voltage or drain-source voltage needs to be applied. This could damage the nanowire chip as the breakdown voltage through the 400 nm BOX layer was $\approx$ 10 V.

In the subsequent paper [138], the thickness of the device layer was varied to produce three wafers giving nanowire heights of 105, 65, and 50 nm. These NWs were 1 μm long and 2 μm wide. They demonstrated that the devices can be operated both in inversion as well as accumulation mode, and the increased sensitivity to accumulation was explained in terms of the distance between the channel and the receptor-coated surface with respect to the Debye screening length. An inversion layer (electrons) formed near the BOX surface at positive gate voltages. For large negative back-gate voltages, an accumulation layer (holes) formed near the thermal oxide interface. For positive gate voltages, the electrons were distributed under the inversion layer in a depletion layer 4 nm thick (the Debye length).

The devices used for the bioaffinity test were treated with an oxygen plasma and then functionalized with APTES and biotin. A voltage $V_{DS} = 2.0$ V was applied. A low ionic strength with a Debye length of 2.3 nm was used during the measurements while the streptavidin was added. Subpicomolar sensitivities for streptavidin attachment were demonstrated for the thinnest Si layers (45 nm) with the inversion layer at the BOX surface.

For detection of streptavidin, the response is larger in the accumulation mode than in the inversion mode, but this did not lead to a higher detection sensitivity because of increased noise. The effect was attributed to the location of the conducting channel, which for holes is closer to the screened surface charges of the biomolecules. Furthermore, the response increased for decreasing silicon thickness in both the accumulation mode and the inversion mode [136].

Finally, it was verified that the simulated conductance ratios were larger in the accumulation mode as observed, but larger than were obtained experimentally. The authors speculated that screening, both by holes in the channel, as well as from charged molecules in the liquid, may be less effective due to the close proximity of the channel to the sensing surface [139].

This remains to be explained. One possible reason is that a plasma treatment, as was used prior to functionalization for these demonstrations, can completely depassivate the Si dangling bonds at the thermal oxide interface if it is not protected by a polysilicon or metal layer. It is possible that lines of defects created by the potentials created by the oxygen plasma or ion bombardment had some effect. In any case, the simulations and experiments demonstrated the necessity of taking the possible nonuniformity of the fixed charge at the oxide interface into account.

6.4.5 ELIMINATING THE OXIDE LAYER

Another effort has recently focused on functionalizing SiNWs very close to the silicon surface using ultrathin oxides or functionalizing the Si<100> surfaces directly [109]. It showed that the detection sensitivity could be increased dramatically in a 0.165 M electrolyte if the probes were bonded by electrostatic forces to an amine-terminated surface. This meant that the electrolyte concentration was high enough to study the target conjugation on the sensor in real time and obtain the association and dissociation rate constants.

The SiNWs were made using the SNAP process (described later) to form an array 2 mm long of 400 SiNWs patterned with a 35 nm pitch on SOI wafers. The NWs were 20 nm high and 20 nm wide and were both p- and n-doped to a level 8×10^{18} cm^{-3}. After a lift-off process to form the metal contacts and leads, the wafer was annealed in forming gas (95% N2, 5% H$_2$ at 475°C for 5 min). Finally, it was covered with silicon nitride, and openings in the nitride were etched to expose the NWs. Two functionalization processes were compared. In the first, which is comparable to the process used on the VLS devices described earlier, the nanowires were cleaned in piranha solution, leaving a 0.5–1 nm thick chemical silicon oxide, and this was functionalized using (3-aminopropyl) dimethylethoxy silane in toluene. For the second process, the oxide was removed with dilute HF, hydrogen-terminated with NH$_4$F, and then the photochemical method was used to form an amine-modified surface covalently bonded with Si-C bonds to the Si surface [140,141].* It is selective to the bare silicon surface so the BOX surface is not functionalized, eliminating signals from the fringes of the NWs.

The bioaffinity data were presented for the p-doped samples. Single-stranded 10 μM DNA (16 mer) in 1× SSC buffer (15 mM Na Citrate, 150 mM NaCl, pH 7.5, ionic strength 0.165 M) flowed through a microchannel over the arrays for 1 h and were allowed to attach by electrostatic forces to the amine-terminated surface of the SiNWs. The solution potential was maintained at a constant value by controlling the current to a metal electrode with a feedback loop. The same solutions were used in the SPR measurements.

When compared to the devices with chemical oxide, the SiNWs without the oxide had improved field transistor characteristics in solution, with a significantly enhanced sensitivity to DNA conjugation and a dynamic range ~100 times greater. They also showed a response that varied with the log of the salt concentration.

Apart from demonstrating that the covalent bonding of the amine directly to the Si surface enhanced the detection sensitivity, the measurements demonstrated other significant results. Since the Debye length is calculated to be ~0.6 nm in the 0.165 M buffer used for the measurements, the results suggest that the DNA must somehow confine itself very tightly to the surface of the nanowire with the covalently bonded amine, or otherwise the DNA charge would not modify the transistor current. The details of the binding and thickness of the DNA membrane are not clear; however, the results show that conjugation can be more effectively monitored by using a combination of the electrostatic binding of the probe molecules and the direct covalent bonding of the amine-terminated alkyl linker molecules to the Si surface.

The authors also presented the results of a statistical analysis based on the Langmuir model that determined k_{on}, k_{off}, and the binding affinity constant K_A. (see Equation 6.37 above), These were found to be in the same range as the values obtained by SPR, leading the authors to conclude that the resistance changes of the nanowires could be used for quantitative determination of binding kinetics but with a higher sensitivity than SPR.

Subsequently, another group demonstrated better sensitivity for hybridization with PNA probes using covalent bonding of the probes, rather than the electrostatic attachment just described [142]. They also used deep UV lithography, more compatible with production than e-beam lithography or the SNAP method discussed later, to produce 63±5 nm square nanowires. The same photochemical process* was used to form the covalently bonded amine-terminated surface to the NWs. This was followed by covalent attachment of amine-terminated PNA probes using the bifunctional linker glutaraldehyde, which can couple two amine groups. This allowed for a higher density of PNA attachment that was also selectively attached to the SiNWs. Measurements were carried out in 0.01 × SSC buffer (100 times lower than above) with 22 mer targets. Conjugation was detected with 10 fM solutions of 22 mer targets, and a 1 bp mismatch could be distinguished. Measurements during the conjugation were not reported.

6.4.6 Alternate Nanowire Fabrication Methods

Electron beam lithography patterning techniques have a resolution limit on the order of 15 nm [143]. A patterning method called superlattice nanowire pattern transfer, or SNAP [144], has demonstrated the fabrication of large arrays of silicon nanowires on SOI substrates, with wire widths and pitches of 10–20 and 40–50 nm, respectively. Briefly, the SNAP method transfers the pattern made by cleaving the edge of a GaAs crystal, which has had a GaAs/AlGaAs superlattice deposited on it. After treating the wafer with a suitable etchant, the cleaved edge has a series of nano-trenches on it. Pt is deposited at an oblique angle onto this edge tool. Then the Pt on the surface is transferred

---

* Hydrogen-terminated Si surfaces are covalently linked to 10-N-Boc-Amino-dec-1-ene with an ultraviolet initiated reaction. Removal of the t-BOC protecting group leaves primary amines covalently linked to the SiNW surfaces.

by contact with an SOI wafer to create an etch mask. With careful (low-energy) reactive-ion etching and spin-on glass doping good resistivity characteristics can be obtained in the nanowires even though the doping profile in the nanowires is not uniform [145]. Nanowires 14 nm wide, (5–10) per sensing element, fabricated using the SNAP technique were able to detect the conjugation of DNA at target concentrations <220 attomolar in 0.15 M electrolyte (0.15 M NaCl, 0.15 M sodium citrate, pH 7.0) [146]. Target solutions flowed over the nanosensors through a microfluidics channel. Both p- and n-type nanosensors exhibited sensitivity over a broad dynamic range, and their response scaled logarithmically with concentration.

SiNWs have also been transferred onto insulator-coated silicon wafers using a process that is akin to contact printing [147]. The electrical properties of these nanowires can be controlled by choosing a silicon substrate with the intended doping. After thermal oxidation, 0.8–1 μm wide oxide lines were defined on the substrate by stepper photolithography. A triangular structure was formed, determined by the <111> etch stop plane of silicon, with the nanowires aligned in the (110) direction. Cycles of deep silicon reactive ion etching followed by removal of the oxide created silicon lines with an inverted triangle-shaped cross section supported by narrower silicon pillars. Thermal oxidation of the silicon substrate was then followed by etching of the oxide to produce free SiNWs with a sub-100 nm diameter near the center of the inverted triangle. After transfer, and the formation of source-drain contacts, Al_2O_3 was deposited by atomic layer deposition, followed by the deposition and patterning of the gate metal. Well-ordered silicon nanowires with lengths varying from 5 to 200 μm and diameters varying from 20 to 100 nm were demonstrated to have good electrical characteristics. Variations of this process could be used to fabricate biosensor arrays.

The use of alternatives such as polysilicon (poly) and p-n junction SiNWs have been investigated as a less expensive alternative to SOI substrates. A group working with the poly has recently reported the detection of pathogenic strains (H5 and H7] of avian influenza virus (HPA1) DNA at the pM-fM levels [148]. High-quality device-grade poly fabricated using a standard sidewall spacer technique was used to make an array of NWs 80 nm wide and ≈50 nm thick on silicon nitride. They were functionalized with a standard APTES glutaraldehyde process.

Another poly CMOS process has produced an underlap nanowire geometry where the 900 nm long 20 nm high nano-gap is etched into an oxide layer between a poly gate and silicon substrate [149]. Preliminary measurements with the antigens for the avian flu virus (AIa antigen) were successful at the 10 μg/mL level.

Finally, we mention work that has made biosensors using a Si nanowire on a crystalline silicon layer that forms a p-n junction with the Si bulk substrate [150]. The NWs were 50–150 nm wide, and the leakage current for the reverse-biased junction was reported to be low enough to be neglected. Tests with the PSA antigen were reported to be successful with 10 ng/mL target solutions.

6.4.7 NANOWIRE ELECTRICAL MODELS

The use of 2-D simulations for cross sections of the potentials and carrier densities in the nanowires mentioned above involved the setting of a potential and fixed charge at the nanowire surfaces. The following calculations were done for an ideal MOS device; that is, the charge in the oxide and fast states at the Si/SiO_2 interface were ignored. The simulators use the standard models [14] for the silicon to calculate the hole and electron concentrations in terms of the potential, using Fermi–Dirac statistics and the quasi-Fermi energy established in the semiconductor. These are used with Poisson's equation, and we refer to these as the standard equations for the semiconductor. They are written below for convenience.

In the semiconductor, the equations are given in terms of ε_{Si} the permittivity, N_B the doping concentration, p the hole concentration, n the electron concentration, n_i the intrinsic carrier concentration, and E_F the quasi-Fermi level:

$$-\nabla \cdot (\varepsilon_{Si} \nabla \phi) = q(p - n + N_B)$$

$$p = n_i \exp((E_F + \Delta E_r - V_{GS} - \phi) / \phi_t)$$

$$n = n_i \exp(-(E_F + \Delta E_r - V_{GS} - \phi) / \phi_t) \tag{6.38}$$

Here, V_{GS} is the potential applied between the NW source and a reference electrode immersed in bulk electrolyte, and ΔE_r accounts for the reference electrode potential in solution and work function and electron affinity terms. The latter term was used for the ISFET, but for NWs it is common to use no reference electrode in solution, or sometimes just a metal electrode made of Pt or Au (not polarized) is used to control the solution potential.

In the electrolyte without biomolecules, the Poisson–Boltzmann (PB) equation applies, and the permittivity and the bulk concentration are given by ε and n_o, respectively. The ions are monovalent ($z = 1$) with concentrations of cations n_+ and anions n_-:

$$-\nabla \cdot (\varepsilon \nabla \phi) = q(n_+ - n_-)$$

$$n_\pm = n_0 \exp(\mp \phi / \phi_t). \tag{6.39}$$

6.4.7.1 Cylindrical Membranes

Recently, the ion-permeable membrane model considered in detail in Section 6.2 has been applied to a cylindrical nanowire and upgraded to include ion partition at the membrane–electrolyte interface, the target–probe equilibrium, and the site binding to the oxide [29]. Calculations were done for cylindrical symmetry for the nanowire, and the solution potential was fixed at a reference potential. The standard model was used to describe the conduction in silicon, and the conductance was calculated using a gradual channel approximation for the Si surface potential.

Poisson's equation was used to calculate the potential through the insulator and Helmholtz layer assuming site binding at the outer Helmholtz plane. The PB equation for the insulator with permittivity ε_{ox} and thickness d_{ox} is given in terms of σ_s the surface charge density:

$$-\nabla \cdot (\varepsilon_{ox} \nabla \phi) = \sigma_s. \tag{6.40}$$

The permittivity and site binding appropriate to silicon nitride were used to calculate σ_s, including the silanol (Equation 6.10) and amine (Equation 6.13) contributions [24]. However, the authors assumed the standard Helmholtz capacitance, 20 μF cm^{-2} of SiO$_2$, rather than the 70.8 μF cm^{-2} obtained for silicon nitride [26,27]. A good approximation for the silanol groups described the surface charge in terms of the pI (pH$_{pzc}$ in Section 6.2) instead of pK_a and pK_b.

The PB equation was also used to determine the potential ϕ and ion charge concentrations in the membrane with permittivity ε_m and charge density N_m due to the biomolecules as described for the aforementioned 1-D models (cf. Equation 6.17). The free energy barrier ΔE_m that the mobile ions encounter as they permeate from the solution into the membrane included the Born charge–dielectric interaction energy, in large part due to the permittivity difference between solution and membrane. The ion–solvent and ion–dipole interactions may also make significant contributions [74]. This gives the PB equation in the membrane:

$$-\nabla \cdot (\varepsilon_m \nabla \phi) = q(n_+ - n_-) + qN_m$$

$$\text{with } n_\pm(r) = n_0 \exp(-\Delta E_m \mp \phi / \phi_t) \tag{6.41}$$

A generalized Langmuir–Freundlich isotherm [151] was employed to describe the surface concentration N_{ds} of DNA hybridized to DNA or PNA probes, in terms of C, the target concentration in solution, N_p the total density of probes, and K_A the binding affinity:

$$N_{ds} = N_p \frac{(K_A C)^\nu}{1 + (K_A C)^\nu} \qquad (6.42)$$

where ν is a parameter ($0 < \nu \leq 1$) related to the heterogeneity of surface binding energies (cf. Equation 6.35 for the Langmuir isotherm).

Two schemes represented the fixed charge from 16 mer strands of DNA; scheme A for a 5.5-nm-thick membrane for DNA standing on the surface with $\varepsilon = 80\varepsilon_0$ and scheme B for a 2 nm thickness of DNA parallel to the surface with $\varepsilon = 20\varepsilon_0$. The ion partition was only considered for the latter, the denser membrane.

Self-consistent calculations of the potentials and charge distributions in the radial direction were done for an array of E_F values using Equations 6.38–6.42 and the site-binding terms. These numerical solutions were integrated in the radial direction and then along the length using a gradual channel approximation to calculate the NW conductance for a small potential V_{DS} applied between the contacts.

The conductances G_{ss} before attachment and G_{ds} after attachment were calculated for increasing gate potential, the conductance increasing from accumulation through flatbands to depletion and inversion, characteristic of a MOS device. The "subthreshold regime" includes the accumulation of carriers (holes) below flatbands and their depletion above, similar to the changes in current described for the ISFET in Section 6.2. The charged molecule attachment causes shifts of the characteristic curves, and these were presented as a shift of the voltage near flatbands (in the subthreshold regime) by a quantity ΔV_0 after probe attachment and ΔV_{hyb} after hybridization.

The calculations show that the shift ΔV_{hyb} is 5 mV for scheme A and 5 mV for B; however, scheme A showed a much larger shift ΔV_0 on probe attachment, 42 mV versus 7 mV for scheme B. For scheme B, the Donnan potential is not reached in the membrane, resulting in a lower shift with probe attachment. The energy barrier was responsible for a reduction in the effective ionic strength in the membrane, thereby significantly increasing the Donnan potential. The energy barrier at the membrane–electrolyte interface, not carefully discussed in previous work, has an important impact on the design of BioFETs with optimal sensitivity.

The quantity describing the sensitivity to hybridization, $\Delta G / G \equiv (G_{ds} - G_{ss}) / G_s$, was calculated. This had the highest value for gate potentials in inversion, where it is $\Delta G / G = \exp(-\Delta V_{hyb} / \phi_t) - 1$. However, the largest change of $\Delta G / G$ with gate potential was in the subthreshold region (around flatbands). This is where the NW is best able to distinguish the change of conductance due to the addition of biomolecular charge from the background conductance.

Other conclusions reinforced those reached previously for planar BioFETs, that both the mobile electrolyte ions and the site binding contribute to the screening of the biomolecular charge, that this results in a nonlinear response to the attachment, that PNA probes enhance the sensitivity, that the Donnan potential could be reached in the membrane for normal probe densities encountered in practice, that reaching the Donnan potential reduced the sensitivity, and that better sensitivity could be obtained if the DNA was laid down.

The calculations were compared to experimental data for the nanowires fabricated by the SNAP method discussed in Section 6.4.5 [109]. Another important conclusion is that the heterogeneity of the surface represented by the parameter ν in the Langmuir–Freundlich isotherm is critical in determining the NW input dynamic range (the difference between the minimum and maximum conductance), as deduced from comparing devices with and without oxide. The superior input dynamic range in the NWs without oxide had a lower value of ν ($= 1/2$) than those with the native oxide.

This is a paper well worth reading in some detail because it is the most complete analysis of the electrical detection of DNA hybridization to date.

6.4.7.2 Determination of Nanowire Capacitance and Surface Potential

The PB equation (Poisson equation with Boltzmann statistics for the carriers) in silicon and the PB equation in the electrolyte and the oxide layer were used to calculate the potential profiles for a circular cross section of the cylinder in another recent paper. It is discussed in this section because it also provides a simple picture of the detection mechanism in a NW [121].

For hole densities p in silicon, the Debye length $\lambda_{Si} = \sqrt{\varepsilon_{Si}\phi_t / pq}$. For the uniform doping density N_B of these devices ($\approx 10^{18}$–10^{19} cm^{-3}) the Debye length for the majority carriers (holes) should be ≈ 1–1.5 nm (see Reference 14, p. 78). For no net charge in the NW, the hole density is considered to be p_0 throughout, equal to the boron doping density N_B. Calculations were compared to the results obtained on NWs with length $L = 2$ μm and radius $R = 5$ nm with a 1-nm-thick native oxide, probably cleaned with an oxygen plasma. Measurements were performed on an APTES-functionalized surface at pH in the range 4–9 in 10 mM phosphate buffer solution.

For the ideal MOS device with no attached charge, the potential across the NW $\phi_{Si}(r) = 0$ and the response to a variable magnitude of charge at the oxide surface simulated the effect of charged biomolecules or changes in pH. The applied voltage between the contacts was assumed to be negligible, so the conductance was calculated without considering the variation along its length. The charge shifts the surface potential ϕ_{Si} at the perimeter by the quantity $\delta\phi_{Si}$ as a result of the band bending associated with the shift of the quasi-Fermi energy level in the semiconductor. This changes the carrier (hole) concentration in the semiconductor by δp, thereby changing the screening (depletion) length λ_{Si}. Because of the cylindrical symmetry, the potential in the nanowire changes from the value $\delta\phi_{Si}$ at the perimeter, reaching a minimum or a maximum value in the center as shown in Figure 6.19.

The calculations show that for a small hole concentration (in depletion or subthreshold regime), the screening length is approximately equal to the Debye length and $\lambda_{Si} \gg R$, resulting in an almost uniform hole density change given by $\delta p(r) = p \exp(-\delta\phi_{Si}/\phi_t)$ at all points along the radius. The potential falls almost exponentially from the NW surface described by the screening length λ_{Si}.

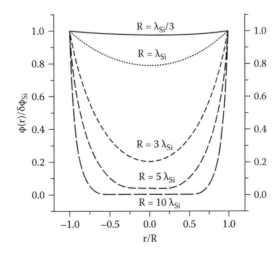

FIGURE 6.19 Normalized potential $\phi(r)/\delta\phi_{Si}$ as a function of normalized radial position r/R inside a SiNW with radius R for different screening lengths λ_{Si}: λ_{Si} = 3R, R, R/3, R/5, and R/10. (From X. P. A. Gao et al., *Nano*, **10**, 547–552, 2010 Supplementary.)

For $\lambda_{Si} \gg R$, changes in carrier densities with r are small, and the change in conductance of the nanowire is given by

$$G = q\mu \int_0^R 2\pi r \,\delta p(r)\,dr = q\mu \pi R^2 p \exp(-\delta\phi_{Si}/\phi_t) \tag{6.43}$$

where μ is the carrier mobility. When $\phi_{Si} = 0$, the conductance is given by $G_0 = q\mu\pi R^2 p$ and the maximum sensitivity, $\Delta G/G_0 = (G - G_0)/G_0$, can be achieved by operating in the subthreshold regime (hole depletion) regime, where the volume of the NW is fully utilized.

For comparison, measurements were done as a function of pH and V_{GS}, the potential on the Au pad. These NWs are close to inversion with the gate potential $V_{GS} = 0$ V, and in the subthreshold regime for $V_{GS} = 0.2$–0.5 V. To calculate a change in conductance δG, a parameter ξ was defined (the "gate coupling efficiency") to fit the values of G obtained as a function of V_{GS}:

$$G = q\mu\pi R^2 p \exp(-\xi\delta\phi_{Si}/\phi_t) \tag{6.44}$$

This gives $(\delta G/\delta\phi_{Si})/G = \delta\ln G/\delta\phi_{Si} = -\xi/\phi_t$, and the value of $\delta\phi_{Si}$ to change G by one decade is $\delta\phi_{Si} = -\phi_t\ln(10)/\xi$. A value of $S_s \approx 0.33$ was found in the subthreshold regime for an APTES-treated oxide at pH 4.

The measured conductance was used to determine the value of $\delta\phi_{Si}$ at pH 4 from the transconductance $\delta G/\delta V_{GS}$ (≈ 700 nS/V) fit to the data in the linear regime; that is, $\delta G(\text{pH } 4) = (\delta G/\delta V_{GS})\delta\phi_{Si}(\text{pH } 4)$, which determined a $\delta\phi_{Si}/\delta\text{pH} = -30$ mV/ pH. This can be compared to the $\delta\phi_{Si}/\delta\text{pH} = \approx -46$ mV/ pH obtained for an ISFET with a high-quality thermal oxide [16]. However, it is known that a lower-quality oxide can form a hydrogel at its surface [152], resulting in a higher $\delta\phi_{Si}/\delta\text{pH}$, and this could be even higher at higher pH [16,44]. In addition, the effect of the APTES has an effect on pH sensitivity. Any fast states at the Si–SiO$_2$ interface will also change the subthreshold slope [14].

Useful estimates were made for the nanowire capacitance C_{NW} and the double-layer capacitance C_{DL} for a cylinder using good approximations of the PB equation for the electrolyte and semiconductor. For $\lambda_{Si} \gg R$, the nanowire capacitance $C_{NW} \approx 2\pi p q R \lambda_{Si} L/\phi_t$ and for $\lambda_{Si} \gg R$, $C_{NW} \approx \pi p q R^2 \lambda L/\phi_t$. The electrolyte Debye length has been given in terms of the permittivity ε and the Debye length λ_b above (Equation 6.5). For the NW cylinder in low salt concentrations, $\lambda_b \gg R$, $C_{DL} = \pi\varepsilon L/\ln(\lambda_b/R)$.

Using the double cylinder capacitance formula $C_{ox} = 2\pi\varepsilon_{ox}L/\ln(1 + t_{ox}/R)$ where t_{ox} and ε_{ox} are the oxide thickness and permittivity, respectively, the capacitance was determined to be $C_{ox}/L = 1.4 \times 10^{-15}$ F/μm, which results in a dielectric constant $\varepsilon_{ox}/\varepsilon_0 \approx 28$. This suggests that the oxide must be quite permeable to electrolyte. However, this is open to interpretation because the authors did not include the capacitance C_a due to site binding in their analysis and used a questionable equivalent circuit.

An equivalent circuit presented to describe the NW, oxide, and double layer shows the double-layer capacitance in parallel with the combination of oxide and nanowire capacitances in series: $C = 1/(1/C_{ox} + 1/C_{NW}) + C_{DL}$. This relation is not correct because it violates Gauss' law between the nanowire center and the bulk electrolyte [39,40] and does not describe all the potential drops between the nanowire and the gate [153]. The equivalent circuit for the nanowire should be identical to that of the planar BioFET referenced in Section 6.2 and should include the capacitance combination $C_a + C_{DL}$ in series with C_{DL} and C_{ox}: $C^{-1} = C_{NW}^{-1} + C_{ox}^{-1} + (C_a + C_{DL})^{-1}$. Nevertheless, the limiting NW capacitances and description of the operation of the NW are quite useful.

6.4.7.3   3-D Models

Recently, major efforts have been made to calculate the potentials and charges in the semiconductor and solution phases self-consistently [154]. Quantum effects have been considered in this analysis, but may not be significant at a nanowire radius above 10 nm. These simulations can describe most of the phenomena associated with the conjugation of biomolecules at the nanowire surfaces. Here, we describe one effort that shows the practical information that can be deduced from these calculations. The simulations generally still have difficulty obtaining exact fits to the experimental data, but they are very instructive.

Nair used a 3-D PB solver capable of simulating the electrostatics of biomolecules on arbitrarily shaped conducting surfaces in the electrolyte [155]. In the vicinity of the molecule attachment, a cylindrical NW was approximated as an equipotential surface. The oxide layer was considered to be a perfect dielectric without site binding. The calculated electric flux on the NW surface due to the biomolecules in the electrolyte solution was used as the boundary condition for the PB and drift-diffusion equations in the semiconductor. The structure of the biomolecules was obtained from the Protein Data Bank [156], and sophisticated profiles of their charge distributions were mapped from the force field parameters of molecular dynamics simulators [157]. Electrolyte ions were excluded from the biomolecules, but a Helmholtz layer on the molecules was not considered.

With the simulations, they were able to describe experimental results and confirm the conclusions of previous simulations. Important nanowire design issues were presented, starting with a discussion of some general issues. They presented an estimate of the statistical fluctuations of dopant density in the semiconductor at low concentrations due to a distribution of lines of charges on the electrolyte near the NW surface and argued that this would result in serious wire sensitivity variations at low doping levels so that it would become difficult to detect small numbers of molecules at the sensor surface. This could be alleviated if the initial conductivity of each nanowire were stored in a memory before detection of the targets and statistical averages over many nanowires were averaged to obtain the sensitivity for each measurement.

Some observations were made that would help compare the sensitivities for a NW in air S_{air} and after it was put in water S_{water}. Comparing an NW in accumulation and depletion, the simulations showed S_{water} (NW in depletion) $\leq S_{air}$ (NW in depletion) and S_{water} (NW in accumulation) $\geq S_{air}$ (NW in accumulation). Thus, the observed sensitivity in air would depend on the doping type. They found that the sensitivity in air is an order of magnitude higher when the NW is in depletion than in accumulation, whereas in water, the difference is significantly smaller. This was related to the large dielectric constant of water. At the same time, the fields would spread out significantly further away from the nanowire outside the layer of attached biomolecules.

The calculations show that for negative surface charge, n-type NWs are expected to be more sensitive than p-type, and for positive charge, p-type NWs are more sensitive, as has been borne out by tests [118,146]. They show that sensitivity scales inversely with length and increases with decreasing NW radius as shown [126,131].

The simulations show results consistent with experience indicating that the probability of differentiating between DNA strands of the same length, but with different base pair sequences, would be very low for DNA probes but higher with PNA. They showed that proteins are better detected away from their pI where they carry significant charge. This is expected, and is consistent with recent measurements on nanowires [122,131].

This work did not include site binding, and the authors pointed out that this could be particularly important for proteins because their charge will be determined by shifts in surface pH, and it could easily be included in the calculations. This work did also not include the Helmholtz (Stern) layer around the biomolecules. In work on planar sensors described in Section 6.2 [28] and for the membrane model applied to the foregoing cylinder [29], this contribution was considered to be important.

An analysis of the effect of site binding and fixed charge on the performance of BioFETs in inversion, weak inversion, and the subthreshold region has also been considered in other recent work [158,159].

6.4.8 SLOW RESPONSE DUE TO DIFFUSION

With the advent of nanowire sensors, the first efforts were made to determine how the slow diffusion of the relatively large biomolecules limits the ultimate practical sensitivity that can be obtained in solution if the response is limited by diffusion [160]. For the lower concentrations being investigated, the question arises: What is the density of conjugated molecules on the sensor surface if one waits a practical time to record the change in signal if targets are simply allowed to diffuse to the sensor surface [161–163]. This analysis considers the diffusion equation describing the time dependence of the analyte concentration C and requires that Equation 6.35 be modified to consider the time dependence. Recently, modeling of the transient response has considered the effect of site-binding and target detachment ($k_{off} = 0$) [161].

The geometry of the sensor surface is treated by a fractal analysis [164] and considers the minimum detectable concentration $C(t = 0) \equiv C_0$ and its relation to the settling time t_s available for incubation. This results in a scaling relationship $C_0 \approx k_D t_s^{1/Df}$, where D_F is the fractal dimension of the surface, and k_D is related to the physical dimensions of the sensor and the diffusion coefficient of the analyte. The work was summarized recently [164]. Figure 6.20 shows the minimum detectable concentration C_0 (C at time $t = 0$) and its relation to the settling time t_s available for incubation. It clearly shows the superior time response of NWs and arrays of nanowires (including "Nanonets") over planar sensors for target concentrations. Further work on the statistical analysis of the arrival of individual targets has also been discussed [165]. Even for the nanowires, it is clear that techniques need to be found to overcome the diffusion limit, which applies also to targets applied through laminar flow channels.

6.4.9 CONCLUSIONS

The performance and repeatability of nanowire sensors will no doubt improve as engineers and scientists perfect the fabrication techniques with the help of the wealth of data and the good models now at hand.

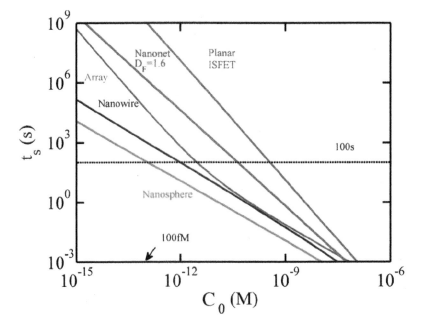

FIGURE 6.20 Plot of time settling time t_s versus initial target concentration C_0 based on a fractal analysis of several nanowire geometries: cylindrical nanowire, nanosphere, planar ISFET, nanonet, array of nanowires. (From M. Alam and P. Nair, Proc. 9th ULIS, pp. 40–47, © 2008, IEEE.)

There are some obvious areas for improvement and some questions. Is there an alternative to the relatively poor but very high-capacitance native (chemical) oxides that are often used? How thin can a much better insulator layer get and can it be eliminated? Perhaps the use of Si<110> wafer to produce FinFET [129] structures with vertical Si<111> walls on each side of the nanowire could produce better results and help eliminate the oxide layer altogether.

There are opportunities to eliminate the fixed charge at the Si–SiO$_2$ interface that plague some devices. It is well known to CMOS engineers that plasma treatments of a SiO$_2$–Si interface unprotected by poly or metal gates completely depassivates the Si-dangling bonds. Our own experience is that even plasma cleaning equipment promoted as using "remote" or "gentle" plasmas can leave interface state densities $>5 \times 10^{12}$ eV^{-1}cm^{-2}. This level could dramatically decrease the sensitivity of the best nanowire. UV-ozone treatment or an atomic oxygen beam not in line of sight with a remote plasma have proved effective in removing carbon contamination while maintaining an interface state density $<2 \times 10^{11}$ eV^{-1}cm^{-2} without reannealing in forming gas [166]. Is there a fully reliable wet process that can remove the adventitious carbon or processing residue prior to linker molecule attachment? Can cleaning with liquids such as powerful alkaline phosphate detergents commonly used to clean delicate optical components be used or will more aggressive solutions [167] be required to remove the carbon? Ozonated water does not seem to have been tested with nanowires.

The only functionalization that has proved to allow the electrical measurement during conjugation at high salt concentrations uses electrostatic binding of the probes [109]. Can better electrostatic binding layers or membranes be fabricated?

Some work on the effect of the discontinuity in dielectric constant at the membrane–electrolyte interface and the water–air interface have been referenced earlier [29,37] but more details await to be explored. Furthermore, there is a large dielectric constant offset at the gate–membrane interface, and this should be further considered, particularly for proteins and lipid layers [168,169].

All of the efforts to make the most comprehensive models described in this chapter are some way from predicting the sensitivity of real devices even if the simple models for the Helmholtz layer and site binding were included. That will remain so until the details of all the interfacial reactions are better understood.

Finally, it is clear that turbulence, advection, electrophoresis, electrokinetic, or other processes will need to be employed in the target applicators to get reasonable response times at low target concentrations.

REFERENCES

1. C. Stagni, C. Guiducci, L. Benini, B. Riccò, S. Carrara, B. Samorí, C. Paulus, M. Schienle, M. Augustyniak, and R. Thewes, *IEEE J. Solid-State Circuits*, **41**, 2956–2963 (2006).
2. R. Thewes, C. Paulus, M. Schienle, F. Hofmann, A. Frey, R. Brederlow, P. Schindler-Bauer, M. Augustyniak, M. Atzesberger, B. Holzapfl, M. Jenkner, B. Eversmann, G. Beer, M. Fritz, T. Haneder, and H.-C. Hanke, *Proceeding of the 30th European Solid-State Circuits Conference and 34th European Solid-State Device Research Conference*, Leuven, Belgium, September 21–23, 2004, ed. R. P. Mertens C. L. Claeys, M. Steyaert, and C. L. Claeys, pp. 19–28 (2004).
3. P. Bergveld, *IEEE Trans. Biomed. Eng.*, **17**, 70–71 (1970).
4. P. Bergveld, *IEEE Trans. Biomed. Eng.*, **19**, 342–351 (1972).
5. P. Bergveld, *Sens, Actuat. B*, **88**, 1–20 (2003).
6. S. D. Moss, J. Janata, and C. C. Johnson, *Anal. Chem.*, **47**, 2238–2243 (1975).
7. J. Janata, *Electroanalysis,* **16**, 1831–1835 (2004).
8. H.-S. Wong, H. White, *IEEE Trans. Electron Dev.*, **36**, 479–487 (1989).
9. T. C. W. Yeow, M. R. Haskard, D. E. Mulcahy, H. I. Seo, and D. H. Kwon, *Sens. Actuat. B*, **44**, 434–440 (1997).
10. A. K. Wanekaya, W. Chen, N. V. Myung, and A. Mulchandani, *Electroanalysis*, **18**, 533–550 (2006).
11. Y. Tsividis, *Operation and Modeling of the MOS Transistor*, 2nd ed. (New York, McGraw-Hill, 1999).
12. E. H. Nicollian and J. R. Brews, *MOS (Metal Oxide Semiconductor) Physics and Technology*, 2nd ed. (New York: Wiley, 1982).

13. D. W. Greve, *Field effect Devices and Applications: Devices for Portable, Low-Power, and Imaging Systems* (Upper Saddle River, NJ, Prentice Hall, 1998).

14. S. M. Sze, *Physics of Semiconductor Devices*, 2nd ed. (New York, John Wiley & Sons, 1981).

15. D. A. McQuarrie, *Statistical Mechanics* (New York, Harper and Row, 1976).

16. R. E. G. van Hal, J. C. T. Eijkel, and P. Bergveld, *Sens. Actuat. B: Chem.*, **24–25**, 201–205 (1995).

17. D. E. Yates, S. Levine, and T. W. Healy, *J. Chem. Soc. Faraday Trans.* 1, **70**, 1807–1818 (1974).

18. R. B. M. Schasfoort, P. Bergveld, R. P. H. Kooyman, and J. Greve, *Anal. Chim. Acta*, **238**, 323–329 (1990).

19. L. Bousse and J. D. Meindl, in *Geochemical Processes at Mineral Surfaces*, ed. J. A. David and K. F. Hayes (American Chemical Society, Washington, 1986), p. 79.

20. L. Bousse, N. F. de Rooij, and P. Bergveld, *IEEE Trans. Electron Dev.*, **30**, 1263–1270 (1983).

21. L. Bousse, H. H. van den Vlekkert, and N. F. de Rooij, *Sens. Actuat. B*, **2**, 103–110 (1990).

22. M. Klein and M. Kuisl, *VDI-Ber.*, **509**, 275–279 (1984).

23. T. Mikolajick, R. Kühnhold, and H. Ryssel, *Sens. Actuat. B*, **44**, 62–267 (1997).

24. D. L. Harame, L. J. Bousse, J. D. Shott, and J. D. Meindl, *IEEE Trans. Electron Devices*, **34**, 1700–1706 (1987).

25. A. H. M. Smets and M. C. M. van de Sanden, *Phys. Rev. B*, **76**, 073202 (2007).

26. S. Martinoia and G. Massobrio, *Sens. Actuat. B*, **62**, 182–189 (2000).

27. S. Martinoia, G. Massobrio, and L. Lorenzelli, *Sens. Actuat. B*, **105**, 14–15 (2005).

28. S. Uno., M. Iio, H. Ozawa, and K. Nakazato, *Jpn. J. Appl. Phys.*, **49**, 01AG07 (2010).

29. Y. Liu and R. W. Dutton, *J. Appl. Phys.*, **106**, 014701 (2009).

30. M. Stelzle, G. Weissmueller, and E. Sackmann, *J. Phys. Chem.*, **97**, 2974–2981 (1993).

31. O. S. Lee and G. C. Schatz, *J. Phys. Chem. C*, **113**, 15941–15947 (2009).

32. S. Miklavic and S. Marcelja, *J. Phys. Chem.*, **92**, 6718–6722 (1998).

33. S. Misra, S. Varanasi, and P. P. Varanasi, *Macromolecules*, **22**, 4173–4179 (1989).

34. T. M. Herne and M. J. Tarlov, *J. Am. Chem. Soc.*, **119**, 8916–8920 (1997).

35. A. W. Peterson, R. J. Heaton, and R. M. Georgiadis, *Nucl. Acid. Res.*, **29**, 5163–5168 (2001).

36. P. Gong and R. Levicky, *Proc Natl. Acad. Sci.*, **105**, 5301–5306 (2008).

37. B. W. Ninham and V. Yaminsky, *Langmuir*, **13**, 2097–2108 (1997).

38. W. Kunz, P. Lo Nostro, and B. W. Ninham, *Curr. Opin. Coll. Interf. Sci.*, **9**, 1–18 (2004).

39. D. Landheer, D., G. Aers, W. R. McKinnon, M. J. Deen, and J. C. Ranuarez, *J. Appl. Phys.*, **98**, 044701/1–15 (2005).

40. D. Landheer, W. R. McKinnon, G. Aers, W. Jiang, M. J. Deen, and M. W. Shinwari, *IEEE Sens. J.*, **7**, 1233–1242 (2007).

41. W. R. McKinnon and D. Langheer, *J. Appl. Phys.*, **104**, 124701 (2009).

42. D. Landheer, W. R. McKinnon, W. H. Jiang, and G. Aers, *Appl. Phys. Lett.*, **92**, 253901 (2008).

43. B. K. Wunderlich, P. A. Neff, and A. R. Bausch, *Appl. Phys. Lett.*, **91**, 083904 (2007).

44. R. E. G. van Hal, J. C. T. Eijkel, and P. Bergveld, *Adv. Coll. Interf. Sci.*, **69**, 31–62 (1996).

45. P. Bergveld, *Biosens. Bioelectron.*, **6**, 55–72 (1991).

46. W. H. Jiang, D. Landheer, G. Lopinski, W. R. McKinnon, A. Rankin, E. Ghias-Begloo, R. Griffin, N. G. Tarr, N. Tait, J. Liu, and W. N. Lennard, *ECS Trans.*, **16**, 441–450 (2008).

47. M. Kosmulski, *Chemical Properties of Material Surfaces* (New York, Dekker, 2001).

48. W. R. McKinnon and D. Landheer, *J. Appl. Phys.*, **100**, 054703 (2006).

49. M. Egholm, O. Buchardt, L. Christensen, C. Behrens, S. M. Freler, D. A. Driver, R. H. Berg, S. K. Kim, B. Norden, and P. E. Nielsen, *Nature*, **365**, 566–568 (1993).

50. S. Tomac, S. Sarkar, T. Ratilainen, P. Wittung, P. E. Nielsen, B. Nordén, and A. Gräslund, *J. Am. Chem. Soc.*, **118**, 5544–5552 (1996).

51. J. Summerton, in *Discoveries in Antisense Nucleic Acids*, C. Brake, Ed.; *Advances in Applied Biotechnology*; (Portfolio Publishing Co., The Woodlands, TX, 1989); pp. 71–80.

52. N. Tercero, K. Wang, P. Gong, and R. Levicky, *J. Am. Chem. Soc.*, **131**, 4953–4961 (2008).

53. G. F. Blackburn, Chemically sensitive field effect transistors, *Biosensors: Fundamentals and Applications*, ed. A. P. F Turner, I. Karube, and G. S. Wilson (Oxford, Oxford University Press, 1987), pp. 481–530.

54. G. S. Manning, *J. Chem. Phys.*, **51**, 924–933 (1969).

55. M. Deserno and C. Holm, "Cell model and Poisson-Boltzmann theory: A brief introduction," in *Electrostatic Effects in Soft Matter and Biophysics,* eds. C. Holm, P. K'ekicheff, and R. Podgornik, *NATO Science Series II—Mathematics, Physics and Chemistry*, Vol. 46 (Kluwer, Dordrecht, 2001), p. 27.

56. A. Poghossian, A. Cherstvy, S. Ingebrandt, A. Offenhäusser, and M. J. Schöning, *Sens. Actuat. B: Chem.*, **111–112**, 470–480 (2005).

57. D. Cahen, R. Naaman, and Z. Vager, *Adv. Funct. Mater.*, **15**, 1571–1578 (2005).

58. O. De Sagazan, M. Harnois, A. Girard, F. LeBihan, A. C. Salaun, S. Crand, and T. Mohammed-Brahim, *ECS Trans.*, **14**, 3–10 (2008).
59. E. Souteyrand, J.P Cloarec, J. R. Martin, C. Wilson, I. Lawrence, S. Mikkelson, and M. F. Lawrence, *J. Phys. Chem. B*, **101**, 2980–2985 (1997).
60. J. P. Cloarec, J. R. Martin, P. Polychronakos, I. Lawrence, M. F. Lawrence, and E. Souteyrand, *Sens. Actuat. B*, **58**, 394–398 (1999).
61. P. Estrela, P. Migliorato, H. Takiguchi, H. Fukushima, and S. Nebashi, *Biosens. Bioelectron.*, **20**, 1580 (2005).
62. J. Fritz, E. B. Cooper, S. Gaudet, P. K. Sorger, and S. R. Manalis, *Proc. Natl. Acad. Sci.*, **99**, 14142–14146 (2002).
63. D.-S. Kim, H.-J. Park, H.-M. Jung, J.-K. Shin, P. Choi, J.-H. Lee, and G. Lim, *Jpn. J. Appl. Phys. B*, **43**, 3855 (2004).
64. F. Uslu, S. Ingebrandt, D. Mayer, S. Bocker-Meffert, M. Odenthal, and A. Offenhausser, *Biosens. Bioelectron.*, **19**, 1723–1731 (2004).
65. F. Pouthas, C. Gentil, D. Côte, and U. Bockelmann, *Appl. Phys. Lett.*, **84**, 1594–1596 (2004).
66. T. Sakata, M. Kamahori, and Y. Miyahara, *Mater. Sci. Eng. C*, **24**, 827–832 (2004).
67. T. Sakata and Y. Miyahara, *ChemBioChem*, **6**, 703–710 (2005).
68. T. Sakata and Y. Miyahara, *Biosens. Bioelectron.*, **22**, 1311–1316 (2007).
69. S. Ingebrandt and A. Offenhäusser, *Phys. Stat. Sol.* (a), **203**, 3399–3411 (2006).
70. E. Souteyrand, J. P Cloarec, J. R. Martin, C. Wilson, I. Lawrence, S. Mikkelson, and M. F. Lawrence, *J. Phys. Chem. B*, **101**, 2980–2985 (1997).
71. J. P. Cloarec, J. R. Martin, P. Polychronakos, I. Lawrence, M. F. Lawrence, and E. Souteyrand, *Sens. Actuat. B*, **58**, 394–398 (1999).
72. J. P. Cloarec, N. Deligianis, J. R. Martin, I. Lawrence, E. Souteyrand, C. Polychronakos, and M. F. Lawrence, *Biosens. Bioelectron.*, **17**, 405–412 (2002).
73. R. L. Smith, J. Janata, and R. J. Huber, *J. Electrochem. Soc.*, **127**, 1599–1603 (1980).
74. A. J. Bard and L. R. Faulkner, *Electrochemical Methods: Fundamentals and Applications* (Wiley, New York, 1980).
75. E. Katz and I. Willner, *Electroanalysis*, **15**, 913–946 (2003).
76. J. E. B. Randles, *Discuss Faraday Soc.* 1, 11 (1947).
77. B. V. Ershler, *Discuss. Faraday Soc.*, **1**, 269 (1947).
78. L. Bousse and P. Bergveld, *J. Electroanal. Chem.*, **152**, 25–39 (1983).
79. T. Matsuo and M. Esashi, Extended Abstracts of the Electrochemical Society Spring Meeting (1978), Seattle, WA., pp. 202–203.
80. T. Matsuo and H. Nakajima, Characteristics of reference electrodes using a polymer gate ISFET, *Sens. Actuat.* **5**, 293–305 (1984).
81. P. Bergveld, A. Van den Berg, P. D. Van der Wal, M. Skowronska-Ptasinska, E. J. R. Sudholter, and D. N. Reinhoudt, *Sens. Actuat.*, **18**, 309–327 (1989).
82. H. Nakajima, T. Matsuo, and M. Esashi, *J. Electrochem. Soc.*, **129**, 141–143 (1982).
83. A. Errachid, J. Bausells, and N. Jaffrezic-Renault, *Sens. Actuat. B*, **60**, 43–48 (1999).
84. A. Errachid, N. Zine, J. Samitier, and J. Bausells, *Electroanalysis*, **16**, 1843–1851 (2004).
85. M. M. G. Antonisse, B. H. M. Snellink-Ruël, R. J. W. Lugtenberg, J. F. J. Engbersen, A. van den Berg, and D. N. Reinhoudt, *Anal. Chem.*, **72**, 343–348 (2000).
86. J. Kruise, J. G. Ripens, P. Bergveld, F. J. B. Kremer, J. R. Starmans, J. Haak, J. Feijen, and D. N. Reinhoudt, *Sens. Actuat. B*, **6**, 101–105 (1992).
87. J. M. Chovelon, N. Jaffrezic-Renault, Y. Gros, J. J. Fombon, and D. Pedone, *Sens. Actuat. B.*, **3**, 43–50 (1991).
88. E. Juhasz, K. N. Marsh, and T. Plebanski, contributors to Section "Electrolytic Materials" in *Recommended Reference Materials for Realization of Physicochemical Properties*, Ed. K. N. Marsh (Pergamon Press Ltd., IUPAC, 1981), *Pure Appl. Chem.*, **53**, 1841–1845 (1981).
89. T. C. Chen and G. Hefter, *J. Phys. Chem. A*, **107**, 4025–4031 (2003).
90. R. Gulich, M. Köhler, P. Lunkenheimer, and A. Loidl, *Radiat. Environ. Biophys.*, **48**, 107–114 (2009).
91. M. W. Denhoff, *J. Phys. D: Appl. Phys.*, **39**, 1761–1765 (2006).
92. A. Friebe, F. Lisdat, and W. Moritz, *Sens. Mater.*, **5**, 065–082 (1993).
93. A. Demoz, E. M. J. Verpoorte, and D. J. Harrison, *J. Electroanal. Chem.*, **389**, 71–78 (1995).
94. R. D. Armstrong, and G. Horvai, *Electrochim. Acta*, **35**, 1–7 (1990).
95. H. Maupas, C. Saby, C. Martelet, N. Jaffrezic-Renault, A. P. Soldatkin, M. H. Charles, T. Delair, and B. J. Mandrand, *Electroanal. Chem.*, **406**, 53–58 (1996).

96. C. Schyberg, C. Plossu, D. Barbier, N. Jaffrezic-Renault, C. Martelet, H. Maupas, E. Souteyrand, M. H. Charles, T. Delair, and B. Mandrand, *Sens. Actuat. B*, **27**, 457–460 (1995).

97. C. J. van Oss, R. J. Good, and M. K. Chaudhury, *J. Chromatogr.*, **376**, 111–119 (1986).

98. B. Prasad and R. Lal, *Meas. Sci. Technol.*, **10**, 1097–1104 (1999).

99. F. Patolsky, M. Zayats, E. Katz, E., and I. Willner, *Anal. Chem.*, **71**, 3171–3180 (1999).

100. A. B. Kharitonov, M. Zayats, A. Lichtenstein, E. Katz, and I. Willne, *Sens. Actuat. B*, **70**, 222–231 (2000).

101. X. Li, Y. Zhou, T. C. Sutherland, B. Baker, J. S. Lee, and H.-B. Kraatz, *Anal. Chem*, **77**, 5766–5769 (2005).

102. S. Ingebrandt, Y. Han, F. Nakamura, A. Poghossian, M. J. Schöning, and A. Offenhäusser, *Biosens. Bioelectron.*, **22**, 2834–2840 (2007).

103. Y. Han, A. Offenhäusser, and S. Ingebrandt, *Surf. Interface Anal.*, **38**, 176–181 (2006).

104. R. GhoshMoulick, X. T. Vu, S. Gilles, D. Mayer, A. Offenhäusser1, and S. Ingebrandt, *Phys. Status Solidi A*, 206, 417–425 (2009).

105. S. Schäfer, S. Eick, T. Hoffman, T. Dufaux, R. Stockmann, G. Wrobel, A. Offenhäusser, and S. Ingebrandt, *Biosens. Bioelectron.*, **24**, 1201–1208 (2009).

106. S. Iijima, *Nature*, **354**, 56–58 (1991).

107. A. M. Morales and C. M. Lieber, *Science*, **279**, 208–211 (1998).

108. D. P. Yu, Z. G. Bai, Y. Ding, Q. L. Hang, H. Z. Zhang, J. J. Wang, Y. H. Zou, W. Qian, G. C. Xiong, H. T. Zhou and S. Q. Feng., *Appl. Phys. Lett.*, **72**, 3458–3460 (1998).

109. Y. L. Bunimovich, Y. S. Shin, W. Yeo, M. Amori, G. Kwong, and J. R. Heath, *J. Am. Chem. Soc.*, **128**, 16323–16331 (2006).

110. J. Wang, D. Palecek, P. Nielsen, G. Rivas, X. Cai, H. Shiraishi, N. Dontha, D. Luo, P. A. M. Farias, *J. Am. Chem. Soc.*, **118**, 7667–7670 (1996).

111. J. Liu, L. Tiefenauer, S. Tian, P. E. Nielsen, and W. Knoll, *Anal. Chem.*, **78**, 470–476 (2006).

112. T. E. Creighton, *Proteins—Structures and Molecular Properties*, 2nd ed. (W. H. Freeman and Company, New York, 1993).

113. E. D. Minot, A. M. Janssens, I. Heller, H. A. Heering, C. Dekker, and S. G. Lemay, *Appl. Phys. Lett.*, **91**, 093507 (2007).

114. Y. Cui, Q. Wei, H. Park, and C. M. Lieber, *Science*, **293**, 1289–1292 (2001).

115. W. U. Wang, C. Chen, K.-H, Lin, Y. Fang, and C. M. Lieber, *PNAS*, **102**, 3208–3212 (2005).

116. F. Patolsky, G. Zheng, O. Hayden, M. Lakamamyali, X. Zhuang, C. M. Lieber, *PNAS*, **101**, 14017 (2004).

117. J. -I Hahm and C. M. Lieber, *Nano Lett.*, **4**, 51–54 (2004).

118. G. Zheng, F. Patolsky, Y. Cui, W. U. Wang, and C. M. Lieber, *Nat. Biotechnol.*, **23**, 1294–1301 (2005).

119. G. Zheng, F. Patolsky, and C. M. Lieber, *Mater. Res. Soc. Symp. Proc.*, Vol. 900E, 0900-O07-04, 1–6 pages.

120. F. Patolsky, G. Zheng, and C. M. Lieber, *Nat. Protoc.*, **1**, 1711–1724 (2006).

121. X. P. A. Gao, G. Zheng, and C. M. Lieber, *Nano Lett.*, **10**, 547–552 (2010).

122. A. Kim, C. S. Ah, H. Y. Yu, J.-H. Yang, I.-B. Baek, C.-G. Ahn, C. W. Park, and M. S. Jun, *Appl. Phys. Lett.*, **91**,103 901 (2007).

123. T. A. Stamey, Z. Chen, and A. F. Prestigiacomo, *Clin. Biochem.*, **31**, 475–481 (1998).

124. M. J. Deen, M. W. Shinwari, and J. C. Ranuarez, Noise considerations in field-effect biosensors, *J. Appl. Phys.*, **100**, 074703 (2006).

125. J.-H. Choi, Y.-J. Park, and H.-S. Min, *IEEE Electron Dev. Lett.*, 16, 527–529 (1995).

126. Z. Li, Y. Chen, X. Li, T. I. Kamins, K. Nauka, and R. S. Williams, *Nano Lett.*, **4**, 245–248 (2004).

127. Z. Li, B. Rajendran, T. I. Kamins, X. Li, Y. Chen, and R. S. Williams, *Appl. Phys. A: Solids Surf.*, **80**, 1257–1263 (2005).

128. W. J. Royea, A. Juang, and N. S. Lewis, *Appl. Phys. Lett.*, **77**, 1988 (2000); and references therein.

129. Y. Liu, K. Ishii, T. Tsutsumi, M. Masahara, and E. Suzuki, *IEEE Electron Dev. Lett.* **24**, 484–486 (2003).

130. Y. L. Bunimovich, G. Ge, K. C. Beverly, R. S. Ries, L. Hood, and J. R. Heath, *Langmuir*, **20**, 10 630–10 638, (2004).

131. E. Stern, J. F. Klemic, D. A. Routenberg, P. N. Wyrembak, D. B. Turner-Evans, A. D. Hamilton, D. A. LaVan, T. M. Fahmy, and M. A. Reed, *Nature*, **445**, 519–522 (2007).

132. S. C. Sun, and J. D. Plummer, *IEEE J. Solid-State Circuits*, **15**, 562–573 (1980).

133. E. Stern, R. Wagner, F. J. Sigworth, R. Breaker, T. M. Fahmy, and M. A. Reed, *Nano Lett.* **7**, 3405–3409 (2007).

134. E. Stern, A. Vacic, and M. A. Reed, *IEEE Trans. Electron Dev.*, **55**, 3119–3130 (2008).

135. E. Stern, A. Vacic, N. K. Rajan, J. M. Criscione, J. Park, B. R. Ilic, D. J. Mooney, M. A. Reed, and T. M. Fahmy, *Nat. Nano.*, **353**, 1–5 (2009).

136. N. Elfström, R. Juhasz, I. Sychugov, and T. Engfeld, *Nano Lett.*, **9**, 2608–2612 (2007).

137. N. Elfström and J. Linnros, *Appl. Phys. Lett.*, **91**, 103502 (2007).
138. N. Elfström, A. W. Karlström, and J. Linnros, *Nano Lett.*, **8**, 945–949 (2008).
139. N. Elfström and J. Linnros, *Nanotechnology*, **19**, 235201 (2008).
140. J. A. Streifer, H. Kim, B. M. Nichols, and R. J. Hamers, *Nanotechnology*, **16**, 1868–1873 (2005).
141. T. Strother, R. J. Hamers, and L. M. Smith, *Nucl. Acid. Res.*, **28**, 3535–3541 (2000).
142. G.-J. Zhang, J. H. Chua, R.-E. Chee, A. Agarwal, S. M. Wong, K. D. Buddharaju, and N. Balasubramanian, *Biosens. Bioelectron.*, **23**, 1701–1707 (2008).
143. C. Vieu, F. Carcenac, A. Pepin, Y. Chen, M. Mejias, A. Lebib, L. Manin Ferlazzo, L. Couraud, and H. Launois, *Appl. Surf. Sci.*, **164**,111–117 (2000).
144. N. A. Melosh, A. Boukai, F. Diana, B. Gerardot, A. Badolato, P. M. Petroff, and J. R. Heath., *Science*, **300**, 112–115 (2003).
145. R. A. Beckman, E. Johnston-Halperin, Y. Luo, N. Melosh, J. Green, and J. R. Heath, *J. Appl. Phys.*, **96**, 5921–5923 (2004).
146. M. M.-C. Cheng, G. Cuda, Y. L. Bunimovich, M. Gaspari, J. R. Heath, H. D. Hill, C. A. Mirkin, A. J. Nijdam, R. Terracciano, T. Thundat, and M. Ferrari, *Curr. Opin. Chem. Biol.*, **10**, 11–19 (2006).
147. K.-N. Lee, S.-W. Jung, W.-H. Kim, M.-H. Lee, K.-S. Shin, and W.-K. Seong, *Nanotechnology*, **18**, 445302 (2007).
148. C.-H. Lin, C.-H. Hung, C.-Y. Hsiao, H.-C. Lin, F.-H. Ko, and Y.-S. Yang, *Biosens. Bioelectron.*, **24**, 3019–3024 (2009).
149. K.-W. Lee, S.-J. Choi, J.-H. Ahn, D.-I. Moon, T. J. Park, S. Y. Lee, and Y.-K. Choi, *Appl. Phys. Lett.*, **96**, 033703 (2010).
150. C.-G. Ahn, C. W. Park, J. -H. Yang, C. S. Ah, A. Kim, T.-Y. Kim, H. Y. Yu, M. Jang, S. -H. Kim, I.-B. Baek, S. Lee, and G. Y. Sung, *J. Appl. Phys.*, **106**, 114701 (2009).
151. R. Sips, *J. Chem. Phys.*, **16**, 490–495 (1948).
152. J. Janata, *Analyst*, **119**, 2275–2278 (1974).
153. L. Bousse, *J. Chem. Phys.*, **76**, 5128–5133 (1982).
154. C. Heitzinger and G. Klimeck, *J. Comput. Electron.*, **6**, 387–390 (2007).
155. P. R. Nair and M. A. Alam, *IEEE Trans. Electron Dev.*, **54**, 3400–3408 (2007).
156. H. M. Berman, J. Westbrook, Z. Feng, G. Gilliland, T. N. Bhat, H. Weissig, I. N. Shindyalov, and P. E. Bourne, *Nucl. Acid. Res.*, **28**, 235–242 (2000).
157. W. F. van Gunsteren and H. J. C. Berendsen (Biomos BV Gromos-87 Manual, 1987).
158. K. Lee, P. R. Nair, H. H. Park, D. Y. Zemlyanov, A. Ivanisevic, M. A. Alam, and D. B. Janes, *J. Appl. Phys.*, **103**, 114510 (2008).
159. K. Lee, P. R. Nair, A. Scott, M. A. Alam, and D. B. Janes, *J. Appl. Phys.*, 105, 102046 (2009).
160. P. E. Sheehan and L. J. Whitman, *Nano Lett.*, **5**, 803–807 (2005).
161. P. R. Nair and M. A. Alam, *Nano Lett.*, **8**, 1281–1285 (2008).
162. P. R. Nair and M. A. Alam, *Phys. Rev. Lett.*, **99**, 256101 (2007).
163. P. R. Nair and M. A. Alam, *Appl. Phys. Lett.*, **88**, 233120 (2006).
164. M. A. Alam and P. R. Nair, *Proc. 9th ULIS*, pp. 40–47, (2008).
165. J. Go and M. A. Alam, *Appl. Phys. Lett.*, **95**, 033110 (2009).
166. D. Landheer and M. Denhoff, unpublished.
167. Y. Han, D. Mayer, A. Offenhäusser, and S. Ingebrandt, *Thin Solid Films*, **510**, 175–180 (2006).
168. R. R. Netz, *Phys. Rev. E*, **60**, 3174–3182 (1999).
169. S. Q. Lud, M. G. Nikolaides, I. Haase, M. Fischer, and A. R. Bausch, *ChemPhysChem*, **7**, 379–384 (2006).

7 MEMS-Based Optical Chemical Sensors

Huikai Xie
Department of Electrical & Computer Engineering,
University of Florida, Gainesville, FL 32611

Zhi-Mei Qi
State Key Laboratory of Transducer Technology, Institute of
electronics, Chinese Academy of Sciences, Beijing 100190

CONTENTS

7.1 MEMS INTRODUCTION

7.1.1 MEMS OVERVIEW

Integrated circuits (ICs) have dramatically changed the world in the past few decades by facilitating technologies such as computers and cell phones. The drivers of IC innovations have been cost and functionality. Following this trend toward smaller size, lower cost, higher performance and more functions, a new field, microelectromechanical systems, or MEMS, gradually developed in the 1980s. The best-known MEMS device in the early days of the MEMS field was the micromotor. Though they did not live up to the initial expectations, MEMS micromotors inspired numerous researchers to work on MEMS.

MEMS is called Microsystems in Europe and Micromachines in Japan. Despite the terminology, MEMS employs existing IC fabrication technologies to manufacture various miniature sensors and actuators. Almost all disciplines in engineering, science, and medicine are involved, including microelectronics, electronics, mechanical, chemical, physics, biology, medicine, chemistry, aerospace, etc. With the price going down, MEMS devices are finding their way into many consumer electronics products, including laptops and cell phones. On account of their small sizes, MEMS microsensors have many advantages over traditional sensors, such as portability, high speed, high resolution, and low power. MEMS sensors are also much cheaper because of batch fabrication. Furthermore, most MEMS fabrication processes are IC compatible, so MEMS can be integrated with IC electronics to provide "smart" sensors or "smart" microsystems.

MEMS is widely used even in our daily life. The airbags in our cars have Analog Devices Inc. (ADI)'s microaccelerometers embedded. Our cars may also have Robert Bosch Corporation (Bosch)'s MEMS gyroscopes for electronic stability program (ESP) and Freescale Semiconductors (Freescale)'s MEMS tire pressure sensors. We use Texas Instrument's (TI's) digital micromirror devices (DMDs)-enabled portable projectors to deliver presentations. We also use HP's MEMS inkjet printers. Moreover, our new cell phones may be installed with ST Microelectronics (ST)'s gyroscopes and accelerometers. Of course, these are just a few MEMS application examples. In fact, the worldwide MEMS market has already reached to 6.5 billion USD.

Due to the wide range of materials used and the difficulty of handling fragile microstructures, MEMS started with dedicated "in-house" fabrication. For instance, the major MEMS manufacturers, such as ADI, Bosch, ST, Freescale, and TI, all had their own MEMS production fabs. As we have learned from IC fabrication, the "in-house" fab model is not cost-effective and it cannot foster technological innovations. The industry has realized this problem and pushed hard to establish MEMS foundries. Currently, the MEMS foundry sector is still not mature, but a number of MEMS foundries, such as Dalsa Semiconductor, Micralyne, Silex Microsystems, Asia Pacific Microsystems (APM), IMT, Tronics Microsystems, MEMSCAP, and X-FAB, are already capable of providing reliable MEMS fabrication services. Another piece of good news is that the CMOS foundry giant, TSMC, has also become an MEMS foundry.

With those MEMS foundries available, the number of MEMS design houses or fabless companies keeps increasing in the past few years. Those fabless companies include HP and Lexmark (inkjet printers); Invensense, Melexis, Qualtre, and Senodia (gyroscopes); Knowles and Akustica (microphones); SiTime and Silicon Clocks (oscillators); and many others.

7.1.2 MEMS FABRICATION

MEMS fabrication, though developed from IC fabrication, has its own unique fabrication techniques because of the need for creating movable microstructures in most cases. The key MEMS fabrication techniques include surface micromachining and bulk micromachining.

Surface micromachining creates microstructures on top of the substrate. At the micron scale, it is difficult to form suspended microstructures. So the surface micromachining technique was developed based on the "sacrificial" concept. As shown in Figure 7.1a, the MEMS structure is built up layer by layer on a solid substrate, similar to IC fabrication. The main difference for MEMS fabrication is that at least one of the thin-film layers is selectively removed to form suspended MEMS structures, as shown in Figure 7.1b. The selectively removed layer is called a *sacrificial layer*. This process step is called "release." Many materials, including dielectrics, metals, and polymers, can be used as sacrificial layers as long as there is enough selectivity to the structural material. The commonly used sacrificial materials are silicon dioxide, aluminum, porous silicon, and photoresist. The scanning electron micrograph (SEM) in Figure 7.1c shows microgears made by Sandia National Lab.

Bulk micromachining refers to processes that remove part of "bulk" substrate to form trenches, cavities, through-wafer holes, etc. A classic bulk micromachining process is anisotropic silicon etch by KOH. As shown in Figure 7.2a, the etch rate in <100> directions is much higher than that in <111> directions, leaving almost unattacked (111) planes. This well-defined slope, 54.7° exactly, is ideal for inkjet nozzles. It is also used for making V-grooves (Figure 7.2b) for aligning optical fibers. The process is simple and has an extremely low cost and high yield.

Another import bulk micromachining process is deep reactive ion etch (DRIE). This process was first developed by Bosch. Now DRIE has become a standard process for MEMS fabrication. It has a high etching rate, which is attributed to the high-density inductively coupled plasma (ICP). The high aspect ratio is achieved by alternatively performing etching and passivation, as shown in Figure 7.3. Both etching and passivation are done in the same processing chamber, and the change from one

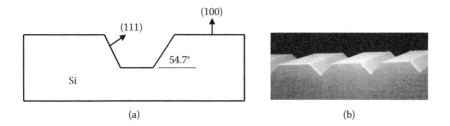

FIGURE 7.1 Surface micromachining.

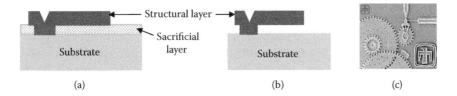

FIGURE 7.2 Anisotropic wet etch of silicon.

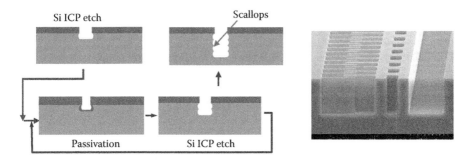

FIGURE 7.3 DRIE silicon etch.

process to the other is accomplished by simply switching the gas channels between SF6 and CHF3. Because of the repetitive cycles, scallops form on the sidewalls. In order to achieve smooth sidewalls, additional hydrogen anneal may be used. The typical aspect ratios are around 25:1, but aspect ratios as high as 100:1 are also reported.

In addition to surface and bulk micromachining, there are several other MEMS fabrication techniques. For example, researchers may use LIGA (German for Lithographie, Galvanoformung, Abformung; or lithography, electrodeposition, and molding in English) for producing ultra-high-aspect-ratio (>100:1) microstructures, but LIGA is very expensive, and its availability is very limited due to the need for synchrotron radiation sources. Wafer-to-wafer bonding is another popular micromachining process that is often used to create single-crystal, three-dimensional microstructures. There are several wafer-to-wafer bonding techniques. The commonly used fusion bonding has high yield but requires a high temperature at about 1000°C. Bonding temperature can be lowered to about 450°C by using glass frit bonding or using anodic bonding but at a high bias voltage of hundreds of volts. Bonding temperature can be further reduced to 100°C–400°C by using eutectic bonding or epoxy bonding.

7.1.3 Chapter Outline

In Section 7.1.1, we only mentioned MEMS accelerometers, gyroscopes, microphones, and pressure sensors. There are a large number of other MEMS sensors. All these sensors can be divided into several categories: mechanical, chemical, optical, magnetic, thermal, and biological sensors. The focus of this chapter in the next few sections is optical chemical sensors. In Section 7.2, optical sensing principles are introduced. Absorption, polarization, phase, spectroscopy, interference, and birefringence can all be used for optical sensing. In this chapter, Fourier transform spectroscopy (FTS) and optical-waveguide-based sensing are presented. Section 7.3 describes MEMS FTS, and Section 7.4 discusses MEMS waveguide chemical sensors.

7.2 OPTICAL SENSING PRINCIPLES

Light has a dual wave–particle nature, and it exhibits many unique properties. The following one-dimensional wave equation describes a plane wave propagating in the z-direction in a nonabsorbing media:

$$E = E_0 \sin(kz - \omega t), \tag{7.1}$$

where k is the propagation constant and equal to $2\pi/\lambda$, λ is the wavelength, ω is the angular frequency and is equal to $2\pi v$, and v is the frequency. The velocity of the wave is given by

$$v = v \times \lambda = c/n, \tag{7.2}$$

where c is the speed of light in a vacuum and n is the refractive index of the media. When two coherent waves reach a photodetector, interference occurs, which can be described by

$$|E|^2 = |E_{0,1}|^2 + |E_{0,2}|^2 + 2E_{0,1}E_{0,2} \cos\left[\frac{2\pi}{\lambda_0}\left(\int_0^{l_1} n_1(l)dl - \int_0^{l_2} n_2(l)dl\right)\right] \tag{7.3}$$

where λ_0 is the free-space wavelength, n_1 and n_2 are respectively the refractive indices along the two optical paths, and l_1 and l_2 are respectively the physical lengths of the two optical paths.

Equation 7.3 suggests several optical detection methods based on interference: (1) *l*-related changes: displacement, strain, thermal expansion, etc.; and (2) *n*-related changes: stress, temperature, absorption, concentration, composition, electric field, magnetic field, etc.

The particle nature of light can be easily described by photons. A photon has zero mass, but it does have energy, W_p, and momentum, p, which are respectively given by

$$W_p = h\nu = hc / \lambda_0 \tag{7.4}$$

$$p = \frac{h\nu}{c} = \frac{h}{\lambda_0} \tag{7.5}$$

where h is the Planck's constant.

When a light is incident on matter (gas, solid, or liquid), the light may be reflected, absorbed, or transmitted, depending on the energy state structure of the matter and the photon energy of the light. The energy states can be represented by a Jablonski diagram, as shown in Figure 7.4, where S_i and T_i (I is a positive integer), respectively, denote singlet states and triplet states. If the photon energy is greater than the energy gap between the ground state and S_1, the photon will be absorbed, and the photon energy is transferred to an excited electron. The excited electron descends to S_1 quickly through thermal relaxation. Then a new photon with an energy slightly less than the absorbed photon is emitted. This emission from S_1 to S_0 is called *fluorescence*. A portion of the excited electrons at S_1 are thermally relaxed to T_2 and then T_1. The emission from T_1 to S_0 is called *phosphorescence*. Since both fluorescence and phosphorescence are strongly related to the energy states, they can be used to detect various species.

When light passes through an absorbing medium, the light intensity will be attenuated, and the attenuation can be described by the following equation:

$$P(\lambda) = P_0(\lambda) \cdot 10^{-\alpha(\lambda)bM_c} \tag{7.6}$$

where $\varepsilon(\lambda)$ is the molar extinction coefficient in L (liters) mol^{-1} cm^{-1}, b is the optical path length (in cm) in the absorbing medium, and M_c is the molar concentration (mol/L). The extinction coefficient is wavelength dependent. Thus, this relation can be used for direct measurement of the concentration. More importantly, absorption spectroscopy can be used to simultaneously measure

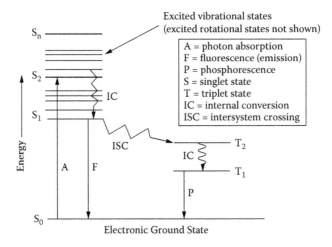

FIGURE 7.4 The Jablonski diagram of photon–electron interaction.

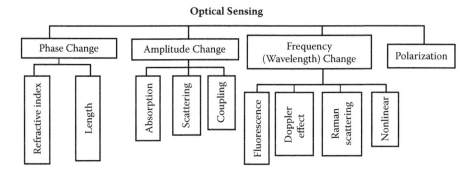

FIGURE 7.5 Optical sensing schemes.

multiple species and pinpoint a specific species as well because the extinction coefficient is wavelength dependent.

Raman scattering shifts the wavelength of the scattered photon. The wavelength shift is determined by the vibrational or rotational energy of a molecule. Thus, Raman spectroscopy can be used for material analysis and chemical sensing. Raman spectroscopy is especially powerful for obtaining vibrational spectra in an aqueous medium because water shows very weak Raman scattering but strong IR absorption. Furthermore, polarization, birefringence, nonlinear optical effects, evanescent wave coupling, and surface plasmon resonance can also be used for chemical sensing.

Figure 7.5 summarizes the optical sensing principles. The advantages of optical sensing include remote sensing; multiple channel/multiparameter detection; isolation from electromagnetic interference; minimally invasive for in vivo measurement; high selectivity and specificity; and good safety and biocompatibility.

7.3 MEMS-BASED FOURIER TRANSFORM SPECTROSCOPY FOR CHEMICAL SENSING

7.3.1 QUANTIZED STATES IN MOLECULES

Spectroscopy is an important chemical or material analysis method because every molecule has its unique energy level structure. The interactions between electrons and nucleuses form many energy levels, or states, that can be modeled as molecular orbitals. The lowest energy state is the ground state, S_0. Above S_0, there are singlet states S_1, S_2, and so on, and triplets T_1, T_2, and so on, as shown in Figure 7.6. These are the electronic states of a molecule. The atoms of a molecule can generate different modes of vibration; these modes have different energy levels, which are called *vibrational states of a molecule*. For molecules with a net dipole moment, there also exist rotational states. The required energies for obtaining excited states in these three types are largely different. As conceptually sketched in Figure 7.6, the electronic transition requires the highest energy and rotational transitions the least. Electronic transitions involve the spectral range from deep UV to near IR; vibrational transitions correspond to the mid-IR to far-IR range; and rotational transitions are in the microwave frequency range.

The combination of electronic and vibrational transitions spans almost the entire optical spectrum. Figure 7.7 summarizes various optical spectroscopy methods, such as fluorescence spectroscopy, Raman spectroscopy, and IR spectroscopy, which can all be used for chemical and biological detection.

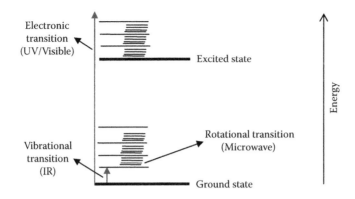

FIGURE 7.6 Quantized states of a molecule.

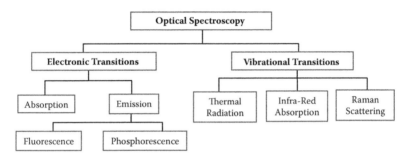

FIGURE 7.7 Various mechanisms in optical spectroscopy.

7.3.2 Optical Spectroscopy

Rapid detection of chemical and biological agents is very important for national border security, battlefields, and antiterrorism efforts [1]. Terrorism has become an increasing threat to the United States since 2001. Explosives have been extended from traditional TNT to improvised explosive devices (IEDs) made of raw industrial chemicals such as nitric acid, ammonium nitrate, diesel fuel, and sugar, so traditional explosive detection technologies are far from sufficient. Many terrorist organizations may use chemical weapons since they are very cheap to make and easy to access and transport. Moreover, there is an increasing threat of biowarfare agents and bioterrorism [2]. However, the requirement of time-consuming laboratory testing and analysis may cause catastrophic results. On the civilian side, public safety and a healthy working environment also require immediate identification of toxic chemicals, toxic materials, and numerous other hazardous substances. Disposing of chemical waste safely and properly is a growing challenge to the world. On-site assessment usually is not sufficient, and further tests need to be conducted off-site or in a nearby laboratory [3]. This can add a tremendous amount of time and expense to the identification and clean-up process. Furthermore, chemical/biosensors can play an extremely important role in medicine to reduce health care cost, improve quality of life, and save lives, but most chemical sensors and biosensors cannot be used at the point of care because of their poor selectivity and stability and/or poor portability.

Optical spectroscopy, combined with a spectral library, can be used to quickly identify the presence of particular chemical or biological agents [4,5]. The chemical bonds and functional groups of an object are unique and thus generate a unique absorbance spectral pattern. In addition to functional groups, hydrogen bonding, molecular conformations, and even chemical reactions can be determined by analyzing these absorption spectral patterns. Thus, optical spectroscopy has high selectivity and specificity. Furthermore, optical spectroscopy can be obtained without contact or even remotely [6], which is very important for the safety of the on-site workers and first responders or the point-of-care medical personnel.

To obtain an optical spectrum, either a tunable light source or a broadband light source may be used. Using tunable light sources simplifies the detection system as no spectrometers are needed, but such light sources are expensive and have small spectral bandwidth and limited availability of spectral coverage. Broadband light-source-based systems are less expensive and can work in a wide spectral range, but they do need spectrometers. A spectrometer may be a tunable Fabry–Perot etalon or a dispersive component such as a prism or a grating to individually detect the wavelengths of the transmitted or reflected light. However, there is a serious trade-off between the substance concentration resolution and the spectral resolution for all these methods because the separation of the spectral components reduces the signal power of each spectral component.

7.3.3   FOURIER TRANSFORM SPECTROSCOPY

Fourier transform spectroscopy (FTS) does not require dispersive components, and it uses only a single photodetector that collects the optical signal of the entire spectral range [7]. In contrast, non-FTS uses a photodetector array, and each photodetector only receives the optical signal in a small fraction of the entire spectrum. Therefore, FTS has much higher signal-to-noise ratio (SNR), and much more affordable photodetectors can be used.

To obtain FTS, a Michelson interferometer is typically used. As shown in Figure 7.8, the light is split into two beams that are respectively directed to two mirrors. One mirror is fixed, and the other mirror is moveable. The two optical beams from the beam splitter (BS) are reflected back by both mirrors, and then the merged optical beam passes through the sample before it reaches the photodetector.

The sample may absorb specific wavelengths and the photodetector records the interferogram signal, $I(z)$, which can be expressed as

$$I(z) = \tfrac{1}{2}I(0) + \tfrac{1}{2}\int_{-\infty}^{\infty} G(k)e^{ikz}\,dk \tag{7.7}$$

where z is the position of the movable mirror, $I(0)$ is the intensity for zero path difference, $k\,(=2\pi/\lambda)$ is the angular wave number, and $G(k)$ is the spectral power density. Of interest are the wavelengths at which those "intensity peaks" occur. As can be seen in Equation 7.7, $I(z)$ and $G(k)$ are a Fourier transform pair. Thus, the spectral density can be decoded by a reverse Fourier transform, that is,

$$G(k) = \frac{1}{\sqrt{2\pi}}\int_{-\infty}^{\infty}\left[2 \cdot I(z) - I(0)\right]e^{-ikz}\,dz \tag{7.8}$$

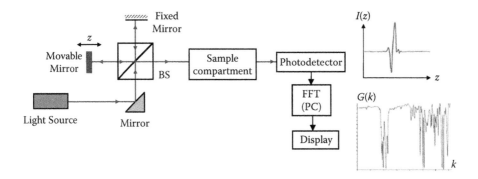

FIGURE 7.8 (See color insert).   Simplified block diagram of an FTS system. BS: Beam splitter.

7.3.4 FOURIER TRANSFORM INFRARED SPECTROSCOPY

FTS in the infrared (IR) range, or FTIR, is very powerful in chemical and biological sensing since almost all vibrational transitions and some rotational transitions of molecules that carry molecular fingerprints fall into the IR range. Also, optical components, especially dispersive spectrometers for IR, are expensive. So FTIR is widely used for molecular identification and sensing.

FTIR spectroscopy is sensitive and fast with high resolution and high specificity. The resolution of the Nicolet 4700 by Thermo Scientific is 0.4 cm^{-1}. However, conventional FTIR spectrometers are expensive, large, and not portable. For instance, the world's smallest FTIR spectrometer ALPHA claimed by Bruker Optics still weighs 15 lb [8]. Thus, portable, inexpensive, miniaturized FTIR instruments are urgently needed for field, on-site, and point-of-care applications. The key challenges for FTIR miniaturization include the bulky scanning mechanisms and the bulky and expensive cooled IR detectors.

7.3.5 MEMS FTIR

Traditional FTIR spectrometers are table-top systems, which deliver very accurate and high performance results. However, they are bulky and expensive to buy and to maintain, due to their precision mirror-scanning mechanism. The advent of MEMS has opened the door to a new realm of miniaturized spectrometers, using micromirrors and microactuators. Several research groups have reported miniaturized FTIR spectrometers based on the Michelson interferometer approach. The main difference between those FT spectrometer designs is the actuation method for the micromirror. There are several MEMS actuation mechanisms, namely, electrostatic, electrothermal, electromagnetic, and piezoelectric actuators. The principles of the actuation mechanisms are introduced before the miniaturized FTS systems are reviewed.

7.3.5.1 MEMS Actuators

7.3.5.1.1 Electrostatic Actuation

Electrostatic actuation is the most popular actuation mechanism in the MEMS world because of its high speed, low power consumption, and easy fabrication. As shown in Figure 7.9, the actuator may consist of either parallel plates or interdigitated beams. The parallel plate actuators can generate motions in plane or out of plane (vertical), as shown in Figure 7.9a. Interdigitated beams, or comb fingers, can also generate in-plane or vertical motions. There are three different operation modes for a comb drive: transverse, longitudinal, and vertical, as respectively shown in Figures 7.9b, 7.9c, and 7.9d.

All the configurations shown in Figure 7.9 can be modeled by a simple parallel plate actuator. Assuming the gap between the parallel plates is g and the area of one plate is A, then the parallel plate capacitance can be expressed as

$$C(z) = \frac{\varepsilon_0 A}{g(z)} = \frac{\varepsilon_0 A}{g_0 - z} = \frac{\varepsilon_0 A}{g_0}\left(1 - \frac{z}{g_0}\right)^{-1} = C_0\left(1 - \frac{z}{g_0}\right)^{-1} \tag{7.9}$$

where g_0 is the gap at rest, z is the displacement of the movable plate, $C_0 = \varepsilon_0 A/g_0$ is the capacitance at rest, and ε_0 is the permittivity of air. If a voltage V is applied to the parallel plates, an electrostatic force perpendicular (or transverse) to the plate is generated, which is given by

$$F(z,V) = \frac{1}{2}\frac{dC(z)}{dz}V^2 \tag{7.10}$$

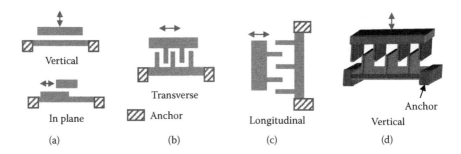

FIGURE 7.9   Various configurations of electrostatic actuation.

Substituting Equation 7.9 into 7.10 yields

$$F(z,V) = \frac{C_0 V^2}{2g}\left(1 - \frac{z}{g_0}\right)^{-2} \tag{7.11}$$

When the displacement is much smaller than the gap, the transverse force is constant, that is,

$$F(z,V) = \frac{C_0 V^2}{2g} \tag{7.12}$$

The applied voltage may also generate an electrostatic force parallel to the plates if the two plates are not overlapped perfectly and the overlap area is variable. This parallel or longitudinal electrostatic force can be expressed as

$$F(x,V) = \frac{1}{2}\frac{dC(A)}{dx}V^2 = \frac{\varepsilon_0 C_0 V^2}{2g}\frac{dA}{dx} \tag{7.13}$$

where the gap stays constant and the fringing capacitance is neglected. This force is constant if the plates are rectangular and the motion is perpendicular to one of the edges.

7.3.5.1.2 Magnetic Actuation

When a current, I, flows through a wire that is perpendicular to an external magnetic field B, the wire will experience a magnetic force. According to Laplace's law, the magnetic force, sometimes also called the *Lorenz force*, is given by

$$F_{mag} = BI\ell \tag{7.14}$$

where ℓ is the length of the wire exposed to the magnetic field.

Magnetic force can be used to generate either in-plane or out-of-plane linear motion or rotation. The magnetic field may come from a permanent magnet or a coil. Figure 7.10 shows magnetic actuators with magnetic materials deposited on the movable parts. The external magnetic field induces magnetic poles in the magnetic material region. The interaction of the induced field with the external field generates a force that tends to align the magnetic-coated movable part with the external magnetic field. The actuation can be either in plane or out of plane depending on the direction of the external magnetic field.

Coils can generate variable magnetic fields by varying the injected current. Figure 7.11 shows coil-based magnetic actuation methods. Either piston motion or rotation can be generated depending on the spring design and the direction of the external magnetic field.

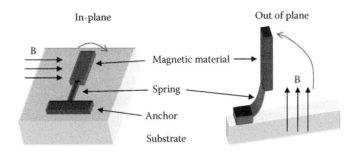

FIGURE 7.10 Magnetic actuators using magnetic materials.

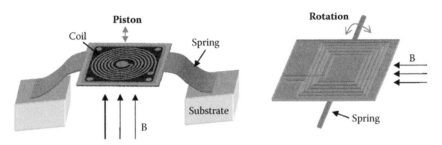

FIGURE 7.11 (See color insert). Magnetic actuators using coils.

7.3.5.1.3 Electrothermal Actuation

Most electrothermal actuators are based on thermal bimorph actuation. A bimorph structure consists of two layers that have different temperature coefficients of expansion (TCEs). When there is a temperature change, ΔT, a stress is induced due to the TCE difference, resulting in a bending of the bimorph beam. The temperature change can be caused by Joule heating or infrared absorption. A bimorph structure is illustrated in Figure 7.12, which consists of two layers with TCEs of α_1 and α_2, respectively. The tangential angle at the tip of the bimorph beam is given by

$$\theta_T = L\beta_r(\alpha_1 - \alpha_2)\Delta T \tag{7.15}$$

where L is the length of the bimorph, and β_r is a constant related to the thicknesses and Young's moduli of the two layers.

Rotation can be used to generate a large motion by using angular amplification. As shown in Figure 7.13a, a rigid frame is attached to the end of the bimorph. If the frame is long, even a small angular change by the bimorph actuator will generate a large motion at the tip of the frame. However, with this simple bimorph design, a large tilt and large lateral shift will be generated at the same time, greatly limiting its applications. The tilting problem can be overcome by adding a second set of bimorph and frame. As shown in Figure 7.13b, the second set is folded back, so its rotation is the same as but opposite in direction to the first bimorph set, resulting in zero tilting if the lengths of the two bimorphs and two frames are chosen properly. But this improved design still has a large lateral shift.

In order to compensate for the lateral shift at the same time, a third set of bimorph and frame is added, as shown in Figure 7.13c. The lateral shift can be canceled if the following relations are satisfied:

$$l_1 = l_3 = \frac{1}{2}l_2 \text{ and } L_1 = L_2, \tag{7.16}$$

where l_1, l_2, and l_3 are the lengths of the first, second, and third bimorph, respectively, and L_1 and L_2 are the lengths of the two frames. Then both the tilt and the lateral shift are compensated for. Thus, it is called the *tilt- and lateral shift-free* (TLSF) bimorph design.

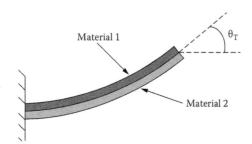

FIGURE 7.12   Bimorph structure.

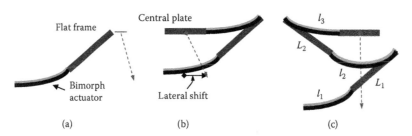

FIGURE 7.13   Bimorph beam designs. (a) Simple bimorph with a flat frame, (b) tilt-compensated bimorph set, (c) tilt-lateral-shift-free (TLSF) bimorph design.

7.3.5.2   Miniature FTIR Based on Electrostatic Micromirrors

7.3.5.2.1   Electrostatic Micromirrors with Sidewall as the Reflective Mirror Surface

Large out-of-plane motion is generally very difficult to generate. Thus, researchers use the large stroke of in-plane actuation to move mirrors with reflective sidewall surfaces.

Manzardo et al. [9] reported an MEMS-based FTIR system in which an electrostatic comb drive actuator was used to move the MEMS mirror [9]. Figure 7.14a shows the layout design of the electrostatic comb drive. A micromirror is attached to the actuator. In order to maintain linear motion of the mirror, two identical comb drive actuators, A and B, are placed opposite each other. When the two out-of-phase voltage signals with the same amplitude V_{ac} are applied to A and B, respectively, the displacement of the mirror, Δx, is given by

$$\Delta x = 2 \frac{\varepsilon_0 Nh}{g} V_{ac} V_{dc} / k \tag{7.17}$$

where k is the spring constant of the suspension springs, h is the height of the comb fingers, g is the gap between the comb fingers, N is the number of comb fingers on the movable comb, ε_0 is the permittivity in air, and V_{dc} is the DC bias voltage.

The mirror surface was a silicon sidewall formed by silicon DRIE, as shown in Figure 7.14b. The mirror size was 75×500 μm. With $V_{ac} = 10$ V and $V_{dc} = 15$ V, the maximum displacement measured was $\delta_{max} = 77$ μm, leading to a maximum optical path difference of 154 μm. The estimated spectroscopic resolution at 633 nm was $\Delta\lambda = \lambda^2/\delta_{max} \approx 5.2$ nm, which matched the experimental result shown in Figure 7.15a. A measured interferogram of an 800 nm luminescence diode is shown in Figure 7.15b.

Yu et al. [10] reported another sidewall MEMS mirror for FTIR [10]. In this FTIR, the silicon beam splitter, MEMS micromirrors, and fiber U-grooves were all integrated on a single chip. All these components were made of the (110) device layer of SOI wafers. The fabrication process combined the flexible definition capability of DRIE with the KOH wet etching. KOH etching does provide smooth vertical sidewalls. However, that requires the (110) device layer, which makes the mirrors oriented not at 45° but at an awkward angle of 70.53°. A portion of the fabricated device is shown in Figure 7.16a.

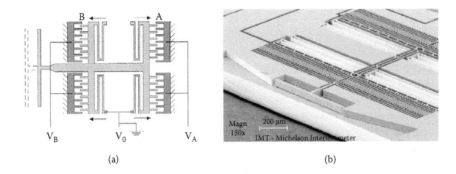

(a) (b)

FIGURE 7.14 Schematic of the actuator used for FT spectroscopy.

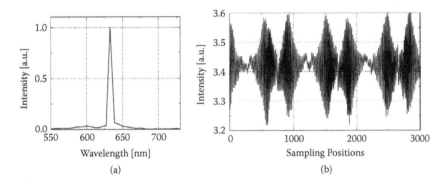

(a) (b)

FIGURE 7.15 (a) The spectrum of a He–Ne laser obtained at $\delta_{max} = 77$ µm, (b) interferogram of a luminescence diode measured over several scans.

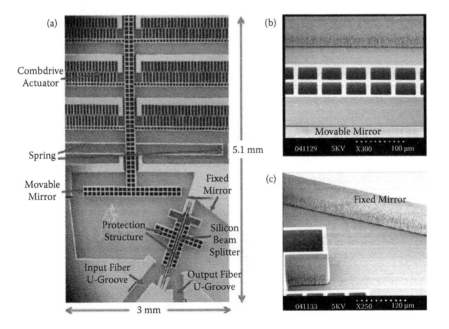

FIGURE 7.16 (a) An SEM image of the micromachined Fourier transform spectrometer, (b) SEM of the movable micromirror with the sidewall obtained by KOH etching, (c) SEM of the fixed mirror with the sidewall defined by DRIE.

The entire device size was 4×8 mm. The closed-up views of the fixed mirror and movable mirror are shown in Figure 7.16b and Figure 7.16c, respectively. The movable mirror is much smoother than the fixed mirror.

The maximum displacement of the mirror was about 25 μm, corresponding to a drive voltage of 150 V. So the maximum achievable OPD was 50 μm, leading to a spectral resolution of 45 nm for 1500 nm near-IR light. Figure 7.17a shows measured interferograms of three different wavelengths (1500, 1520, and 1580 nm). Due to the nonconstant scan velocity, the interferograms need to be resampled. Figure 7.17b shows the spectra obtained from the FTIR spectrometer after resampling and apodization to remove nonlinear effects.

The second sidewall micromirror improves the mirror surface smoothness but at the price of the inefficient Michelson interferometer design, and the mirror size is still limited to the thickness of the device layer, which is typically less than 100 μm. Furthermore, the scan range is only about 25 μm even at 150 V, which greatly limits the spectral resolution.

7.3.5.2.2 Electrostatic Micromirrors with Large Out-of-Plane Displacement

In order to overcome the shortcomings of sidewall micromirrors, Sandner et al. [11] proposed a miniaturized MEMS-based FTIR spectrometer using an electrostatic micromirror that moves out of plane [11]. The layout of the device is shown in Figure 7.18a, which is 1.8×9 mm in size. The

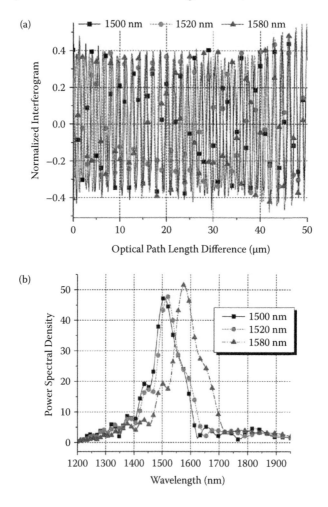

FIGURE 7.17 Measured spectra after resampling of the interferograms.

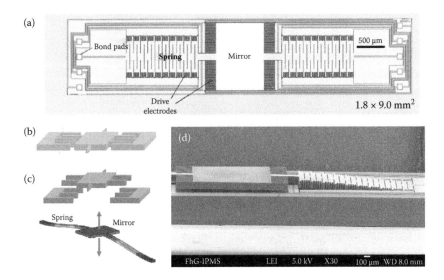

FIGURE 7.18 (See color insert).   Translatory electrostatic micromirror.

device has a central mirror plate, a long spring beam, and comb drives. The basic idea is to excite the out-of-plane mode (Figure 7.18c) of a regular in-plane comb drive (Figure 7.18b). In order to achieve large out-of-plane translatory motion, the device must have high Q. Sander et al. [11] were able to achieve a maximum displacement of 200 μm at the 5 kHz resonance at 40 V by placing the mirror in a vacuum chamber of 100 Pa.

The 400 μm OPD produces a spectral resolution of 25 cm^{-1} theoretically. The mirror plate is 1.5 × 1.1 mm. This larger mirror aperture significantly reduces coupling loss. Figure 7.19 shows an open view of a vacuum-packaged FTIR system. The MEMS mirror is operated at 100 Pa in a vacuum chamber, and a Peltier-cooled IR detector is used for IR detection. The reference interferometer monitors the position and velocity of the mirror in real time. Figure 7.20 shows the absorption spectra of 0.2 mM Acetone and 0.2 mM Ethyl alcohol.

This out-of-plane micromirror has a large optical aperture and large scan range. However, the large scan range is achieved by resonance operation in a vacuum. Vacuum packaging is expensive, and it increases the overall packaging size dramatically. Furthermore, stability and reliability in a high vacuum are additional concerns.

7.3.5.3 Miniature FTIR Based on Electromagnetic Micromirrors

Using sidewall mirrors has the advantage of planar integration. The two main limitations are the small mirror size and small scan range. Solf et al. [12] proposed to use electromagnetic actuation to obtain a large scan range [12]. Meanwhile, a deep sidewall mirror created by LIGA technology was used for easy integration. As shown in Figure 7.21, all parts such as the collimation optics, the detectors, the actuator, and the mirrors were assembled on a chip. The optical bench and the mirrors were made of 380-μm-thick electroplated permalloy. The chip footprint is 11.5 × 9.4 mm. The light coupled to the beam splitter via the optical fibers is split into two beams: one traveling to the fixed mirror on the left and the other reaching the movable mirror on the right. The movable mirror is connected to and can be displaced by the electromagnetic actuator. The actuator has two assembled microcoils. Both optical beams are reflected back to the beam splitter and then directed to the photodiode, where the interference is detected. The actuator is a linear reluctance motor made from a soft magnetic FeNi alloy. The movable plunger is suspended by a set of four folded cantilever beams. An actuation coil is placed at either end of the plunger, so the plunger can be pulled in both directions, which doubles the travel range.

FIGURE 7.19 (See color insert).   Photograph of a prototype FTIR system.

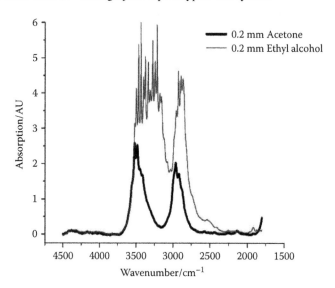

FIGURE 7.20   Absorbance spectra of acetone and ethyl alcohol in a 0.2 mm flow cell.

The generated magnetic force depends on the number of windings in the coils. With 300 windings, the maximum displacement is about 110 µm at 12 mW. If the input power exceeds 12 mW, the actuator becomes instable, showing pull-in behavior. Figure 7.22a shows the interferogram of a 1540 nm laser. In this experiment, the displacement of the actuator was about 54 µm, and an 850 nm laser was used to measure the mirror displacement. Figure 7.22b shows the derived spectrum after Fourier analysis; the resolution was 24.5 nm at 1540 nm.

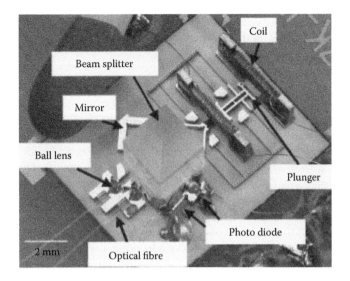

FIGURE 7.21 Top view of the LIGA-enabled FTIR.

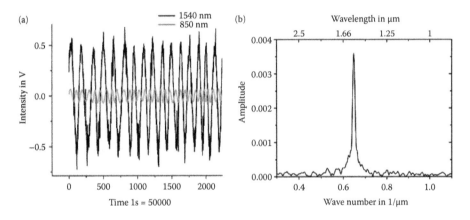

FIGURE 7.22 (a) An interferogram of a 1540 nm laser. An 850 nm laser was used for position measurement. (b) Measured spectrum of the 1540 nm laser.

In summary, this LIGA-based MEMS FTIR design achieved a large scan range and large optical aperture. However, due to the pull-in instability, only a fraction of the full scan range is usable. Also, the LIGA process is very expensive and has very limited access.

7.3.5.4 Miniature FTIR Based on Electrothermal Micromirrors

As discussed earlier, both electrostatic and electromagnetic MEMS mirrors have been used to miniaturize FTS systems. However, these actuation methods can generate only about tens of microns mirror scan range, so the spectral resolution is low. A maximal scan range of 200 µm by an electrostatic actuator was reported more recently [10], but the large scan range was achieved at resonance and vacuum packaging was required, sacrificing the cost efficiency and compactness of the system.

For FTS, the spectral resolution is inversely proportional to the travel range of the movable mirror, that is,

$$\Delta\sigma = 1/\left(2 \cdot \Delta z_{max}\right) \tag{7.18}$$

where σ is the spectroscopic wave number ($\sigma = k/2\pi = 1/\lambda$), $\Delta\sigma$ is the spectral resolution in wave number (units: cm^{-1}), and Δz_{max} is the maximum displacement of the mirror. We may convert the wave number into a wavelength, that is,

$$\Delta\lambda = \lambda^2 \cdot \Delta\sigma = \lambda^2 / (2 \cdot z_{max}) \tag{7.19}$$

So, if the maximum displacement is 50 μm, the wave number resolution is 100 cm^{-1}. If the light source has a center wavelength of 1.5 μm, the wavelength resolution will be 22.5 nm. Note that the wavelength resolution is increased with the square of the wavelength. So for longer wavelengths (e.g., midwave or long-wave IR), the wavelength resolution deteriorates quickly. Therefore, MEMS mirrors with large displacements are essential for FTIR. For instance, for a 1.0 μm wavelength light, a 1 nm (or 1 cm^{-1}) resolution requires the mirror to travel 0.5 mm. For a 2 μm wavelength light, the mirror has to travel 2 mm for the same resolution.

Electrothermal micromirrors, on the other hand, can generate a large actuation range of several hundreds of microns without operating at resonance by using a unique large-vertical-displacement (LVD) bimorph actuator design [13], and a recently developed lateral-shift-free (LSF) LVD actuator design achieves large vertical actuation of over half a millimeter with very small lateral shift and tilting [14].

As shown in Figure 7.23b, the LSF-LVD actuator consists of two rigid frames and three Al/SiO$_2$ bimorph beams connected in series for tilting and lateral shift compensation. The mirror plate (Al coating on a Si layer) is supported by two LVD actuators at the opposite edges. Each actuator consists of two pairs of LSF-LVD actuators that connect with each other by a rigid frame and share the third bimorph beams. The third bimorph beams connect all along the mirror edges to minimize the tilting during the vertical actuation. The bimorph beams curl upward after release, resulting in an initially elevated mirror rest position. With electrothermal actuation on the embedded Pt heater, the bimorphs bend downward due to differential thermal expansions of Al and SiO$_2$, resulting in a net vertical displacement of the mirror plate. The MEMS mirror is fabricated using a similar process presented in Reference 13. Figure 7.23a shows an SEM of a fabricated device with a mirror size of 1.8×1.6 mm.

The MEMS mirror demonstrated a maximum vertical displacement of about 1 mm at a 4 V DC voltage applied on both actuators, as in Figure 7.24a. A maximum tilting angle (MTA) of ~2.5° has been observed. The tilting is mainly due to fabrication variations of the heater electrical resistances and the two actuator structures. To minimize the tilting, the two actuators need to be driven separately and an optimum voltage ratio between the two driving voltages needs to be experimentally

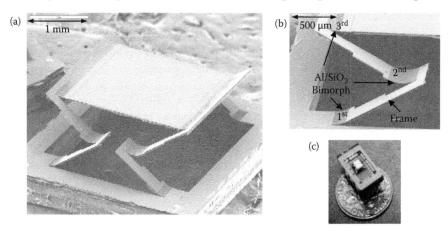

FIGURE 7.23 (a) SEM pictures of a fabricated LVD micromirror. (b) LSF-LVD actuator. (c) A packaged device with a U.S. dime.

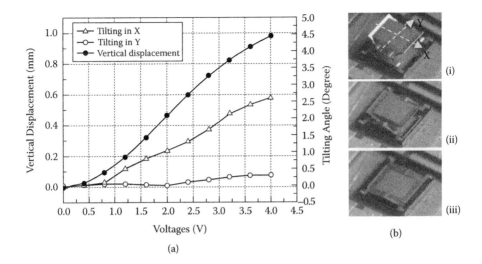

(a)

(b)

FIGURE 7.24 (a) Experimental results of vertical displacement and tilting angle versus actuation voltage, (b) microscopic pictures of a packaged device actuated at (1) 0 V, (2) 2 V, and (3) 3.5 V.

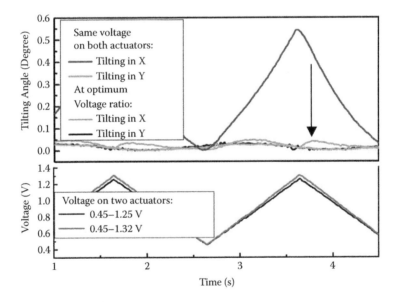

FIGURE 7.25 Tilting angles measured by PSD with same voltage on both actuators and minimized tilting angle at optimum voltage ratio.

characterized. A position-sensitive detector (PSD) is employed to measure the tilting angle. During the mirror actuation, the PSD detects the shift of the reflected light beam that is normally incident on the MEMS mirror (after being reflected by the beam splitter), and the tilting angle can be calculated. Figure 7.25 shows an MTA of 0.55° with a ramp driving voltage of 0.45–1.32 V at 0.5 Hz on both actuators. With one actuation voltage being adjusted to 0.45–1.25 V, the MTA is minimized to 0.06°.

The LSF micromirror was placed in an FTS setup as the scanning mirror. A He-Ne laser was used for the spectroscopy measurement experiment. Figure 7.26 shows the interferograms obtained by applying two ramp waveforms of 0.45–1.25 V and 0.45–1.32 V on the two actuators, respectively. The obtained interferogram gives the interference signal in the time domain. However, direct

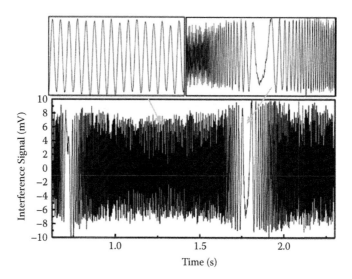

FIGURE 7.26 Interferogram of a He-Ne laser by actuation voltages of 0.45–1.25 V and 0.45–1.32 V on the two actuators, respectively. Insets showing the mirror's deceleration/ acceleration range and a uniform mirror velocity range.

correlation of the interferogram in the time domain to the optical path difference (OPD) is not applicable due to the nonuniformity of the mirror velocity for the entire actuation range. Therefore, the data has to be resampled to account for the scanning nonuniformity. Phase correction, adaptive digital filtering, and polynomial interpolation methods have been used previously to resample the data linearly with OPD. Our approach is slightly different. The raw interferogram data is first over-sampled using interpolation. The fringe maxima and minima are then found, and equally spaced sample points are selected between each fringe maximum and minimum. Since the spacing of the consecutive points between each maximum and minimum is equal, the mirror velocity nonuniformity is corrected by this method. These sample points can be used to calibrate the spectrometer, by using a light source with a known wavelength, for example, a He-Ne laser. The calibration results can be applied to an interferogram for an unknown wavelength by resampling it with the calibration sample points.

As shown in Figure 7.27, spectral resolutions of 71.4 and 19.2 cm^{-1} were obtained by MEMS mirror actuation ranges of about 70 μm and 261 μm (with OPD of 140 μm and 522 μm), respectively.

7.3.5.5 A Mirror-Tilt-Insensitive (MTI) Fourier Transform Spectrometer [15]

The other challenge of FTS miniaturization is to control the tilting of the scanning mirror. As shown in Figure 7.28a, undesired tilting of the scan mirror can easily cause misalignment of the two returning beams and thus impair the interference signal used for the Fourier transform. The stability of the movable mirror during scanning is very important [16,17]. As shown in Figure 7.28b, if the movable mirror tilts at an angle θ and the distance from the movable mirror to the photodetector is l, then the two light beams reflected back from the two mirrors will be separated by $\Delta y = \tan\theta \times l$. Assume that θ = 1°, or 0.0175 rad, and l = 10 mm, yielding Δy = 175 μm. This means that the two reflected light beams will be offset by 175 μm. More importantly, due to the tilting of the movable mirror, the wavefronts of the light beams from the two mirrors have different phase differences at different locations of the photodetector. Thus, multiple fringes if any will be detected by the photodetector at the same time. Use the foregoing assumption that the mirror aperture size d = 1 mm. Then the maximum optical path difference at the photodetector surface is $d \times \tan\theta$ = 17.5 μm, which is about 12λ for 1.5 μm light. This large range of phase difference will cause serious distortion of the interferogram, which in turn yields inaccurate spectra.

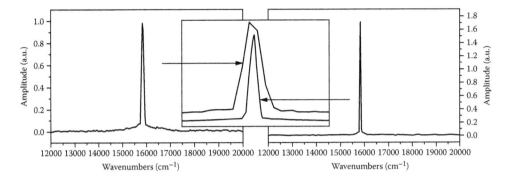

FIGURE 7.27 Spectrum of He-Ne laser with spectral resolutions of 71.4 and 19.2 cm⁻¹.

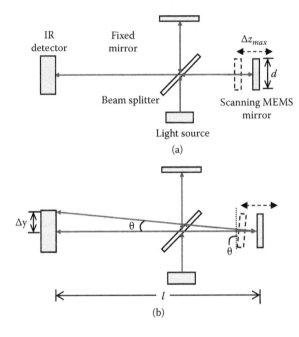

FIGURE 7.28 (a) Schematic of a conventional FTIR spectrometer, (b) schematic showing the beam shift by the mirror tilting.

For MEMS mirrors, tilting during large-range scanning is almost inevitable. Therefore, a mirror-tilt-insensitive (MTI) FTS has been proposed using the dual-reflective LVD MEMS mirror and a corner-cube retroreflector to compensate for the small tilting of the MEMS mirror [15]. Figure 7.29a shows the MTI-FTS concept. Both light beams from the beam splitter are directed to the dual-reflective MEMS mirror. There are one corner-cube retroreflector and one fixed mirror in the two arms of the interferometer, respectively. At the rest position, the MEMS mirror is placed at the zero OPD point. With the actuation of the MEMS mirror, the OPD of the two subarms can be generated, and interferograms can be detected by a photodetector. The tilting of the scanning mirror is compensated by taking advantage of the dual reflections of the MEMS mirror and the retroreflector, whose reflected light beam is always parallel to the incident beam. When there is a tilting of the MEMS mirror in any direction, both light beams will bounce off the MEMS mirror with the equal but opposite tilting. Due to the right-angle arrangement of the corner cube, the returning light beam will be shifted to the opposite direction. Therefore, the two returning light beams will have the same amount of shift at the same direction when propagating back to the beam splitter. Another

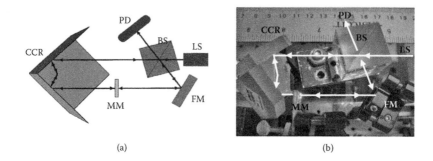

FIGURE 7.29   (a) FTS system with dual-reflective MEMS mirror, (b) a photo of the MTI-FTS demonstration setup. CCR: corner-cube retroreflector, PD: photodetector, BS: beam splitter, MM: MEMS micromirror, FM: fixed mirror, LS: light source.

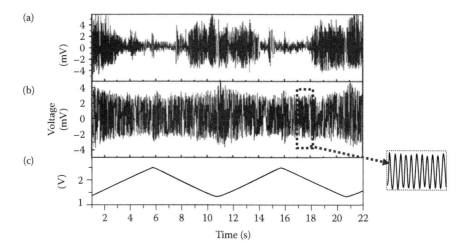

FIGURE 7.30   Interferograms of the He-Ne laser by (a) a conventional FTS setup, (b) MTI-FTS setup, (c) with micromirror actuated at a ramp voltage from 1.2 V to 2.5 V at 0.1 Hz.

advantage of this MTI FTS is that the equivalent OPD is four times the physical scan range of the MEMS mirror, which is doubled compared to a conventional FTS so that the spectral resolution can be further enhanced [18].

Figure 7.29b shows a picture of the experimental setup, which measures about $12 \times 5 \times 5$ cm^3. A much smaller size can be obtained by further reducing the sizes of components such as the beam splitter and retroreflector. FTS experiments were performed using a He-Ne laser on a conventional Michelson-interferometer-based FTS and the MTI-FTS by the same MEMS mirror. The MEMS mirror was scanning vertically by 0.4 mm at 0.1 Hz. Figure 7.30a shows the interferogram obtained from a conventional FTS, which was greatly impaired by the tilting of the MEMS mirror. The interference signal was detected only at small driving voltages of each driving cycle, where the tilting angle is very small. As a comparison, the MTI-FTS is much less sensitive to the mirror tilting and gives good interference signals in the entire cycle by the same mirror actuation as in Figure 7.30b.

Figure 7.31 shows the spectrums of a He-Ne laser obtained after the calibration of the nonuniform mirror scan at different mirror scan ranges. Spectral resolutions of 19.2 and 8.1 cm^{-1} were obtained by the mirror actuation ranges of about 131 and 308 μm (with the OPD of 522 μm and 1.23 mm), respectively.

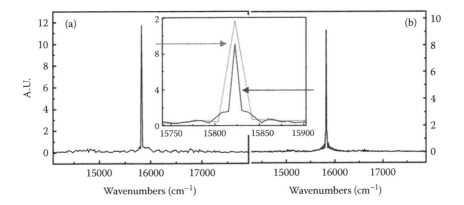

FIGURE 7.31 Spectrum of He-Ne laser with different resolutions obtained by the micromirror being actuated at ramp waveforms of (a) 1.2 to 2 V and (b) 1.2 to 2.5 V, respectively. Inset compares the resolutions of 19.2 and 8.1 cm⁻¹.

7.3.6 SUMMARY

As discussed in this section, electrostatic, electromagnetic, and electrothermal micromirrors all have been used for miniature FT spectroscopy. Electrostatic micromirrors have high speed and low power consumption, but the scan range is small, typically less than 100 μm. One electrostatic actuator design was reported to have a maximum scan of 200 μm, but it requires vacuum packaging and resonance operation, which significantly increases cost and overall packaging size. Electromagnetic actuation can achieve about 400 μm, but the pull-in behavior limits the usable range to only 54 μm. And the high-cost LIGA technology is needed for fabrication. Electrothermal bimorph micromirrors can reach about 1 mm scan range with low-cost batch fabrication. Promising results have been obtained from electrothermal MEMS mirror-based FTS systems. Repeatability and reliability will continue to be the main hurdles for the commercialization of MEMS FTS systems.

7.4 INTERFEROMETRIC MEMS CHEMICAL AND BIOCHEMICAL SENSORS

7.4.1 INTRODUCTION

Optical waveguide (OWG)-based chemical and biochemical sensors have many intrinsic advantages over their electrical counterparts, including high sensitivity, label-free detection capability, resistance to electromagnetic disturbance, absence of electric shock and spark, good safety, and high reliability. Owing to these advantages, OWG-based chemical and biochemical sensors have received widespread applications in various fields. With the current rapid development of optical MEMS technology, OWG sensors are attracting increasing interest because they are well compatible with the MEMS processing techniques. OWGs have the slab and stripe geometries that have been applied to chemical and biochemical sensors. The stripe OWGs are a typical MEMS component, and the slab OWGs also keep MEMS characteristic in one dimension. The stripe-geometrical OWG sensors have been currently fabricated by using the MEMS process combined with the SION technology on silicon substrate. The next-generation OWG sensors would be single-silicon-chip devices containing the sensing elements, LED sources, Si-photodiode detectors, and CMOS data processor. Moreover, it is anticipated that OWG-based lab-on-chip systems can be realized by integrating or stacking the MEMS sample-preprocessing and microfluidic components onto the IO sensor chip.

As shown in Figure 7.32, optical chemical and biochemical sensors use functional thin films rather than the corresponding bulk materials to interact with the analyte for rapid detection. Because the film thickness is small, ranging from nanometers to micrometers, the bulk optical sensors with

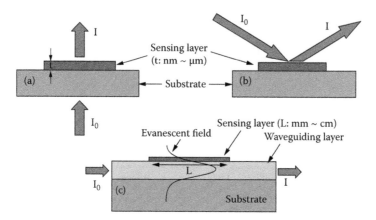

FIGURE 7.32 Schematic diagrams of bulk optical sensors with (a) transmission and (b) reflection measurements; (c) schematic of an OWG sensor.

either transmission or reflection measurement have a poor sensitivity that is not enough for trace detection. OWG-based chemical and biochemical sensors are quite different from the bulk optical devices, which make use of the evanescent field to interrogate the sensing layer immobilized to the waveguide surface, translating changes in the physical or chemical properties of the sensing layer due to the interaction with analytes into a measurable signal that can be intensity, phase, resonant angle, or resonant wavelength. The evanescent field can interact with the sensing layer in a long path length ranging from millimeters to centimeters; thus, the sensitivity of OWG sensors can be several orders of magnitude higher than that of bulk optical sensors. OWG sensors are also superior to bulk optical devices in resistance to environmental interference. In the big family of OWG-based chemical and biochemical sensors, one of the most widely studied sensors is integrated optical (IO) interferometer sensors [19–53]. This is because the phase of a guided mode is extremely sensitive to changes in the refractive index and/or thickness of the sensing layer, and these changes can easily occur with almost all of the surface interactions. An OWG interferometer without immobilizing a sensing layer on its surface is an effective and versatile transducer. It can be applied to a large variety of chemical and biochemical sensors as long as the waveguide surface is functionalized with different molecule-recognition materials.

From the structural point of view, OWG interferometers mainly include the Mach–Zehnder interferometer (MZI) [19–33], polarimetric (or difference) interferometer (PI or DI) [34–38], Young interferometers (YIs) that contain the slab and stripe structures [39–52], and Fabry–Perot and Michelson interferometers [53,55]. The first three types are most often used for chemical and biochemical sensing applications. According to the works reported so far, the MZI and YI sensors generally have high sensitivity. Compared with the MZIs and YIs, PI sensors have a simple structure but a low sensitivity. The Fabry–Perot interferometer (FPI) has a very limited application as chemical and biochemical sensors [53]. The Michelson interferometer is mainly used for displacement measurement, and it is seldom applied to chemical and biochemical sensing [54]. In this section, the OWG interferometers as versatile transducer elements—not specific chemical or biochemical sensors—were summarized.

7.4.2 INTEGRATED OPTICAL MACH–ZEHNDER INTERFEROMETER (MZI) SENSORS

MZI sensors are typical IO devices, consisting of three-dimensional single-mode waveguides with the geometry shown in Figure 7.33. The two Y-shaped waveguide structures serve as the power splitter and recombiner, respectively, which are connected with each other via the two parallel arms. The MZI chips are generally covered with a buffer layer of low-index materials such as SiO_2 and MgF_2, leaving a window in the region of one of the two arms for chemical and biochemical sensing.

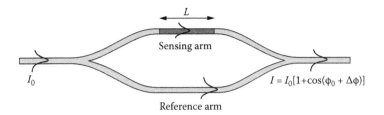

FIGURE 7.33 Schematic of an integrated optical MZI sensor.

The other arm with the covered layer acts as the reference due to its evanescent-wave insensitivity. The length (L) of the window ranges from millimeters to centimeters. For an MZI sensor operating at a wavelength λ, the output light intensity (I) changes with the phase change ($\Delta\varphi$) for the guided mode in the sensing arm according to Equation 7.20. Equation 7.21 indicates that $\Delta\varphi$ is related to a variation in the modal index (N_{eff}) induced by the evanescent field interaction with the surface analyte. The sensitivity (S) to refractive index of bulk liquid (n_C) per unit path length is defined as a ratio of $\Delta\varphi$ to Δn_C, which can be written as Equation 7.22.

$$I = I_0\left[1 + \gamma\cos\left(\phi_0 + \Delta\phi\right)\right] \tag{7.20}$$

$$\Delta\phi = \left(\frac{2\pi L}{\lambda}\right)\Delta N_{eff} \tag{7.21}$$

$$S = \frac{\Delta\phi}{\Delta n_C} = \left(\frac{2\pi L}{\lambda}\right)\left(\frac{\Delta N_{eff}}{\Delta n_C}\right) \tag{7.22}$$

where $I_0 = (I_{max} + I_{min})/2$ (I_{max} and I_{min} are the peak and trough intensities in the interference pattern), the so-called fringe contrast $\gamma = (I_{max} - I_{min})/(I_{max} + I_{min})$, ϕ_0 is the initial phase of the guided mode in the sensing arm with a length L.

The parameters γ, ϕ_0, and S are important to IO interferometric sensors, which together determine the sensors' detection limit. The greater the value of γ, the larger the resolution in $\Delta\phi$ and the lower the detection limit; to detect a very small concentration of an analyte, ϕ_0 must have a value that will force the MZI to work in the linear response region. For IO MZI sensors, γ is mainly determined by the performance of the Y-shaped power splitter. An ideal Y-shaped power splitter can provide equal intensities for the sensing and reference arms, leading to the highest fringe contrast. It is, however, practically difficult to fabricate 3dB Y-junction power splitters. Therefore, IO MZI sensors generally have a fringe contrast of $\gamma < 1$.

Several substrate materials including glass sheet, silicon wafer, LiNbO$_3$, and GaAs crystals have been used to fabricate MZIs for chemical and biochemical sensor applications [19–33]. Silicon-based MZIs, which are currently of great interest, include two structures: the conventional waveguide structure based on total internal reflection (TIR) [23–26] and the antiresonant reflecting optical waveguide (ARROW) structure [30–33]. The two structures consist of Si/SiO$_2$/Si$_3$N$_4$/SiO$_2$ multilayer ridge-line waveguides that are made using the MEMS process combined with SiON technology. However, the waveguiding layer is the Si$_3$N$_4$ film for the TIR structure and the upper SiO$_2$ film for the ARROW structure. In 1999, Heideman and the coworker Lambeck [23] realized a monolithic TIR-type MZI sensor on silicon substrate [23]. This sophisticated device contains a ZnO electro-optic phase modulator, a polarizer, a sensing window of $L = 5$ mm and two fiber couplers. Figure 7.34a and 7.34b show the top view of the device structure and the longitudinal cross section along one branch of the device. Figure 7.34c displays a photopicture of the real MZI device. With

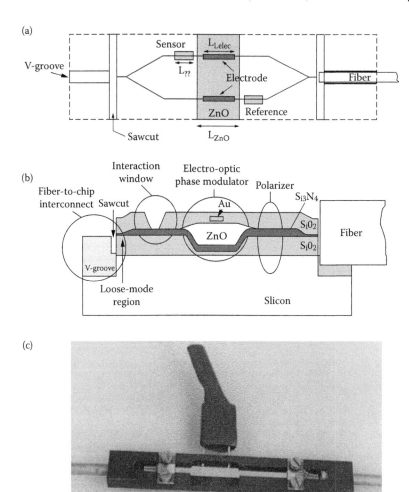

FIGURE 7.34 (a) Top view of the structure of a fiber-coupled MZI with a ZnO modulator, (b) longitudinal cross section along one branch of the device, (c) photograph of the real device. (From R. G. Heideman, et al., Sen. Actuat. B61 (1999), 672–676)

the phase modulator, the initial phase of the MZI sensor can be adjusted to make it operate in the linear response region. Their device works at $\lambda = 850$ nm, and has the following performance data: Resolution of N_{eff} is $\Delta N_{eff} = 5 \times 10^{-9}$; resolution of n_C is $\Delta n_C = 2 \times 10^{-8}$; resolution of the chemically induced effective increase of the layer thickness is about 2×10^{-5} nm, corresponding to a mass coverage of 0.01 pg/mm^2; drift of N_{eff} under laboratory conditions is smaller than 5×10^{-8}/h; fiber insertion loss is 10dB. With $\Delta N_{eff} = 5 \times 10^{-9}$ and $\Delta n_C = 2 \times 10^{-8}$, the sensitivity per unit path length was determined as $\Delta N_{eff}/\Delta n_c = 0.25$. This value is high due to a large difference in refractive index between the lower SiO$_2$ cladding layer and the Si$_3$N$_4$ core layer.

Because glass-OWG-based MZIs are cheap, low-loss, resistant to acidic/alkaline corrosion, relatively easy to fabricate, and compatible with the single-mode fiber coupling, they have been widely used as chemical and biochemical sensors [19–21]. Glass-OWG-based MZIs are made by the standard photolithographic process combined with the ion exchange method. The ion-exchange glass waveguides with a graded-index profile have a very weak evanescent field that results in a small value of $\Delta N_{eff}/\Delta n_c$ and consequently give rise to a small sensitivity per unit length of the sensing arm. In 1991, Boiarski et al. [19] prepared a glass-based MZI by the potassium ion exchange method [19].

With the device, they measured that the value of $\Delta N_{eff}/\Delta n_c$ ranges from 0.0018 to 0.002, being 100 times smaller than that for the Si_3N_4-core-layer MZI mentioned earlier. To improve the sensitivity of the glass-based MZI sensors, Qi et al. [28] designed a channel-planar composite optical waveguide (COWG) to serve as the sensing arm [28]. As shown in Figure 7.35a, a channel-planar COWG is composed of a single-mode glass channel waveguide locally overlaid with a tapered thin film of high-index materials such as TiO_2 and Ta_2O_5. Adiabatic transition of the guided mode in the channel-planar COWG based on the tapered velocity coupling theory results in a significantly enhanced evanescent field as compared with that of a bare glass channel waveguide. Combination of the calculated and experimental data demonstrated that use of the channel-planar COWG as the sensing arm of a glass-OWG-based MZI sensor can make the sensitivity increase more than 70 times. Because of a significant difference between the COWG sensing arm and the bare channel reference arm, the low-index buffer layer is no longer required for the glass-OWG-based MZI sensors.

7.4.3 COMPOSITE WAVEGUIDE-BASED POLARIMETRIC INTERFEROMETER (PI) SENSORS

In addition to the channel-planar COWG, a simpler COWG with a planar-planar structure can be readily prepared by using masked sputtering to fabricate a tapered layer of TiO_2 or Ta_2O_5 onto the single-mode slab glass waveguide. Figure 7.36a shows the photopicture of the planar-planar COWG chips, and the bright stripes are the tapered thin films of TiO_2. Single-mode slab and channel glass waveguides are low-index and polarization-insensitive OWGs, allowing the fundamental transverse electric (TE_0) and magnetic (TM_0) modes to be simultaneously excited with a single beam of linearly polarized laser light under the prism-coupling condition. Single-mode glass waveguides are also low loss and poorly sensitive to surface changes, enabling the TE_0 and TM_0 modes to propagate a long distance and permitting attachment of a measuring chamber to it without perturbing the mode propagation. On the other hand, single-mode thin-film waveguides made of high-index materials such as TiO_2 and Ta_2O_5 generally have a large propagation loss, a large modal birefringence, and a great sensitivity to surface interaction. Therefore, the COWG is an optimized combination of a low-loss, nonbirefrigent, and poorly sensitive glass waveguide and a high-loss, significantly birefringent, and highly sensitive thin-film waveguide. The two tapers of the thin film ensure the adiabatic modal transition between the uncovered and covered regions of the single-mode glass waveguide, effectively inhibiting the intermode conversion. For chemical and biochemical sensor applications, the COWG offers not only a local sensing area but also two inactive areas for placing both the prism couplers and the measuring chamber.

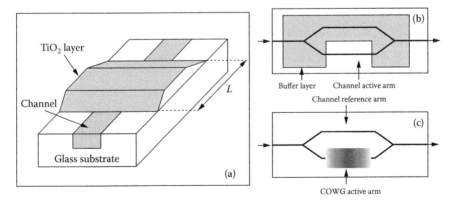

FIGURE 7.35 (a) Structure of a channel-planar composite optical waveguide (COWG), (b) conventional structure of glass-based MZI sensors with both a channel waveguide sensing arm and a low-index buffer layer, (c) modified structure of glass-based MZI sensors with a COWG sensing arm and without having the buffer layer.

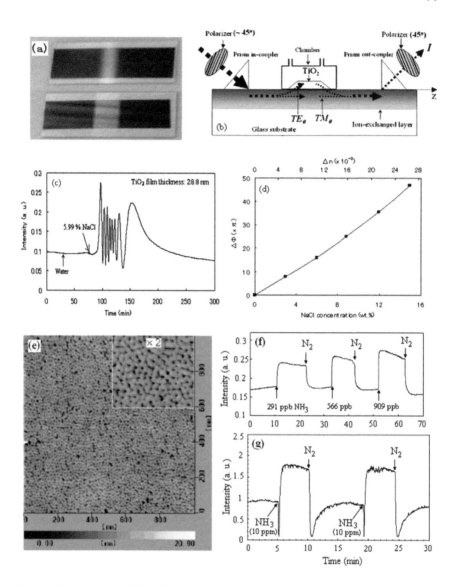

FIGURE 7.36 (a) Photograph of COWG chips, (b) schematic configuration of the COWG-based PI sensor, (c) temporal response of the sensor to a change of refractive index of liquid, and (d) the phase change of the PI sensor versus the concentration of aqueous NaCl solution, (e) AFM image of the mesoporous TiO$_2$ film coated on the COWG chip, (f) and (g) temporal response to ammonia gas of the COWG-based PI sensor coated with the mesoporous TiO$_2$ film.

Both the planar-planar and channel-planar COWGs can be directly used as a very simple yet highly sensitive polarimetric interferometer (PI) sensor based on the spatial separation of the TE$_0$ and TM$_0$ modes in the waveguide region covered with the tapered thin film of TiO$_2$ [34–36]. The spatial separation leads to a great difference between the two modes: the evanescent field of the TE$_0$ mode is much stronger than that of the TM$_0$ mode. Therefore, chemical and physical changes occurring within 100 nm over the TiO$_2$ film can cause a large variation of the effective refractive index of the TE$_0$ mode (N$_{TE}$) relative to that of the TM$_0$ mode (N$_{TM}$). Figure 7.36b illustrates the structure of the COWG-based PI sensor, which is as simple as a conventional OWG absorption sensor. A pair of glass prism couplers are attached to the glass waveguide with a refractive-index-matching liquid (methylene iodide), and a measuring chamber is mounted onto the waveguide to shield the TiO$_2$

film. By using a linearly polarized He-Ne laser beam to irradiate a prism coupler at a given angle of incidence, the TE_0 and TM_0 modes are simultaneously excited in the glass waveguide. With the first TiO_2 taper, the TE_0 and TM_0 modes propagate through the film-covered region, where the two modes spontaneously separate from each other. The second TiO_2 taper makes the two modes recombine in the glass waveguiding layer. The output light beam was passed through a 45° polarization analyzer, to cause interference between the TE and TM components. The interference fringes are monitored in real time with a photodiode detector. The induced changes ($\Delta\varphi$) in the phase difference between the TE_0 and TM_0 modes can be expressed as follows:

$$\Delta\phi = \frac{2\pi}{\lambda}\int_0^L \left(\Delta N_{TE} - \Delta N_{TM}\right)dz \qquad (7.23)$$

where λ is the vacuum wavelength, L is the length of the tapered thin TiO_2 film, and ΔN_{TE} and ΔN_{TM} are variations in N_{TE} and N_{TM}.

Figure 7.36c displays a temporal interference pattern of the PI sensor induced by a refractive index change in the measuring chamber. The phase-difference change ($\Delta\varphi$) versus the concentration and refractive index of aqueous NaCl solution is plotted in Figure 7.36d, indicating that $\Delta\varphi = 1°$ corresponds to $\Delta n_C \approx 1 \times 10^{-6}$. After the tapered layer of high-index materials was covered with a nanoporous thin film, the COWG-based PI sensor can detect ppb-level ammonia gas at room temperature [36]. Figure 7.36e shows atomic force microscopre (AFM) image of the sol-gel templated mesoporous TiO_2 thin film dip-coated on the sputtered TiO_2 layer of the COWG. Figures 7.36f and 7.36g display the responses to different concentrations of ammonia in nitrogen measured at room temperature using the COWG-based PI sensor coated with the mesoporous TiO_2 film. The other PI sensors based on stepindex planar waveguides were also reported [37,38], for which the sensitivity per unit path length are generally small as indicated by Lambeck in Reference 24.

7.4.4 INTEGRATED OPTICAL YOUNG INTERFEROMETER (YI) SENSORS

7.4.4.1 Sensing Principle of Integrated Optical YI

Figure 7.37a shows a schematic of an OWG-based YI sensor that contains two closely adjacent slab or stripe waveguides and a linear CCD detector. The output light from a single waveguide is diffracted in the x direction, and the diffracted light intensity $I_S(x)$ can be expressed by Equation 7.24. The two divergent light beams overlap to yield a spatial interference pattern, and the interfered light intensity $I(x)$ in the x direction can be written as Equation 7.25.

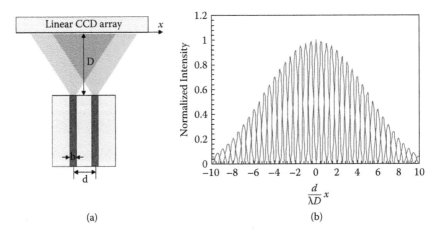

FIGURE 7.37 (a) Schematic of an OWG-based YI device, (b) the simulated interference patterns.

$$I_S(x) = I_0 \sin^2\left(\frac{\pi b x}{\lambda\sqrt{D^2 + x^2}}\right)\left(\frac{\pi b x}{\lambda\sqrt{D^2 + x^2}}\right)^{-2} \tag{7.24}$$

$$I(x) = 2I_S(x)\left[1 + \cos\left(\frac{2\pi d}{\lambda D} x + \phi_0 + \Delta\phi\right)\right] \tag{7.25}$$

$$= 2I_S(x)\left[1 + \cos\left(2\pi \quad fx \quad + \phi_0 + \Delta\phi\right)\right]$$

$$\phi_0 + \Delta\phi = \arctan\left[\frac{\displaystyle\int_{-\infty}^{+\infty} I(x)\sin\left(2\pi \quad fx \quad\right)dx}{\displaystyle\int_{-\infty}^{+\infty} I(x)\cos\left(2\pi \quad fx \quad\right)dx}\right] \tag{7.26}$$

where I_0 is light intensity in a single waveguide, b is the core thickness for slab waveguide or the core width for stripe waveguide, d is the core-to-core spacing, D is the distance from the end face of the OWG chip to the CCD detector, $f = d/(\lambda D)$, ϕ_0 is the initial phase difference between the two waveguides, and $\Delta\phi$ is the phase-difference change induced by surface interaction between the evanescent field and analyte. According to Equation 7.26, the value of $(\phi_0 + \Delta\phi)$ can be determined with the Fourier transform method. Figure 7.37b displays the two interference patterns simulated with $\lambda = 0.633$ μm, $b = 4$ μm, $d = 50$ μm, and $D = 10$ mm. The phase difference between the two patterns is 180°.

7.4.4.2 Slab-Type YI Sensors

Slab OWG Young interferometer (YI) sensors include two types: one is based on a multilayer slab OWG chip made by SION technology, which consists of two planar core layers separated by a buffer layer. As shown in Figure 7.38a, the upper core layer of the slab OWG is used for chemical or biochemical sensing, and the lower core layer serves as the reference; this type of slab OWG YI sensor has been commercialized as a bioanalyzer by the Farfield Sensor Ltd., U.K. [40–43]. Figure 7.38b displays the configuration of the slab YI bioanalyzer. The bioanalyzer is called the *dual-polarization interferometer* (DPI) because the phase changes for the TE and TM modes ($\Delta\phi_{TE}$ and $\Delta\phi_{TM}$) can be simultaneously measured. Figure 7.38c shows a photopicture of the commercial DPI instrument. With $\Delta\phi_{TE}$ and $\Delta\phi_{TM}$ measured at the same time, the effective refractive indexes of the TE and TM modes (N_{TE} and N_{TM}) can be determined according to the following equations:

$$N_{TE} = N_{TE}^0 + \frac{\lambda}{2\pi L}\Delta\phi_{TE} \tag{7.27}$$

$$N_{TM} = N_{TM}^0 + \frac{\lambda}{2\pi L}\Delta\phi_{TM} \tag{7.28}$$

where N_{TE}^0 and N_{TM}^0 are effective refractive indexes of the TE_0 and TM_0 modes prior to measurement, and they are given for each specific OWG chip. Consequently, both the thickness and refractive index of the analyte layer formed on the OWG surface can be obtained by the best fitting of the measured N_{TE} and N_{TM} with the eigenvalue equations for a planar optical waveguide. As an example, Figure 7.39 shows a simulated result for a protein adlayer formed on the waveguide with the core layer of 200 nm and refractive indexes of 1.52, 1.77, and 1.33 for the substrate, core, and clad, respectively. The waveguide N_{TE}^0 and N_{TM}^0 are determined as 1.56861 and 1.60521, respectively,

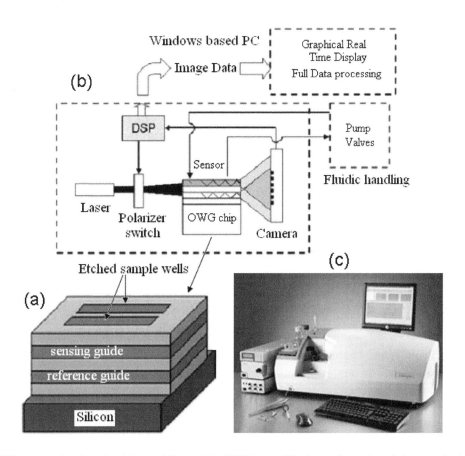

FIGURE 7.38 (a) Principle of the multilayer slab OWG based YI, (b) configuration of the so-called dual-polarization interferometer bioanalyzer with the multilayer slab OWG chip.

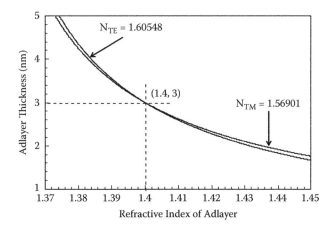

FIGURE 7.39 Simulation for determining thickness and refractive index of the analyte layer on the OWG.

at $\lambda = 633$ nm. Simulation with $\Delta N_{TE} = 0.00027$ and $\Delta N_{TM} = 0.00039$ indicates that the refractive index and thickness of the protein adlayer are $n_{ad} = 1.4$ and $t_{ad} = 3$ nm. Fixing ΔN_{TM} to 0.00039 but changing ΔN_{TE} to 0.0029 leads to $n_{ad} = 1.4667$ and $t_{ad} = 1.5565$ nm.

Another slab OWG YI sensor is schematically shown in Figure 7.40, which uses a double-slit element to divide the single-beam incident light into the two beams that were simultaneously coupled

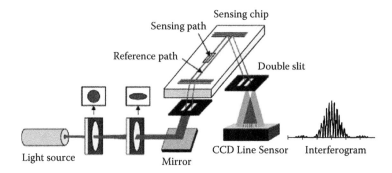

FIGURE 7.40   Optical setup of the slab OWG YI system with a slab waveguide chip containing grating couplers, two double-slit elements, and a linear CCD detector. (From K. Schmitt et al., *Biosens. Bioelectron.* 22 (2007), 2591–2597.)

with an integrated grating into a planar waveguide to yield a sensing and a reference path [26]. The output light beams corresponding to the sensing and reference paths were passed through the second double-slit element to produce a Young interference pattern that was detected with a linear CCD detector. It is seen from Figure 7.40 that the sensor system contains many individual optical elements, making the preparation of the sensor system difficult.

7.4.4.3   Stripe-Type YI Sensors

The stripe waveguide YI sensors generally consist of a sensing and a reference arm as well as a 3dB Y-shaped waveguide splitter that apportions the input optical power to the two arms. The divergent light beams emitting from the arm ends overlap to form a two-dimensional spatial interference pattern. Only the fringe shifts in the transverse direction were detected with a linear CCD detector. Figure 7.41 shows the common structure of a stripe waveguide YI sensor, which is actually half an IO MZI sensor. Figure 7.42 shows an exception of a multichannel IO YI sensor that was developed by Ymeti et al. [49,50]. The multichannel YI sensor contains more than one sensing window, capable of simultaneous detection of multiple analytes. The common features for the existing slab and stripe waveguide YI sensors are as follows: (1) the sensor chips generally consist of step-index OWGs; (2) the butt-coupling method was usually used; and (3) the measurand-induced phase change was determined by means of Fourier analysis of the spatial interference pattern that was detected with the linear CCD detector.

The expensive CCD detector is not conducive to widespread application of integrated YI sensors. As a matter of fact, a simple silicon photodetector with a slit can be readily used to detect the temporal interference pattern of the IO YI sensor instead of the CCD detector. On the basis of their previous works, Qi et al. recently developed a glass-waveguide-based YI sensor with the improved structure [51,52]. As shown in Figure 7.43a, this sensor is different from the aforementioned YI

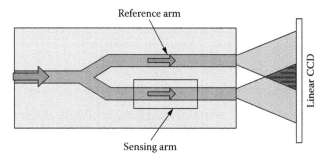

FIGURE 7.41   The common structure of the stripe waveguide YI sensors.

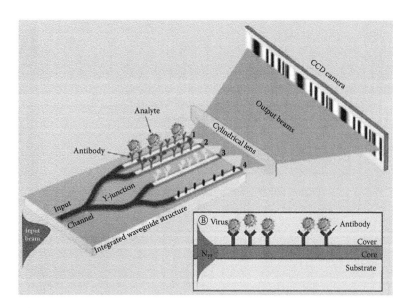

FIGURE 7.42   Schematic diagram of a multichannel waveguide YI sensor. (From A. Ymeti et al., *Nano. Lett.* 7 (2007), 394–397.).

sensors in four respects: (1) the sensor chip is a glass sheet containing several pairs of parallel straight channels but without having a Y-branch power splitter, and the channels are graded-index single-mode waveguides; (2) the sensing arm is a channel-planar COWG that offers the sensor a high sensitivity and enables the uncovered channel waveguide to act as the reference arm; (3) the prism-coupling method was used, to provide the sensor with a high-contrast dotted-line interference pattern in the lateral direction (see Figure 7.43b); and (4) the detector used is not a CCD but a slit-photodiode assembly. The optical power (P) detected with the slit-photodiode assembly can be expressed as Equation 7.29.

$$P = 2P_0 \left[1 + \frac{\lambda D}{\pi da} \sin\left(\frac{\pi da}{\lambda D} \right) \cos\left(\Delta\phi + \phi_0 + \frac{2\pi d}{\lambda D} x_0 \right) \right] \tag{7.29}$$

where λ is the laser wavelength, P_0 is the power of light passing through the slit from one channel, d is the channel-to-channel spacing, a is the slit width, D is the distance from the output prism to the slit. x_0 represents the slit position along the dotted-line interference pattern, ϕ_0 is the phase difference between the two channels prior to measurement, $\Delta\phi$ is the phase-difference change. From Equation 7.29 the fringe contrast of the COWG-based TI sensor can be derived as following equation:

$$\gamma = \frac{P_{\max} - P_{\min}}{P_{\max} + P_{\min}} = \frac{\lambda D}{\pi da} \left| \sin\left(\frac{\pi da}{\lambda D} \right) \right| = \left| \sin c\left(\frac{\pi da}{\lambda D} \right) \right| \tag{7.30}$$

To facilitate replacement of the waveguide chip and to simplify the device operation, we prepared an integrated prism chamber component. With this prism-chamber component, the YI sensor is stable and reliable, capable of real-time detection of very slow physical and chemical processes at the waveguide surface. Figure 7.43c and 7.43d schematically show the front view of the integrated prism chamber component in the close state and the side view in its open state. Figure 7.43e displays a photopicture of an actual prism-chamber component containing a glass waveguide YI chip. Figures 7.43f and 7.43g show the temporal response of the COWG-based YI sensor to unspecific adsorption of proteins.

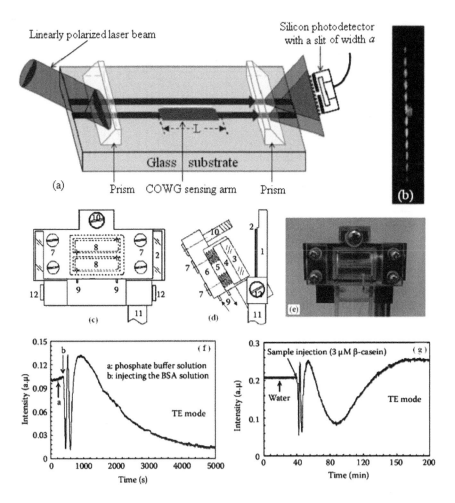

FIGURE 7.43 (a) the front view of the prism chamber integrated system in the close state; (b) the side view in the open state (1. back board, 2. OWG chip, 3. prism coupler, 4. prism holder, 5. springs, 6. front board, 7. slide screws, 8. fluid chamber, 9. inlet and outlet, 10. locking screw, 11. strut, 12. rotation axis, 13. silicone gasket, and 14. temperature sensor); (c) photograph of an actual prism chamber system; (d) response of the YI sensor or BSA adsorption from the PBS solution; (e) the sensor response to unspecific adsorption of β-casein from the aqueous solution.

7.4.5 Integrated Optical Fabry–Perot Interferometer (FPI) Sensor

All the IO interferometer sensors mentioned earlier are based on the evanescent wave interaction with the analyte molecules immobilized on the waveguide surface. For these interferometers, the sensitivity per unit path length heavily relies on the refractive index difference between the substrate (or the lower cladding layer) and the core layer. To increase the integral sensitivity of the device, the evanescent wave sensing pathlength generally ranges from millimeters to centimeters, far beyond the MEMS scale. In 2006, Kinrot [53] demonstrated that the guided-wave interaction with the bulk analyte solution instead of that between the evanescent wave and the surface analyte monolayer allows the sensing path length to be decreased down to submillimeter levels 0. By using MEMS processing and silicon wafer, they fabricated an SU8-waveguide-based MZI in which a periodically segmented waveguide FPI was integrated for bulk sensing of chemical and biochemical substances (see Figure 7.44). Using a 720-μm-long segmented waveguide FPI that is much smaller than those of

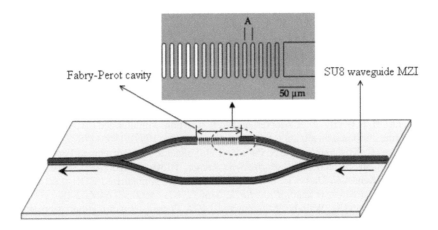

FIGURE 7.44 Schematic diagram of a periodically segmented waveguide Fabry–Perot interferometer integrated in an MZI structure. (From Kinrot and M. Nathan, *J. Lightwave Technol.* 24, (2006), 2139–2145.)

the previously reported devices, the detection limit for the refractive index of a liquid was measured as $\Delta n = 4 \times 10^{-5}$. The high sensitivity of such an integrated FPI sensor results from the multiple-reflection effect induced by Fresnel reflections from the repetitive barrier transitions between the sample and waveguide segments. Kinrot's work opens the way to fabricating MEMS-scale IO chips for interferometry-based chemical and biochemical sensor applications.

7.4.6 Conclusion

IO interferometers as versatile and surface-sensitive transducers were briefly reviewed in this section. For chemical and biochemical sensor application, IO interferometers mainly include MZI, YI, and PI. Monolithic MZI sensors consist of stripe waveguides, generally fabricated by MEMS processing combined with SION technology. Vertically integrated slab YI sensors have become a commercial bioanalyzer. The COWG-based PI sensor is one of the simplest and cheapest IO sensors, but it can provide a required high sensitivity for trace, even ultra-trace, detection. IO interferometric sensor chips generally have a size of several centimeters for convenience of use and for high sensitivity, but the basic components on the chip are the MEMS structures. From this standpoint, IO interferometers can be referred to as interferometric MEMS sensors. Three-dimensional SiON waveguides as typical optical MEMS components offer many opportunities to make small, highly integrated, and sophisticated multifunctional sensor systems. The next-generation IO chemical and biochemical sensors would be single-silicon-chip devices containing OWG sensing elements, LED sources, Si-photodiode detectors, CMOS data processor, and MEMS sample-preprocessing and microfluidic components.

REFERENCES

1. D. H. Ellison, *Handbook of Chemical and Biological Warfare Agents*, 2nd edition, CRC Press, Boca Raton, FL, 2007.
2. Countering Bioterrorism: The Role of Science and Technology, Panel on Biological Issues, Committee on Science and Technology for Countering Terrorism, National Research Council, National Academies Press, 2002.
3. V. M. A. Hakkinen, Gas chromatography/mass spectrometry in on-site analysis of chemicals related to the chemical weapons convention, in *Encyclopedia of Analytical Chemistry* (2000), 1001–1007.
4. J. L. Gottfried, F. C. De Lucia, C. A. Munson, and A. W. Miziolek, Standoff detection of chemical and biological threats using laser-induced breakdown spectroscopy, *Appl. Spectrosc.* 62 (2008), 353–363.

5. N. Gayraud, Ł. W. Kornaszewski, J. M. Stone, J. C. Knight, D. T. Reid, D. P. Hand, and W. N. MacPherson, Mid-infrared gas sensing using a photonic bandgap fiber, *Appl. Opt.* 47 (2008), 1269–1277.

6. D. K. Lynch, M. A. Chatelain, T. K. Tessensohn, and P.M. Adams, Remote identification of in situ atmospheric silicate and carbonate dust by passive infrared spectroscopy, *IEEE Trans. Geoscience and Remote Sensing*, 35(3), (1997) 670–674.

7. E. V. Loewenstein, The history and current status of Fourier transform spectroscopy, *Appl. Opt.* 5 (1966), 845–854.

8. http://www.brukeroptics.com/alpha.html.

9. O. Manzardo, H. P. Herzig, C. R. Marxer, and N. F. de Rooij, Miniaturized time-scanning Fourier transform spectrometer based on silicon technology, *Opt. Lett.* 24 (1999), 1705–1707.

10. K. Yu, D. Lee, U. Krishnamoorthy, N. Park, and O. Solgaard, Micromachined Fourier transform spectrometer on silicon optical bench platform, *Sens. Actuat. A* 130–131 (2006), 523–530.

11. T. Sandner, A. Kenda, C. Drabe, H. Schenk, and W. Scherf, Miniaturized FTIR-spectrometer based on optical MEMS translatory actuator, *Proc. SPIE*, 6466 (2007) 646602-1–646602-12.

12. C. Solf, J. Mohr, and U. Walrabe, Miniaturized LIGA Fourier transformation spectrometer, in *Proc. IEEE*, 2 (2003), pp. 773–776.

13. A. Jain, H. Qu, S. Todd, and H. Xie, A thermal bimorph micromirror with large bi-directional and vertical actuation, *Sens. Actuat. A.* 122 (2005), pp. 9–15.

14. L. Wu, A. Pais, S.R. Samuelson, S. Guo, and H. Xie, A miniature Fourier transform spectrometer by a large-vertical-displacement microelectromechanical mirror, *Proc. of the 2009 OSA Spring Optics and Photonics Congress*, Vancouver, BC, Canada, FWD4, 2009.

15. L. Wu, A. Pais, S. R. Samuelson, S. Guo, and H. Xie, A mirror-tilt-insensitive Fourier transform spectrometer based on a large vertical displacement micromirror with dual reflective surface, *Proc. of the 15th International Conference on Solid-State Sensors, Actuators and Microsystems (Transducers'09)*, June 21–25, 2009, Denver, CO, pp. 2090–2093.

16. B. Saggin, L. Comolli, and V. Formisano, Mechanical disturbances in Fourier spectrometers, *Appl. Opt.* 46 (2007), 5248–5256.

17. R. C. M. Learner, A. P. Thorne, and J. W. Brault, Ghosts and artifacts in Fourier-transform spectrometry, *Appl. Opt.* 35 (1996), 2947–2954.

18. U. Wallrabe, C. Solf, J. Mohr, and J. G. Korvink, Miniaturized Fourier transform spectrometer for the near infrared wavelength regime incorporating an electromagnetic linear actuator, *Sens. Actuat. A: Physical*, 123–124 (2005), pp. 459–467.

19. A. A. Boiarski, R. W. Ridgway, J. R. Busch, G. Turhan-Sayan, and L. S. Miller, *SPIE* 1587 (1991), 114–128.

20. N. Fabricius, G. Gauglitz, and J. Ingenhoff, *Sens. Actuat. B* 7 (1992), 672–676.

21. J. Ingenhoff, B. Drapp, and G. Gauglitz, *Fresenius J. Anal. Chem.* 346 (1993), 580–583.

22. Y. Liu, P. Hering, and M.O. Scully, *Appl. Phys. B* 54 (1992), 18–23.

23. R. G. Heideman and P. V. Lambeck, *Sens. Actuat. B* 61 (1999), 100–127.

24. P. V. Lambeck, *Meas. Sci. Technol.* 17 (2006), R93–R116.

25. F. Brosinger, H. Freimuth, M. Lacher, W. Ehrfeld, E. Gedig, A. Katerkamp, F. Spener, and K. Cammann, *Sens. Actuat. B* 44 (1997), 350–355.

26. S. Busse, M. DePaoli, G. Wenz, and S. Mittler, *Sens. Actuat.* 80 (2001), 116–124.

27. Kunz, R. E. *Sens. Actuat. B* 38 (1997), 13–28.

28. Z. Qi, N Matsuda, K. Itoh, M. Murabayashi, and C. Lavers, *Sens. Actuat. B* 81 (2002), 254–258.

29. P. Hua, B. J. Luff, G. R. Quigley, J. S. Wilkinson, and K. Kawaguchi, *Sens. Actuat. B* 87 (2002), 250–257.

30. D. Jimenez, E. Bartolome, M. Moreno, J. Munoz, and C. Dominguez, *Opt. Commun.* 132 (1996), 437–441.

31. F. Prieto, L.M. Lechuga, A. Calle, A. Liobera, and C. Dominguez, *J. Lightwave Technol.*, 19 (2001), 75–83.

32. F. Prieto, B. Sepulveda, A. Calle, A. Llobera, C. Dominguez, and L.M. Lechuga, *Sens. Actuat. B* 92 (2003), 151–158.

33. S. -H. Hsu and Y. -T. Huang, *Opt. Lett.* 30 (2005), 2897–2879.

34. Z. Qi, K. Itoh, M. Murabayashi, and C. R. Lavers, *Opt. Lett.* 25 (2000), 1427–1429.

35. Z. Qi, K. Itoh, M. Murabayashi, and H. Yanagi, *J. Lightwave Technol.* 18 (2000), 1106–1110.

36. Z. Qi, I. Honma, and H. Zhou, *Anal. Chem.* 2006, 78, 1034–1041.

37. Y. M. Shirshov, S. V. Svechnikov, A. P. Kiyanovskii, Y. V. Ushenin, E. F. Venger, A. V. Samoylov, and R. Merker, *Sens. Actuat. A* 68 (1998), 384–387.

38. C. Samm and W. Lukosz, *Sens. Actuat.* B 31 (1996), 203–207.
39. Y. Ren, P. Mormile, L. Petti, and G.H. Cross, *Sens. Actuat. B* 75 (2001), 76–82.
40. M. J. Swann, L. L. Peel, S. Carrington, and N. J. Freeman, *Anal. Biochem.* 329 (2004), 190.
41. G. H. Cross, Y. Ren, and N. J. Freeman, *J. Appl. Phys.* 86 (1999), 6483–6488.
42. S. Ricard-Blum, L. L. Peel, F. Ruggiero, and N. J. Freeman, *Anal. Biochem.* 352 (2006), 252–259.
43. P. D.Coffey, M. J. Swann, T. A. Waigh, F. Schedin, and J. R. Lu, *Opt. Express* 17 (2009), 10959–10969.
44. K. Schmitt, B. Schirmer, C. Hoffmann, A. Brandenburg, and P. Meyrueis, *Biosens. Bioelectron.* 22 (2007), 2591–2597.
45. D. Hradetzky, C. Mueller, and H. Reinecke, *J. Opt. A: Pure Appl. Opt.* 8 (2006), S360–S364.
46. A. Brandenburg, R. Krauter, C. Künzel, M. Stefan, and H. Schulte, *Appl. Opt.* 39 (2000), 6396–6405.
47. E. Brynda, M. Houska, A. Brandenburg, and A. Wikerstal, *Biosens. Bioelectron.* 17 (2002), 665–675.
48. A. Brandenburg and R. Henninger, Integrated optical Young interferometer, *Appl. Opt.* 33 (1994), 5941–5947.
49. A. Ymeti, J. S. Kanger, J. Greve, P. V. Lambeck, R. Wijn, and R. G. Heideman, *Appl. Opt.* 42 (2003), 5649–5660.
50. A. Ymeti, J. Greve, P. Lambeck, T. Wink, S. van Hövell, T. Beumer, R. Wijn, R. Heideman, V. Subramaniam, and J. Kanger, *Nano. Lett.* 7 (2007), 394–397.
51. Z. Qi, S. Zhao, F. Chen, and S. Xia, *Opt. Lett.* 34 (2009), 2113–2115.
52. Z. Qi, S. Zhao, F. Chen, R. Liu, and S. Xia, *Opt. Express* 18 (2010), 7421–7426.
53. N. Kinrot and M. Nathan, *J. Lightwave Technol.* 24, (2006), 2139–2145.
54. R. Fuest, N. Fabricius, U. Hollenbach, and B. Wolf, *SPIE* 1794 (1993), 352–365.
55. O. G. Helleso, P. Benech, and R. Rimet, *Sens. Actuat. A* 47 (1995), 478–481.

Index

A

B